Fundamentals of Data Science

Part I: Inference and Experiment

Jared M. Maruskin

Fundamentals of Data Science

Part I: Inference and Experiment

Cayenne Canyon Press

Jared M. Maruskin

Published by Cayenne Canyon Press
San José, California

ISBN: 978-1-941043-11-0 (softcover)

10 9 8 7 6 5 4 3 2 1

To Those who dedicate their lives
to the secret world of numbers

Preface

I never took a course in statistics. My background, however, is rich in applied mathematics: from aerospace engineering to general relativity to the geometry of nonholonomic mechanics to tracking space debris. I even wrote a book on dynamical systems and geometric mechanics; see Maruskin [2018]. After my PhD, I served for a few years as a professor before transitioning into industry. It is here where I discovered my love for statistics and machine learning. The content of these pages is my practical crash course in the mathematical prerequisites to data science: statistics, modeling, and machine learning. This volume is not a course, but a curriculum. My process for each chapter is to locate a number of references, read, learn, solve, and write. I try to capture the key results of various disciplines, often adding my own perspective or approach.

I write this book for two reasons. The first reason is I write this for myself. By setting out on this endeavor, I seek to forge a mastery in the subjects presented within these pages. As I go through these topics and learn, I have found that I am constantly shuffling through several cornerstone texts, never really settling on any one approach or notation. In writing this, I seek to collect my thoughts and present the material in a manner that is clear and sensible to my own understanding.

The second (and undoubtedly most important) reason I write is I write this book for you. Many of the texts I began with in data science lacked the rigor that I was accustomed to in my studies of other applied topics. Furthermore, the mathematical clarity was scattered over dozens of aforementioned texts and not distilled to its essence in any single place. I therefore feel that these notes will be of benefit to anyone with a solid mathematical background wishing to transition into data science. I also envision this text as a practical reference for practicing data scientists, who wish to have a companion that captures a wide range of material under one roof.

In short: my goal was to lay a foundation of expertise in statistical modeling and machine learning for myself, while leaving a foundational path for others to follow.

Coding examples throughout are written in Python. I have found that writing simple simulations is an effective tool in learning theory. It also serves as a tool to validate our theoretical formulas. Several times in the process of writing this text, I developed a formula, ran a simulation, and found that it didn't work the way I expected. This led me to discover my errors and, when the simulation finally worked, served to validate that I had arrived at the correct conclusions.

Python Distribution

I recommend obtaining the free Anaconda Python 3 distribution, available from **anaconda.com**. It comes with all of the scientific packages pre-installed. Anaconda is a bundle of applications, so you can use it to launch R-Studio, Jupyter Notebook, or Spyder. I write my code in Spyder. The Sypder interface gives you access to a code editor, an iPython console, and help docs, all in one screen. I find it extremely useful running code in the console while I develop source code in the editor. I have also have several friends who use PyCharm from **jetbrains.com** as an alternative.

Many of the examples throughout this text will make use of various packages. The import statements that are used throughout are collected in code block 1. I will further assume the reader to have familiarity with

```python
import numpy as np
import pandas as pd
import sklearn, scipy
import time, datetime
import matplotlib.pyplot as plt
from mpl_toolkits.mplot3d import Axes3D # For 3D
from pandas import DataFrame, Series
from abc import ABC, abstractmethod
```

Code Block 1: Python Preamble

Python. For those starting out, however, I recommend McKinney [2017] as an excellent place to start.

Finally, we will use several of sklearn's built-in data sets to build examples from throughout the text. The load statements returns a *Bunch* object, which essentially acts like a Python dictionary, with key attributes: data, target, DESCR, and feature_names, and filename. As an example, the Boston house-price dataset is loaded in code block 2.

```
1  boston = sklearn.datasets.load_boston()
2  df = DataFrame(boston['data'], columns=boston['feature_names'])
3  df.head() # shows top 5 rows in DataFrame
4  print( boston['DESCR'] ) # prints description of data set
5  y = boston['target'] # target variable
```

Code Block 2: Loading Datasets

San José, California

Jared M. Maruskin
April 2021

Contents

Part I Inference and Experiment

1	**Basecamp: Statistics**	3
	1.1 Probability	3
	1.2 Random Variables	5
	1.3 The Probability Density Function of a Discrete Random Variable	8
	1.4 Multiple Random Variables	9
	1.4.1 Joint and Marginal Distributions	9
	1.4.2 Independence	11
	1.4.3 Covariance and Correlation	11
	1.4.4 Conditional Distributions and Expectation	12
	1.5 Sampling	14
	1.5.1 Point Estimators and Sampling Distributions	14
	1.5.2 Sample Mean and Sample Standard Deviation	15
	1.5.3 Law of Large Numbers	20
	1.5.4 Bias	20
	1.6 The Empirical Distribution	22
	1.6.1 Estimating Distributions from Data	22
	1.6.2 Bias and Variance of the Empirical Distribution	23
	1.6.3 Plug-in Estimators	24
	1.7 The Bootstrap	27
	1.7.1 Introduction	28
	1.7.2 Bootstrap Samples	28
	1.7.3 Estimating the Variance of a Statistic	30
	1.8 Transformations of Random Variables	31
	1.8.1 Univariate Transformations	32
	1.8.2 Bivariate Transformations	33
	1.8.3 Moment-generating Functions	35
	1.8.4 Location and Scale Families	36
	Problems	37

2   A Menagerie of Distributions 39
  2.1 Bernouilli and Binomial 39
      2.1.1 Bernouilli Trials 39
      2.1.2 Binomial Distribution 40
  2.2 Negative Binomial, Geometric, and Hypergeometric 44
      2.2.1 Negative Binomial Distribution 45
      2.2.2 Geometric Distribution......................... 49
      2.2.3 Hypergeometric Distribution 51
  2.3 Poisson and Exponential 54
      2.3.1 Poisson Distribution 54
      2.3.2 Exponential Distribution 56
      2.3.3 Poisson Process 58
  2.4 The Normal Distribution 61
      2.4.1 Normal Distribution 61
      2.4.2 Sums of Normal Random Variables 65
      2.4.3 The Central Limit Theorem 68
  2.5 Chi-Squared, T, and F 70
      2.5.1 The Chi-Squared Distribution................... 70
      2.5.2 Student's T Distribution 76
      2.5.3 Snedecor's F Distribution 78
  2.6 Beta and Gamma 81
      2.6.1 Beta Distribution 81
      2.6.2 Gamma Distribution........................... 83
      2.6.3 Gamma's Relation to the Poisson Process 86
      2.6.4 Beta-Binomial 88
  2.7 Multivariate Distributions 91
      2.7.1 The Multivariate Normal Distribution............. 91
      2.7.2 The Multinomial Distribution 97
  Problems 106

3   Getting Testy 109
  3.1 Hypothesis Testing 109
      3.1.1 The Null Hypothesis 109
      3.1.2 Types of Errors 111
      3.1.3 The Power Function 112
      3.1.4 Confidence Intervals 117
  3.2 Tests for Estimates 118
      3.2.1 The Mean of a Normal Random Variable 118
      3.2.2 The Wald Test 123
      3.2.3 Test for Proportions 125
      3.2.4 Power and Sample Size........................ 126
  3.3 Tests for Dual Populations............................ 134
      3.3.1 Test for Two Means; Equal Variances 134
      3.3.2 Test for Two Means; Unequal Variances 138
      3.3.3 Test for Two Proportions....................... 141

	3.4	Tests for Categories	143
		3.4.1 Pearson's Chi-squared Test	143
		3.4.2 Test for Homogeneity	147
		3.4.3 Test for Independence	148
	3.5	Tests for Categories II: Power Analysis	149
		3.5.1 Noncentral Chi-squared Distribution	149
		3.5.2 Power Function for Chi-square Statistic	154
		3.5.3 Sampling the Simplex	165
	3.6	Tests for Distributions	168
		3.6.1 Chi-squared Goodness-of-Fit Test	170
	3.7	Analysis of Variance	172
		Problems	177

4 It's Alive! ... 179
	4.1	Principles of the Design of Experiments	179
	4.2	A/B Tests	185
		4.2.1 Design	185
		4.2.2 Analysis	189
		4.2.3 Blocking	195
		4.2.4 Online Randomized Block Design	200
	4.3	Single-Factor Experiments	202
		4.3.1 Design	202
		4.3.2 Analysis	206
		4.3.3 Pairwise Comparisons	211
	4.4	Two-Factor Experiments	217
		4.4.1 Design	217
		4.4.2 Analysis: Two-way ANOVA	218
		4.4.3 Analysis: Single Factor Experiment with Random Block Design	222
	4.5	Bandits	223
		Problems	232

5 Estimator 2: Judgement Day ... 235
	5.1	Maximum Likelihood Estimation	235
		5.1.1 Likelihood and Log Likelihood	235
		5.1.2 Score Statistic and Fisher Information	237
		5.1.3 The Method of Scoring	241
		5.1.4 Asymptotic Properties of the MLE	242
		5.1.5 Example: Gamma Random Variable	245
	5.2	Gradient Descent Method	251
		5.2.1 Basic Gradient Descent	251
		5.2.2 Stochastic Gradient Descent	254
	5.3	Censored and Truncated Data	262
		5.3.1 Overview	262
		5.3.2 Censored Data	263

5.3.3   Truncated Data 265
5.3.4   Hazard Function 268
5.3.5   The Pareto Distribution 269
5.3.6   The Weibull Distribution 269
5.4   Online Processes 272
5.4.1   Stochastic Processes 273
5.4.2   Common Types of Stochastic Processes 274
5.4.3   Aggregated Processes 278
5.4.4   Online Processes 280
Problems 287

References 291

Index 295

Part I

Inference and Experiment

1

Basecamp: Statistics

1.1 Probability

We begin with a brief review of a few of the main definitions in probability theory. Please see Capinksi and Kopp [2005], Ross [2012] or Wasserman [2004] for more details.

Definition 1.1. *The sample space, typically denoted Ω, of an experiment or random trial is the set of possible outcomes of that experiment. An event is a subset of the sample space. Each element $\omega \in \Omega$ is called an* outcome.

Probability theory deals with potential outcomes of whatever phenomenon we are trying to study; the sample space is the set of all possible outcomes. We shall consider the null set to be implicitly included in the sample space, so that the sample space is closed under the operation of complement ($\Omega^c = \emptyset$).

Example 1.1. A coin is flipped N times. By using the encoding

$$\text{"heads"} \longrightarrow 1$$
$$\text{"tails"} \longrightarrow 0,$$

we may represent the sample space by the set of all 2^N binary strings of length N. If $N \geq 2$, we may further define the event E as the set of all outcomes such that the first two coin flips result in a heads, i.e.,

$$E = \{s \in \mathbb{B}^N : s_0 = 1 \text{ and } s_1 = 1\},$$

where $\mathbb{B} = \{0, 1\}$ is the set of binary digits (bits). *Note*: we will enumerate components of vectors starting with zero, not one, to be consistent with Python. Hence, $x \in \mathbb{R}^n$ may be represented as $x = \langle x_0, x_1, \ldots, x_{n-1} \rangle$. Another example of an event is the outcome in which the ith coin flip results in heads:

$$E_i = \{s \in \mathbb{B}^N : s_i = 1\}, \qquad i = 0, 1, \ldots, N - 1.$$

The intersection

$$\bigcap_{i=0}^{N-1} E_i = \{s \in \mathbb{B}^N : s_i = 1 \ \forall \ i \in \{0, \ldots, N - 1\}\}$$

is the event in which *each* coin flip is a heads, whereas the union

$$\bigcup_{i=0}^{N-1} E_i = \{s \in \mathbb{B}^N : \exists \ i \in \{0, \ldots, N - 1\}, \ s.t. \ s_i = 1\}$$

is the event in which there is *at least one* heads in the series of coin flips. ▷

We will further say that the events $E_1, E_2, \ldots$ are *disjoint* or *mutually exclusive* if $E_i \cap E_j = \emptyset$ whenever $i \neq j$. A *partition* of the sample space Ω is a set of disjoint sets $E_1, E_2, \ldots$ that cover the sample space, i.e., such that $\cup_{i=0}^{\infty} E_i = \Omega$. Finally, given an event E, we shall define its indicator function as

$$\mathbb{I}_E(x) = \begin{cases} 1 & \text{if } x \in E, \\ 0 & \text{if } x \notin E. \end{cases}$$

In general, we do not allow events to be *any* subset of the sample space. Rather, the set of possible events should satisfy some basic axioms.

Definition 1.2. *Let Ω be a nonempty set, and let $\mathcal{E}$ be a collection of subsets of Ω; i.e., $E \in \mathcal{E}$ means that $E \subset \Omega$. The set $\mathcal{E}$ is called an* algebra *if*

1. *the empty set $\emptyset \in \mathcal{E}$,*
2. *for any $E_1, E_2 \in \mathcal{E}$, the union $E_1 \cup E_2 \in \mathcal{E}$,*
3. *for any $E \in \mathcal{E}$, the complement $E^c \in \mathcal{E}$.*

An algebra is a σ-algebra if, in addition, it is closed under countable unions; i.e., if $E_1, E_2, \ldots \in \mathcal{E}$, then $\cup_{i=1}^{\infty} E_i \in \mathcal{E}$.

We are now ready to define a probability distribution over our sample space.

Definition 1.3. *A* probability space *is a triple $(\Omega, \mathcal{E}, \mathbb{P})$, where Ω is a nonempty set, called the* sample space, *$\mathcal{E}$ is a σ-algebra over Ω, and $\mathbb{P} : \mathcal{E} \to [0, 1]$ is a function, known as a* probability measure *or* probability distribution, *that satisfies the following axioms:*

1. *$\mathbb{P}(E) \geq 0$ for every event $E \subset \Omega$,*
2. *$\mathbb{P}(\Omega) = 1$,*
3. *If $E_1, E_2, \ldots$ are disjoint, then*

$$\mathbb{P}\left(\bigcup_{i=0}^{\infty} E_i\right) = \sum_{i=0}^{\infty} \mathbb{P}(E_i).$$

If the sample space Ω is finite, the function

$$\mathbb{P}(E) = \frac{|E|}{|\Omega|}$$

constitutes a probability distribution over Ω.

Definition 1.4. *Given a sample space Ω and a probability distribution $\mathbb{P}$, two events E and F are* independent *if*

$$\mathbb{P}(EF) = \mathbb{P}(E)\mathbb{P}(F).$$

Here, we use the shorthand $E \cap F = EF$.

Definition 1.5. *Given a probability distribution $\mathbb{P}$ and an event F with $\mathbb{P}(F) > 0$, The* conditional probability *of E given F is defined by*

$$\mathbb{P}(E|F) = \frac{\mathbb{P}(EF)}{\mathbb{P}(F)}.$$

The intuition behind this definition is that we are restricting our world of possible outcomes, i.e., our sample space, to the subset F. We then define our conditional probability as the fraction of F that is occupied by E. As an exercise, the reader may show that the function $\mathbb{P}(E|F)$ satisfies the three probability axioms for the reduced sample space F.

1.2 Random Variables

Definition 1.6. *Given a probability space $(\Omega, \mathcal{E}, \mathbb{P})$, a* random variable *$X$ is a mapping*

$$X : \Omega \to \mathbb{R}$$

that assigns a real number to each outcome, such that, for any real number c,

$$\{\omega : X(\omega) \le c\} \in \mathcal{E}.$$

A discrete *random variable is one that takes countably many values $\{x_0, x_1, \ldots\}$. A random variable that is not discrete is* continuous.

Example 1.2. Consider a single roll of two six-sided dice. The sample space is the set $\Omega = \{(i, j) : i, j = 1, \ldots, 6\}$. The random variable X is defined as the sum of the number on each die, i.e., $X = i + j$. This is a discrete random variable as its values take on only a countable (and finite) range: X = 2, 3, 4, 5, 6, 7, 8, 9, 10, 11, 12. ▷

Regardless of whether a random variable is continuous or discrete, we may define a related function that acts as a cumulative sum of probability over the values that the random variable takes.

Definition 1.7. *Given a random variable X, its associated* cumulative distribution function *(cdf) is the function $F_X : \mathbb{R} \to [0, 1]$ given by*

$$F_X(x) = \mathbb{P}(X \le x).$$

Next, we may define a *probability mass function* for discrete random variables and a *probability density function* for continuous random variables, as follows. We will show in Section 1.2 that this distinction is a bit of an illusion, and that probability density functions can actually be defined for both continuous and discrete random variables, though this formalism is not a standard approach.

Definition 1.8. *Given a discrete random variable X, its associated* probability mass function *(PMF) f_X is the function defined by*

$$f_X(x) = \mathbb{P}(X = x).$$

Example 1.3. Continuing Example 1.2, the probability mass function may be expressed as

$$f_X(x) = \frac{|\Omega_{X=x}|}{|\Omega|},$$

where $\Omega_{X=x}$ is the subset of the sample space with $X = x$. Since the size of the sample space is 36, the probability mass function takes the following values:

$$f(2) = f(12) = \frac{1}{36}, \qquad f(3) = f(11) = \frac{2}{36}, \qquad f(4) = f(10) = \frac{3}{36},$$

$$f(5) = f(9) = \frac{4}{36}, \qquad f(6) = f(8) = \frac{5}{36}, \qquad f(7) = \frac{6}{36}.$$

One can additionally verify that the probability mass function is correctly normalized:

$$\sum_{x=2}^{12} f(x) = 1.$$

Finally, the associated cumulative distribution may be expressed as

$$F(x) = \sum_{t=2}^{x} f(t).$$

$\triangleright$

Definition 1.9. *Given a* continuous *random variable X, its associated* probability density function *(PDF) f_X is the function that satisfies $f_X(x) \ge 0$ for all x,*

$$\int_{-\infty}^{\infty} f_X(x) \, dx = 1,$$

and

$$\int_a^b f_X(x) \ dx = \mathbb{P}(a < X < b),$$

for all $a < b$.

Note that, for continuous random variables, the cumulative distribution function may be represented as

$$F_X(x) = \int_{-\infty}^x f(t) \ dt.$$

Example 1.4. A continuous random variable X is called a *uniform random variable* on the interval $[a, b]$, denoted $X \sim \text{Unif}(a, b)$, if its PDF may be expressed as

$$f(x) = \frac{1}{b - a}$$

for $a < x < b$. ▷

Definition 1.10. *The* expected value *of a function $g : \mathbb{R} \to \mathbb{R}$ of a random variable X is given by*

$$\mathbb{E}[g(X)] = \sum g(x_i) f(x_i), \tag{1.1}$$

if the random variable is discrete, and by

$$\mathbb{E}[g(X)] = \int g(x) f(x) \ dx \tag{1.2}$$

if the random variable is continuous. In particular, we say that the expected value *of a random variable X is given by $\mathbb{E}[X]$, and we use Equations (1.1) or (1.2) with $g(x)$ replaced with x for the case of a discrete or continuous random variable, respectively.*

Example 1.5. Let $X \sim \text{Unif}(a, b)$. Then the expected value of X is given by

$$\mathbb{E}[X] = \frac{1}{b - a} \int_a^b x \, dx = \frac{b^2 - a^2}{2(b - a)} = \frac{a + b}{2}.$$

▷

Proposition 1.1. *Given random variables X and Y and constants a and b,*

$$\mathbb{E}[aX + bY] = a\mathbb{E}[X] + b\mathbb{E}[Y].$$

This follows immediately from the linearity property of integrals and summations.

Definition 1.11. *The* variance *of a random variable X is defined by*

$$\mathbb{V}(X) = \mathbb{E}\left[(X - \mathbb{E}[X])^2\right]. \tag{1.3}$$

Proposition 1.2. *Given random variables X and Y and constants a, b, and c,*

$$\mathbb{V}(aX + bY + c) = a^2\mathbb{V}(X) + b^2\mathbb{V}(Y).$$

Proposition 1.3. *The variance of a random variable may be expressed as*

$$\mathbb{V}(X) = \mathbb{E}[X^2] - \mathbb{E}[X]^2. \tag{1.4}$$

This follows by expanding the square and applying the linearity property of expectation as given by Proposition 1.1.

1.3 The Probability Density Function of a Discrete Random Variable

When considering discrete random variables, on invariably refers to their *probability mass functions*. However, we will do something surprising: we will show that discrete variables also have probability density functions. This will help us unify notation, so that we may consider both cases with a single stroke of the pen.

The "Dirac delta function," $\delta(t)$ is a generalized function that can be thought of as a singularity at the origin:

$$\delta(x) = \begin{cases} \infty & x = 0, \\ 0 & x \neq 0 \end{cases},$$

such that the resulting graph encloses unit area. One may think of the delta function as the limit of a family of standard normal distributions as $\sigma \to 0$. As a result, the delta function has the following properties:

$$\int_{-\infty}^{\infty} \delta(x)\,dx = 1 \quad \text{and} \quad \int_{-\infty}^{\infty} \delta(x)f(x)\,dx = f(0),$$

for any function $f(x)$ that is continuous at $x = 0$. We may further think of the delta function as the derivative of the Heaviside function, defined by

$$H(x) = \mathbb{I}[x \geq 0] = \begin{cases} 1 & \text{for } x \geq 0 \\ 0 & \text{for } x < 0 \end{cases},$$

where $\mathbb{I}$ is the *indicator function*, defined such that $\mathbb{I}[S] = 1$ if statement S is true, and 0 otherwise, for any logical expression S. Thus, integrating the delta function produces a "unit step" or a "unit response."

Now we can do something unexpected: write down the *probability density function* for a discrete random variable.

Proposition 1.4. *Given a discrete random variable X with probability mass function $p(x_i) = m_i$, defined on the countable set $\{x_0, x_1, \ldots\}$, its associated probability density function and cumulative distribution are given by*

$$f(x) = \sum_{i=0}^{\infty} m_i \delta(x - x_i) \qquad and \qquad F(x) = \sum_{i=0}^{\infty} m_i H(x - x_i). \qquad (1.5)$$

We may refer to the individual terms $m_i \delta(x - x_i)$ as point masses, *and the delta functions $\delta(x - x_i)$ as* unit point masses.

This is a very useful result, as now we may represent the expected value of a function by the formula

$$\mathbb{E}[g(X)] = \int_{-\infty}^{\infty} g(x)dF(x),$$

regardless whether the random variable X is discrete or continuous. This follows since, in the discrete case, we have

$$\mathbb{E}[g(X)] = \int_{-\infty}^{\infty} g(x)dF(x) = \sum_{i=0}^{\infty} \int_{-\infty}^{\infty} m_i g(x) \delta(x - x_i) \, dx = \sum_{i=0}^{\infty} g(x_i)f(x_i).$$

Using this notation, we may further express the expected value of a random variable X, as given by Definition 1.10, in the form

$$\mathbb{E}[X] = \int x \, dF(x) = \begin{cases} \sum x_i f(x_i) & \text{if } X \text{ is discrete} \\ \int x f(x) \, dx & \text{if } X \text{ is continuous} \end{cases}. \qquad (1.6)$$

Note 1.1. A more theoretical development of this framework involves measure theory. For our purposes, however, we find that using delta functions and Heaviside functions to be sufficient to unify the theory of discrete and continuous random variables into a common framework.   ▷

1.4 Multiple Random Variables

1.4.1 Joint and Marginal Distributions

Of course, it is possible to define more than one random variable on a given sample space.

Definition 1.12. *Given a discrete bivariate random vector (X, Y), the* joint probability mass function *(joint pmf) is the function $f : \mathbb{R}^2 \to \mathbb{R}$ defined by*

$$f_{X,Y}(x, y) = \mathbb{P}(X = x, Y = y).$$

When we are at no risk of confusion, we may drop the subscript and refer to this function simply as $f(x, y)$.

Example 1.6. Let us consider again the roll of two dice from Example 1.2. Suppose that the two dice are easily discernible; perhaps the first die is red and the second is blue. Let X represent the outcome of the red die, and Y the outcome of the blue die. Here, X and Y are both discrete random variables, which may take the values $\{1, 2, 3, 4, 5, 6\}$. The probability mass function is given by

$$f(x_i, y_j) = \frac{1}{36}, \qquad \text{for } x_i, y_j = 1, 2, 3, 4, 5, 6.$$

Notice that each die is *independent* of the other die. (We will define this more formally momentarily.) In fact, the probability mass function for X is simply $f_X(x) = 1/6$, and the probability mass function for Y is $f_Y(y) = 1/6$. This captures the intuitive meaning of *independence*, i.e.,

$$f_{X,Y}(x, y) = f_X(x) f_Y(y),$$

and *marginalization*, i.e.,

$$f_X(x) = \sum_{y=1}^{6} f_{X,Y}(x, y) \qquad \text{and} \qquad f_Y(y) = \sum_{x=1}^{6} f_{X,Y}(x, y).$$

▷

Definition 1.13. *Given a continuous bivariate random vector* (X, Y), *a function* $f : \mathbb{R}^2 \to \mathbb{R}$ *is the* joint probability density function *(joint pdf) if, for every* $A \subset \mathbb{R}^2$,

$$\mathbb{P}((X, Y) \in A) = \iint_A f_{X,Y}(x, y)\, dx\, dy.$$

When there is no risk of confusion, we may drop the subscript and refer to this function as $f(x, y)$.

Let us take a step back. Since X and Y are both random variables over some sample space Ω, they must each have their own probability mass or probability density function. Our next definition shows us how to reconstruct these individual probability mass or density functions given only the joint distribution.

Definition 1.14. *Given a bivariate random vector* (X, Y) *and is joint probability mass or probability density function* $f_{X,Y}(x, y)$. *If the random variables are discrete, the* marginal probability mass functions *of* X *and* Y *are given by*

$$f_X(x) = \sum_{y} f_{X,Y}(x, y) \qquad \text{and} \qquad f_Y(y) = \sum_{x} f_{X,Y}(x, y). \qquad (1.7)$$

Similarly, if the random variables are continuous, we define the marginal probability density functions *of X and Y by the equations*

$$f_X(x) = \int f_{X,Y}(x,y)\, dy \qquad and \qquad f_Y(y) = \int f_{X,Y}(x,y)\, dx. \qquad (1.8)$$

We refer to the process of transforming the joint probability mass or probability density function to the univariate probability mass or probability density functions as marginalization.

1.4.2 Independence

Definition 1.15. *Given random variables X and Y, we say that the random variables X and Y are* independent, *denoted $X \perp\!\!\!\perp Y$, if the joint distribution can be factored as*

$$f(x,y) = f_X(x)f_Y(y). \qquad (1.9)$$

Similarly, given a multivariate random vector $(X_1, \ldots, X_n)$ with joint pmf or pdf $f(x_1, \ldots, x_n)$, we say that the random variables $X_1, \ldots, X_n$ are independent if

$$f(x_1, \ldots, x_n) = \prod_{i=1}^{n} f_{X_i}(x_i). \qquad (1.10)$$

Definition 1.16. *We say that a set of random variables $X_1, \ldots, X_n$ are* independent and identically distributed *(or* IID*) if they are independent and if each marginal distribution follows the same form, i.e., $F_{X_i}(x) = F_{X_j}(x) = F(x)$ for all $i, j = 1, \ldots, n$. In such a case, we may also say that the random vectors form a* random sample *of the distribution F, and we denote this $X_1, \ldots, X_n \sim F$.*

Example 1.7. Consider tossing a coin n times. Let $X_i \in \{0, 1\}$ represent the outcome of the ith coin toss. Then $X_1, \ldots, X_n$ are IID as they each share the same distribution

$$F(x) = \frac{1}{2}\left(H(x) + H(x-1)\right).$$

Alternatively, they share a common probability mass function: $f(0) = f(1) = 1/2$. Independence implies that the result of one coin toss does not influence another.   ▷

1.4.3 Covariance and Correlation

Definition 1.17. *Let X and Y be random variables with means μ_X and μ_Y and standard deviations σ_X and σ_Y. The* covariance *between X and Y is defined by*

$$\mathrm{COV}(X,Y) = \mathbb{E}\left[(X - \mu_X)(Y - \mu_Y)\right]. \tag{1.11}$$

The correlation *between X and Y is defined by*

$$\rho = \rho(X,Y) = \frac{\mathrm{COV}(X,Y)}{\sigma_X \sigma_Y}. \tag{1.12}$$

Finally, we say that the random variables X and Y are uncorrelated *if* $\mathrm{COV}(X,Y) = 0$.

Proposition 1.5. *Given random variables X and Y, the covariance satisfies*

$$\mathrm{COV}(X,Y) = \mathbb{E}(XY) - \mathbb{E}(X)\mathbb{E}(Y). \tag{1.13}$$

The correlation satisfies

$$-1 \le \rho(X,Y) \le 1. \tag{1.14}$$

Proposition 1.6. *Any two independent random variables $X \perp\!\!\!\perp Y$ are uncorrelated, i.e.,*

$$\mathbb{E}[XY] = \mathbb{E}[X]\mathbb{E}[Y].$$

The converse it not necessarily true.

Definition 1.18. *Let $X = (X_1, \ldots, X_n) \in \mathbb{R}^n$ be a random vector. Then its* variance–covariance matrix[1] *is defined by*

$$\mathbb{V}(X) = \mathbb{E}[(X - \mathbb{E}[X])(X - \mathbb{E}[X])^T]; \tag{1.15}$$

i.e., the matrix $\mathbb{V}(X)$ whose ijth component is given by $[\mathbb{V}(X)]_{ij} = \mathrm{COV}(X_i, X_j)$.

Proposition 1.7. *Let $X = (X_1, \ldots, X_n)$ be a random vector. Then the variance of X satisfies*

$$\mathbb{V}(X) = \mathbb{E}[XX^T] - \mathbb{E}[X]\mathbb{E}[X]^T. \tag{1.16}$$

Proof. Expanding Equation (1.15), we have

$$\mathbb{V}(X) = \mathbb{E}\left[XX^T - X\mathbb{E}[X]^T - \mathbb{E}[X]X^T + \mathbb{E}[X]\mathbb{E}[X]^T\right].$$

Simplifying, while noting that $\mathbb{E}[X^T] = \mathbb{E}[X]^T$, yields the result.   □

1.4.4 Conditional Distributions and Expectation

In the case of two random variables, we previously saw how one might *marginalize* over one of the random variables in order to obtain the probability function of the other. This is tantamount to taking a probability-weighted average over all possible values of the random variable one is marginalizing over, thereby accounting for each of these possibilities. Next, we consider how one may *condition* over one of the random variables, by which we mean restricting the sample space to a subset consistent with a given value of one of the random variables. More formally, we have the following.

---

[1] Often referred to simply as the *covariance matrix* for X.

Definition 1.19. *Given a bivariate random vector (X, Y) with joint pmf or pdf $f(x, y)$ and associated marginal pmf or pdf $f_X(x)$, we define the conditional pmf or pdf of X given $Y = y$ as the function*

$$f_{X|Y}(x|y) = \frac{f_{X,Y}(x, y)}{f_Y(y)}, \tag{1.17}$$

whenever $f_Y(y) \neq 0$.

Note 1.2. Observe that

$$f_{X|Y}(x|y) = \mathbb{P}(X = x|Y = y) = \frac{\mathbb{P}(X = x, Y = y)}{\mathbb{P}(Y = y)}.$$

Thus, the conditional distribution $f(X|Y = y)$ can be viewed as the resulting (re-normalized) probability distribution on the subset $\Omega_{Y=y} = \{s \in \Omega : Y(s) = y\}$ of the sample space Ω that satisfies $Y = y$. ▷

Note 1.3. If $f_Y(y) = 0$, this means that the subset $\Omega_{Y=y} \subset \Omega$ is either the empty set or occurs with probability zero, i.e., $\mathbb{P}(Y = y) = 0$. Thus, defining a probability distribution for X on this subset is ill defined since the outcome $Y = y$, according to our probability measure, cannot occur. ▷

Definition 1.20. *Given random variables X and Y, the conditional expectation of X given $Y = y$ is the expected value of X relative to $f_{X|Y}(x|y)$, i.e.,*

$$\mathbb{E}[X|Y = y] = \begin{cases} \sum x f_{X|Y}(x|y) & discrete\ case \\ \int x f_{X|Y}(x|y)\, dx & continuous\ case \end{cases}. \tag{1.18}$$

If we allow the random variable Y to vary, the conditional expectation $\mathbb{E}[X|Y]$ may be thought of as a function of the random variable Y, and therefore a random variable itself. This motivates the following result.

Theorem 1.1. *Given two random variables X and Y, we have*

$$\mathbb{E}[X] = \mathbb{E}\left[\mathbb{E}[X|Y]\right]. \tag{1.19}$$

The right-hand side of Equation (1.19) is interpreted as follows: The inner expectation, $\mathbb{E}[X|Y]$ is viewed as a random function of Y (i.e., a function of the random variable Y). Therefore the outer expectation is taken over the random variable Y, i.e.,

$$\mathbb{E}\left[\mathbb{E}[X|Y]\right] = \int \mathbb{E}[X|Y = y] f_Y(y)\, dy.$$

As we gather more and more samples from the distribution F, we can construct a histogram of our point estimator $\hat{\theta}$. This is (an approximation of) the sampling distribution.

In short: since a point estimator $\hat{\theta}_n$ is a function of a number of random variables, the point estimator itself is a random variable, and as such, it has its own distribution.

1.5 Sampling

1.5.1 Point Estimators and Sampling Distributions

Often we do not know the form of a distribution, but are instead given a *sample* of values from the distribution, and are tasked with using our data to draw statistical inferences. In particular, we are often interested in inferring the distribution that generated our data, i.e., given a sample $X_1, \ldots, X_n \sim F$, how do we infer the distribution F?

Definition 1.21. *Given a set of data* $\mathcal{X} = \{X_1, \ldots, X_n\}$, a statistic *is any function of the data.*

Definition 1.22. *Given some unobserved quantity of interest θ—it could be an unknown parameter of the distribution, the distribution itself, our some property of the distribution—a* point estimator $\hat{\theta}_n$ *is any statistic that estimates θ. Sometimes we may denote the point estimator as $\hat{\theta}_n$ to emphasize the number of data points in our sample.*

Suppose we have a data set $X_1, \ldots, X_n \sim F$ that was generated by some unknown distribution F, and we have an estimator $\hat{\theta}_n$ of some unknown quantity of interest θ. Of course we will want to know *how good* our estimator actually is. It is therefore natural to be interested in the *distribution of* $\hat{\theta}_n$.

Definition 1.23. *Given a point estimator $\hat{\theta}_n$ of some quantity of interest θ, the distribution of $\hat{\theta}_n$ is called the* sampling distribution, *and the standard deviation of $\hat{\theta}_n$ is called the* standard error:

$$\mathrm{se}(\hat{\theta}_n) = \sqrt{\mathbb{V}(\hat{\theta}_n)}.$$

But what do we mean by "the distribution of $\hat{\theta}_n$"? Isn't our point estimate $\hat{\theta}_n$ just a number?

As a statistic, our point estimator is (clearly) a function of the data, i.e., $\hat{\theta}_n = \hat{\theta}_n(X_1, \ldots, X_n)$. As such, in practice, we are only able to observe the single value of our statistic. However, let us do a little thought experiment. Suppose we are able to draw a second random sample from our distribution, say $X_{12}, X_{22}, \ldots, X_{n2} \sim F$. We can then obtain a second estimator $\hat{\theta}_n^2$ of the unknown quantity θ. But why stop at two? What if we were able to generate many many samples from our distribution, i.e., suppose

$$X_{1j}, \ldots, X_{nj} \sim F, \qquad \text{for } j = 1, \ldots, M,$$

for some large integer M. Then we could compute $\hat{\theta}_n^j = \hat{\theta}_n(X_{1j}, \ldots, X_{1n})$, for $j = 1, \ldots, M$. (Forgive the slight abuse of notation: by $\hat{\theta}_n^j$ we currently mean the point estimator from the jth sample—not the point estimator to the power j.)

1.5.2 Sample Mean and Sample Standard Deviation

The most common examples of statistical estimators are, of course, the sample mean and sample variance of a set of data.

Definition 1.24. *Given a set of random variables* $X_1, \ldots, X_n$, *we define the* sample mean *by*

$$\overline{X}_n = \frac{1}{n} \sum_{i=1}^{n} X_i. \tag{1.20}$$

Definition 1.25. *Given a set of random variables* $X_1, \ldots, X_n$, *we define the* sample variance *by*

$$S_n^2 = \frac{1}{n-1} \sum_{i=1}^{n} \left(X_i - \overline{X}_n \right)^2. \tag{1.21}$$

Naturally, the sample standard deviation *is* S_n.

If our data is an independent and identically distributed sample from some distribution F, the sample mean and sample variance are estimators of the mean and variance of the originating distribution, as we will see in our next theorem. But first, let us state an important lemma that will be useful later, when proving the theorem.

Lemma 1.1. *Let* X *be a random variable with expected value* $\mathbb{E}[X] = \mu$ *and variance* $\mathbb{V}(X) = \sigma^2$. *Then*

$$\mathbb{E}[X^2] = \sigma^2 + \mu^2. \tag{1.22}$$

Proof. This follows immediately from Proposition 1.3. $\qquad\square$

Theorem 1.2. *Let* $X_1, \ldots, X_n \sim F$ *be* IID *and suppose* $\mathbb{E}[X_i] = \mu$ *and* $\mathbb{V}(X_i) = \sigma^2$. *Then*

$$\mathbb{E}[\overline{X}_n] = \mu, \qquad \mathbb{V}(\overline{X}_n) = \frac{\sigma^2}{n}, \qquad \mathbb{E}[S_n^2] = \sigma^2. \tag{1.23}$$

Proof. The first two parts of Equation (1.23) are easy. By Proposition 1.1, we have

$$\mathbb{E}[\overline{X}_n] = \mathbb{E}\left[\frac{1}{n} \sum_{i=1}^{n} X_i \right] = \frac{1}{n} \sum_{i=1}^{n} \mathbb{E}[X_i] = \frac{1}{n} \sum_{i=1}^{n} \mu = \mu.$$

Similarly, by Proposition 1.2, we have

$$\mathbb{V}(\overline{X}_n) = \frac{1}{n^2} \sum_{i=1}^{n} \mathbb{V}(X_i) = \frac{\sigma^2}{n}.$$

For the third part, let us consider the random variable

$$Y = \sum_{i=1}^{n}(X_i - \overline{X}_n)^2 = \sum_{i=1}^{n}\left(X_i^2 - 2X_i\overline{X}_n + \overline{X}_n^2\right).$$

(This is just the sample variance times the divisor $n - 1$.) Summing over the second and third term, we find this expression to be equivalent to

$$Y = \sum_{i=1}^{n}X_i^2 - n\overline{X}_n^2.$$

From Lemma 1.1, we have $\mathbb{E}[X_i^2] = \sigma^2 + \mu^2$, so that

$$\mathbb{E}\left[\sum_{i=1}^{n}X_i^2\right] = n\sigma^2 + n\mu^2.$$

We may also apply Lemma 1.1, along with the results of the first two parts of this theorem, to the random variable $\overline{X}_n^2$, thereby obtaining

$$\mathbb{E}\left[\overline{X}_n^2\right] = \frac{\sigma^2}{n} + \mu^2.$$

Combining the above three equations, we find

$$\mathbb{E}[Y] = n\sigma^2 + n\mu^2 - \sigma^2 - n\mu^2 = (n-1)\sigma^2,$$

thereby proving the result.   □

Example 1.8. In this example, we use built-in Python functions to generate samples from a given distribution. In particular, we will generate samples from a normal distribution with mean $\mu = 3$ and standard deviation $\sigma = 2$ (variance $\sigma^2 = 4$). We will pretend, however, that these values are unknown, and see how well we can infer them from the data.

First, let us compute the sample mean and sample variance from five samples of various size, as shown in Code Block 1.1.   Line 3 generates

```
np.random.seed(444)
for i in range(1, 6):
    x = np.random.normal(loc=3, scale=2, size=10**i)
    print('sample size:', 10**i, '- sample mean:',
          x.mean().round(3), '- sample variance:', x.var().round(3))
```

Code Block 1.1: Example 1.8

a random sample (from a normal distribution with $\mu = 3$, $\sigma = 2$) of a particular size. Line 4 then prints the sample size along with the sample mean and sample standard variance of the given sample. The output is displayed below.

```
sample size: 10 - sample mean: 3.108 - sample variance: 2.563
sample size: 100 - sample mean: 3.017 - sample variance: 3.969
sample size: 1000 - sample mean: 2.959 - sample variance: 4.261
sample size: 10000 - sample mean: 2.969 - sample variance: 3.951
sample size: 100000 - sample mean: 3.004 -sample variance: 4.032
```

We see that as the sample size n increases, our (single, random) point estimate of the mean and variance improves. It is important to note that the sample variance stabilizes as the sample size increases. For any sample size n, the sample variance is a point estimate of the population (true) variance, and the larger the sample size, the better this estimate is. Similarly, the larger the sample size, the better the sample mean approximates the population mean.

It is important to point out that $\overline{X}_n$ is not *just a number*—it is a random variable, as it depends on the data. If we seed the random number generator (line 2) with a different integer, it will produce different sample means and sample variances. And, in general, if we keep drawing random samples, of a fixed size, from a distribution, the sample mean (and sample variance) will be different for each one.

So how good of an approximation is the sample mean? This depends on the population variance and the sample size, according to the middle part of Equation (1.23). To see this in action, we must consider what is meant by: *the variance of the sample mean.*

Instead of considering a single sample of size n, what if we consider many samples of the same size and compute the sample mean for each one? We could then monitor the behavior as the sample size n increases. Let's

```
1  np.random.seed(444)
2  n_samples = 10000
3  for i in range(1, 6):
4      sample_means = np.zeros(n_samples)
5      for j in range(n_samples):
6          x = np.random.normal(loc=3, scale=2, size=10**i)
7          sample_means[j] = x.mean()
8      print('sample size:', 10**i, '- avgerage:',
              sample_means.mean().round(3), '- variance',
              sample_means.var().round(6))
```

Code Block 1.2: Example 1.8 (cont.)

walk through Code Block 1.2. The index i fixes the sample size $n = 10^i$. For a given sample size (i.e., for each i), we then initialize (line 4) an empty vector of length n_samples $= 10,000$, where n_samples is the number of samples we will generate for each i. We then create an inner for-loop that runs from $j = 1, \ldots, \texttt{n\_samples}$.

The output is recorded, below. (For brevity, "average" means the average, or sample mean, of the sample of sample means, and "variance" means the sample variance of the sample of sample means. The code in Code Block 1.2 should illuminate the meaning of this.)

```
sample size: 10 - average: 2.997 - variance: 0.410871
sample size: 100 - average: 3.003 - variance: 0.040004
sample size: 1000 - average: 3.001 - variance: 0.004055
sample size: 10000 - average: 3.0 - variance: 0.000395
sample size: 100000 - average: 3.0 - variance: 0.00004
```

In Code Block 1.2, we consider five different sample sizes, ranging from 10 to 100,000. For a given sample size, the sample mean $\overline{X}_n$ is a random variable. To sample this random variable , say, 10,000 times, we generate 10,000 samples of size n from the original distribution and compute the sample mean of each one. This results in a sample of 10,000 sample means $\overline{X}_n$, for each sample size n. This gives us a feel for how good our estimator $\overline{X}_n$ is at a given sample size. As expected, the variance of the sample of sample means decreases like $1/n$, in agreement with Equation (1.23).

For a given sample size, we may further plot a histogram of the outputs $\texttt{sample\_means}$, which will aid in our visualization of the sample mean as a random variable. We have done this for the cases $n = 10$ and $n = 100$ in Figure 1.1. As the sample size increases, the range of values obtained for $\overline{X}_n$ in accordance with Theorem 1.2. The code used to generate Figure 1.1 is given in Code Block 1.3.

In practice, of course, we typically only have a single sample from our unknown distribution and, therefore, only a single sample mean. By generating many sample means from a known distribution, however, we gain an insight into how far off any one particular sample mean might be from the true mean. Moreover, we see how this error decreases as the sample size increases.   ▷

Note 1.4. In Figure 1.1, we see that the distribution of the sample mean $\overline{X}_n$ is indistinguishable from a normal distribution, i.e., $\overline{X}_n \sim \mathcal{N}(\mu, \sigma^2/n)$. We will see later that this is *exactly* true whenever the sample $X_1, \ldots, X_n$ arises from a normal distribution. We will further see, when we discuss the central limit theorem, that the result is *approximately* true, regardless of the originating distribution, as the sample size n becomes large. A typical rule of thumb is that we may approximate the distribution of $\overline{X}_n$ with the normal distribution $\mathcal{N}(\mu, \sigma^2/n)$ for $n > 30$.   ▷

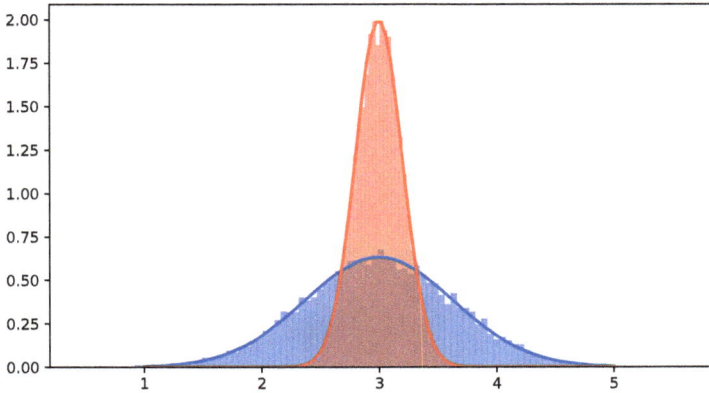

Fig. 1.1: Histogram of sample means for $n = 10$ (blue) and $n = 100$ (red). Solid curves are normal distributions with mean $\mu = 3$ and variance $\sigma^2 = 0.4$ (blue) and $\sigma^2 = 0.04$ (red).

```
plt.figure(figsize=(8, 9/2))
n_samples = 10000
sample_means_10 = np.zeros(n_samples)
sample_means_100 = np.zeros(n_samples)
for j in range(n_samples):
    x = np.random.normal(loc=3, scale=2, size=10)
    sample_means_10[j] = x.mean()
    x = np.random.normal(loc=3, scale=2, size=100)
    sample_means_100[j] = x.mean()
plt.hist(sample_means_10, color='b', bins=100, alpha=0.6,
    normed=True)
plt.hist(sample_means_100, color='r', bins=50, alpha=0.6,
    normed=True)
x = np.linspace(1, 5, num=100)
plt.plot(x, scipy.stats.norm.pdf(x, loc=3, scale=2/np.sqrt(10)),
    'b', linewidth=2)
plt.plot(x, scipy.stats.norm.pdf(x, loc=3, scale=2/np.sqrt(100)),
    'r', linewidth=2)
```

Code Block 1.3: To generate Figure 1.1

1.5.3 Law of Large Numbers

It turns out that, equipped with the appropriate notion of convergence, we can show that the sample mean converges to the population mean as the sample size approaches infinity.

Definition 1.26. *Let $X_1, X_2, \ldots$ be a sequence of random variables, with respective distributions $X_n \sim F_n$, and let $X \sim F$ be a separate random variable.*

1. *We say that the sequence $\{X_n\}$ converges to X in probability if, for every $\varepsilon > 0$, we have*

$$\lim_{n \to \infty} \mathbb{P}(|X_n - X| > \varepsilon) = 0;$$

2. *We say that the sequence $\{X_n\}$ converges to X in distribution if*

$$\lim_{n \to \infty} F_n(x) = F(x),$$

for the set of x at which the distribution F is continuous.

Theorem 1.3 (Law of Large Numbers). *The sample mean $\overline{X}_n$ converges in probability to the population mean $\mathbb{E}[X_i]$ as $n \to \infty$.*

For a proof, see Wasserman [2004]. The following useful theorem is also stated without proof.

Theorem 1.4 (Slutsky's Theorem). *Given two sequences $\{X_n\}_{n=1}^{\infty}$ and $\{Y_n\}_{n=1}^{\infty}$ of random variables with the properties that $X_n \to X$ in distribution and $Y_n \to c$ in probability, then*

(a) $X_n Y_n \to cX$ in distribution, and
(b) $X_n + Y_n \to X + c$ in distribution.

1.5.4 Bias

At first pass, it may appear strange that we use $n-1$ as opposed to n as the divisor in the definition of the sample variance, as given by Definition 1.25. This choice is made, however, so that this estimator is *unbiased*, a concept we shall introduce momentarily.

Definition 1.27. *Given a point estimator $\hat{\theta}_n$ of a quantity θ, estimated from an IID set of data $X_1, \ldots, X_n$, the bias of the estimator is defined by*

$$\text{bias}(\hat{\theta}_n) = \mathbb{E}_\theta[\hat{\theta}_n] - \theta. \tag{1.24}$$

Note 1.5. The expectation $\mathbb{E}_\theta$ is taken with respect to the distribution that generated the data, i.e.,

$$f(x_1, \ldots, x_n; \theta) = \prod_{i=1}^{n} f(x_i; \theta),$$

where θ is a fixed, often unknown, quantity. In other words, given the true distribution (which depends on θ), what is the expected value of the estimator over all possible n-point samples from the distribution? For the following discussion, we shall drop the subscript θ from $\mathbb{E}_\theta$, as there is no danger for ambiguity.   ▷

Note 1.6. From Theorem 1.2, it follows immediately that the sample mean and sample variance, as defined in Definitions 1.24 and 1.25, are unbiased estimators of the population mean and population variance, respectively.   ▷

A useful measure of accuracy is the mean-squared error, defined as follows.

Definition 1.28. *The* mean-squared error *of a point estimator $\hat{\theta}_n$ is given by*

$$\mathrm{mse}(\hat{\theta}_n) = \mathbb{E}[(\theta - \hat{\theta}_n)^2]. \tag{1.25}$$

The definition of bias for a point estimator leads to our first encounter of a result that will be important in broader contexts later on during our discussion of machine learning.

Theorem 1.5 (Bias–Variance Tradeofff). *Given a point estimator $\hat{\theta}_n$ of an unknown, fixed quantity θ, the mean-squared error can be expressed in terms of the bias and variance of the estimator as follows:*

$$\mathrm{mse}(\hat{\theta}_n) = \mathrm{bias}(\hat{\theta}_n)^2 + \mathbb{V}(\hat{\theta}_n)^2. \tag{1.26}$$

Proof. We begin by expanding the quantity $(\theta - \hat{\theta}_n)^2$ as follows:

$$(\theta - \hat{\theta}_n)^2 = (\theta - \mathbb{E}[\hat{\theta}_n] + \mathbb{E}[\hat{\theta}_n] - \hat{\theta}_n)^2$$
$$= (\theta - \mathbb{E}[\hat{\theta}_n])^2 + (\mathbb{E}[\hat{\theta}_n] - \hat{\theta}_n)^2 - 2(\theta - \mathbb{E}[\hat{\theta}_n])(\hat{\theta}_n - \mathbb{E}[\hat{\theta}_n]).$$

Now, since θ and $\mathbb{E}[\hat{\theta}_n]$ are constants, we have

$$\mathbb{E}\left[(\theta - \mathbb{E}[\hat{\theta}_n])(\hat{\theta}_n - \mathbb{E}[\hat{\theta}_n])\right] = (\theta - \mathbb{E}[\hat{\theta}_n])\mathbb{E}\left[(\hat{\theta}_n - \mathbb{E}[\hat{\theta}_n])\right] = 0.$$

Thus, the mean-squared error may be expressed as

$$\mathbb{E}\left[(\theta - \hat{\theta}_n)^2\right] = (\theta - \mathbb{E}[\hat{\theta}_n])^2 + \mathbb{E}\left[(\hat{\theta}_n - \mathbb{E}[\hat{\theta}_n])^2\right],$$

and the result follows from the definitions of bias and variance.   □

Note 1.7. As a result of Theorem 1.5, the mean-squared error of an unbiased estimator is simply equal to its variance, i.e.,

$$\mathrm{mse}(\hat{\theta}_n) = \mathbb{E}[(\hat{\theta}_n - \theta)^2] = \mathbb{E}[(\hat{\theta}_n - \mathbb{E}[\hat{\theta}_n])^2] = \mathbb{V}(\hat{\theta}_n), \qquad \text{if } \mathbb{E}[\hat{\theta}_n] = \theta.$$

▷

1.6 The Empirical Distribution

In this section, we discuss estimating both the distribution function and certain functions of the distribution function from data. For a more details, see Wasserman [2004] and Wasserman [2006].

1.6.1 Estimating Distributions from Data

We next address how one may estimate a distribution given a sample of values generated by the distribution, without making any assumptions as to the underlying form of the distribution.

Definition 1.29. *Let* $X_1, \ldots, X_n$ *be a data set. The* empirical distribution function $\hat{F}_n$ *is the cumulative distribution function over* $\mathbb{R}$ *given by the function*

$$\hat{F}_n(x) = \frac{1}{n} \sum_{i=1}^{n} \mathbb{I}[X_i \leq x]. \tag{1.27}$$

Note 1.8. Suppose the data are IID from a common (oftentimes unknown) distribution, i.e., $X_1, \ldots, X_n \sim F$. In this case, the empirical distribution $\hat{F}_n$ is a point estimator of the original distribution F: it is a statistic, as it is a function of the observed data, and it seeks to estimate an underlying quantity of interest, namely the distribution itself.   ▷

Note 1.9. In the language of Section 1.3, the empirical distribution is simply the distribution obtained by placing an equally weighted *point mass* at each of the observed data points $X_1, \ldots, X_n$.   ▷

We can actually code Definition 1.29 in a single line, as shown in Code Block 1.4. In this code, the variable x is a scalar, whereas data is a vector

```
1   def emp_dist(x, data):
2       # x (float) point to evaluate empirical distribution
3       # data (array) observed data
4       # returns a float
5       return (data <= x).sum() / len(data)
```

Code Block 1.4: Empirical distribution (at a point)

(np.array). The logical evaluation data <= x returns a Boolean vector, i.e., a vector whose components are either True or False. Summing over this vector adds the number of True instances, thereby performing the calculation in Equation (1.27). The problem, however, is that if one were to pass a vector in for the first argument, the function would fail. We would

```
1  def emp_dist(x, data):
2      # x (float or array) point to evaluate empirical distribution
3      # data (array) observed data
4      # returns an array equal in size to x
5      data = data.reshape((len(data), 1))
6      if np.isscalar(x):
7          x = np.array([x])
8      x = x.reshape((len(x), 1))
9      return (data <= x.T).sum(axis=0) / len(data)
```

Code Block 1.5: Empirical distribution (vectorized)

therefore like to take this one step further, by *vectorizing* the function given in Code Block 1.4, as done in Code Block 1.5. By reshaping x and data as column vectors, and then taking the transpose of x, the operation data <= x.T suddenly makes sense, and may further be summed down the rows (axis=0). We invite the reader to dissect this code, playing with the outputs in iPython console, one line at a time. For an np.array object, the shape method is also useful: in particular, one may invoke the call x.shape() before and after the reshape call to see its effect.

Example 1.9. We can use the empirical distribution function from Code Block 1.5 to plot the empirical distribution from scikitlearn's Boston housing price dataset, as shown in Code Block 1.6. The result is shown in Figure 1.2. (We will explain the meaning of the dashed lines later, in Example 1.14.)

```
1  boston = sklearn.datasets.load_boston()
2  df = DataFrame(boston['data'], columns=boston['feature_names'])
3  df.head() # shows top 5 rows in DataFrame
4  print( boston['DESCR'] ) # prints description of data set
5  y = boston['target'] # target variable
6  x = np.linspace(0, 60)
7  F_hat = emp_dist(x, y)
8  plt.plot(x, F_hat)
```

Code Block 1.6: Computing the Empirical Distribution

▷

1.6.2 Bias and Variance of the Empirical Distribution

Since the empirical distribution $\hat{F}_n$, generated from a set of data $X_1, \ldots, X_n$, is a point estimator of the true (often unknown) distribution, it is natural

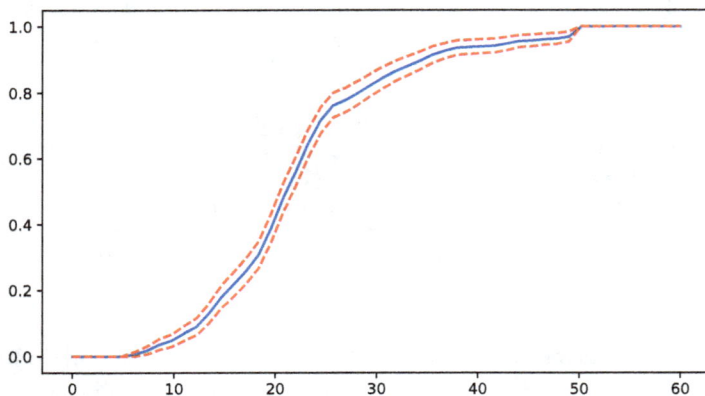

Fig. 1.2: Empirical Distribution of the Boston housing dataset.

to ask what its expected value and variance is, similar to what we achieved with Theorem 1.2.

Theorem 1.6. *Given a set of* IID *data* $X_1, \ldots, X_n \sim F$, *its empirical distribution* $\hat{F}_n$, *and a fixed point* $x \in \mathbb{R}$, *the expectation and variance are given by*

$$\mathbb{E}\left[\hat{F}_n(x)\right] = F(x), \tag{1.28}$$

$$\mathbb{V}\left(\hat{F}_n(x)\right) = \frac{F(x)(1 - F(x))}{n}. \tag{1.29}$$

Equation (1.28) implies that $\hat{F}$ is unbiased. Consequently, from Theorem 1.5, it follows that $\mathrm{mse}(\hat{F}_n(x)) = \mathbb{V}(\hat{F}_n(x))$. We will postpone the proof until our discussion of the binomial random variable.

1.6.3 Plug-in Estimators

One may be interested in many quantities associated with a probability distribution: its mean, variance, quantiles, etc. We next provide a definition of such quantities and then proceed to show how they may be estimated from the empirical distribution.

Definition 1.30. *A* statistical functional *is any real-valued function of a distribution.*

Example 1.10. The expected value and variance of a random variable X is a functional of its probability distribution F, as

$$\mu = \mu(F) = \int x \, dF \quad \text{and} \quad \sigma^2 = \sigma^2(F) = \int (x - \mu(F))^2 \, dF.$$

Hence, the mean and variance may be expressed as a functional of the distribution itself.   ▷

Definition 1.31. *Given a statistical functional* $\theta = T(F)$ *and an* IID *dataset* $X_1, \ldots, X_n \sim F$, *the plug-in estimator of* θ *is defined by*

$$\hat{\theta}_n = T(\hat{F}), \tag{1.30}$$

where $\hat{F}$ *is the empirical distribution generated by the dataset.*

A plug-in estimator is simply the point estimator of a given functional obtained by using the empirical distribution in lieu of the actual distribution, which is often unknown.

Proposition 1.8. *If* $T(F)$ *is a linear functional, i.e., a functional of the form* $\theta = T(F) = \int r(x) \, dF$, *and* $X_1, \ldots, X_n \sim F$ *is an* IID *sample from* F, *then the plug-in estimator is given by*

$$\hat{\theta}_n = \int r(x) \, d\hat{F}_n = \frac{1}{n} \sum_{i=1}^{n} r(x_i). \tag{1.31}$$

Proof. This result follows immediately from our discussion in Section 1.3 and the definition of the empirical distribution. Definition 1.29 is equivalent to the CDF generated by a point mass of strength $1/n$ placed at each datum:

$$\hat{F}_n(x) = \frac{1}{n} \sum_{i=1}^{n} \mathbb{I}[x_i \leq x] = \frac{1}{n} \sum_{i=1}^{n} H(x - x_i).$$

Hence $\hat{f}_n(x) = \hat{F}_n'(x) = \frac{1}{n} \sum \delta(x - x_i)$, and

$$\hat{\theta} = \int r(x) \, d\hat{F}_n = \sum_{i=1}^{n} \int r(x) \delta(x - x_i) \, dx = \frac{1}{n} \sum_{i=1}^{n} r(x_i),$$

and the result follows.   □

Example 1.11. Recalling that the empirical distribution is the distribution that places a point mass of strength $1/n$ at each data point $X_1, \ldots, X_n$, we may express the plug-in estimator for the mean as

$$\hat{\mu}_n = \int x \, d\hat{F}_n = \frac{1}{n} \sum_{i=1}^{n} x_i,$$

which is equivalent to the sample mean. (The calculus of the preceding equation was discussed in Section 1.3 and Proposition 1.8.) As previously discussed, the standard error is $\mathrm{se}(\hat{\mu}) = \sqrt{\mathbb{V}(\overline{X})} = \sigma/\sqrt{n}$. If $\hat{\sigma}$ is an estimate of σ, the corresponding estimated standard error is $\hat{\mathrm{se}} = \hat{\sigma}/\sqrt{n}$.   ▷

Example 1.12. The plug-in estimator of the variance of a probability distribution is given by

$$\hat{\sigma}^2 = \int (x - \hat{\mu})^2 \, d\hat{F}_n = \frac{1}{n} \sum_{i=1}^{n} (x_i - \hat{\mu}_n)^2,$$

which is equivalent to the sample standard deviation S_n scaled by a factor of $(n-1)/n$, i.e., $\hat{\sigma}^2 = (n-1)/nS_n$. Unlike the sample standard deviation, the plug-in estimator of the variance is therefore biased.   ▷

Not all statistical functionals, however, are computed by integrating over a probability distribution. Case in point: quantiles.

Definition 1.32. *Given a distribution function F and a number $p \in (0, 1)$, the pth quantile, Q_p, is defined by*

$$Q_p = F^{-1}(p) = \inf\{x : F(x) \geq p\}. \tag{1.32}$$

The pth quantile is clearly a statistical functional, as it is a function of the distribution function, i.e., $Q_p = Q_p(F)$. The inverse $F^{-1}(p)$ is defined using an infimum[2], to ensure its existence even when no x-value maps directly onto p.

Example 1.13. Consider the probability mass function that places equal weight $1/3$ on the values $x_1 = 1$, $x_2 = 2$, and $x_3 = 3$. The distribution function is

$$F(x) = \begin{cases} 0 & \text{for } x < 1 \\ 1/3 & \text{for } 1 \leq x < 2 \\ 2/3 & \text{for } 2 \leq x < 3 \\ 1 & \text{for } 3 \leq x \end{cases}.$$

Let's compute the 50th percentile ($p = 0.5$). The set $\{x : F(x) \geq 0.5\} = [2, \infty)$. Hence $Q_{0.5} = 2$. A similar calculation shows, for example, that $Q_{0.33} = 1$ and $Q_{0.34} = 2$.   ▷

Proposition 1.9. *Given a set of IID data $X_1, \ldots, X_n \sim F$, the plug-in estimator of the pth quantile Q_p is given by*

$$\hat{Q}_p = \hat{F}_n^{-1}(p) = \inf\{\hat{F}_n(x) \geq p\}. \tag{1.33}$$

We call $\hat{Q}_p$ the pth sample quantile.

---

[2] The infimum is the *greatest lower bound* of a set. So the infimum of the set $(0, 1)$ is $\inf(0, 1) = 0$. An infimum is well defined, even if the minimum value does not exist (as is the case for the open interval $(0, 1)$). Similarly, the *supremum* of a set is its *least upper bound*; e.g., $\sup(0, 1) = 1$.

An interesting application of plug-in estimators is found by approximating the standard error of the cumulative distribution itself. Recall that the empirical distribution function $\hat{F}_n(x)$ is itself a point estimator, with expected value and variance given by Theorem 1.6. Since the standard error $\mathrm{se}(\hat{F}_n(x)) = \sqrt{\mathbb{V}(\hat{F}_n(x))}$ is a function of the distribution, we may compute its plug-in estimator, as done in the following example.

Example 1.14. The plug-in estimator for the standard error of the empirical distribution $\hat{F}_n$ is obtained by replacing F with $\hat{F}_n$ in the right-hand side of Equation (1.29), yielding the expression

$$\hat{\mathrm{se}}(\hat{F}_n(x)) = \sqrt{\frac{\hat{F}_n(x)(1 - \hat{F}_n(x))}{n}}.$$

We may use this to estimate the standard error of the empirical distribution from Example 1.9. We can then plot $\hat{F}_n(x) \pm 2\,\hat{\mathrm{se}}$, as is done in Code Block 1.7. This results in a *confidence interval* for our empirical distribution function, as shown in Figure 1.2.

```
1  se = np.sqrt(F_hat * (1 - F_hat) / len(y))
2  F_up = np.fmin(1, F_hat + 2 * se)
3  F_dn = np.fmax(0, F_hat - 2 * se)
4  plt.plot(x, F_up, 'r--')
5  plt.plot(x, F_dn, 'r--')
```

Code Block 1.7: Computing the standard error (continued from Code Block 1.6).

Note that we cap the upper and lower bounds of our confidence interval at 100% and 0%, respectively, using **numpy**'s built-in **fmin** and **fmax** functions. (In this case, the sample size is large enough so that these caps don't actually affect our bounds.)   ▷

1.7 The Bootstrap

In this section, we build on our discussion of the empirical distribution and devise a method for computing the variance and the distribution of any statistic $T_n = g(X_1, \ldots, X_n)$. Typically we will not have access to the distribution F that actually generated the data.

1.7.1 Introduction

The *bootstrap* is a method for approximating the distribution of any test statistic T_n from a given sample of data. In particular, we are interested in computing the standard error of the test statistic along with a confidence interval (a set of values for T_n which we deem likely to coincide with reality).

The idea of the bootstrap is to replace the variance function $\mathbb{V}_F(T_n)$ with $\mathbb{V}_{\hat{F}}(T_n)$; i.e., when the distribution that generated the data is unknown, we approximate the variance of our statistic using the empirical distribution instead. In most cases, the empirical variance $\mathbb{V}_{\hat{F}}(T_n)$ must be approximated by simulation. Before we discuss this, however, it will be advantageous to perform the computation for a simple case, for which the empirical variance is known.

Example 1.15. Let us consider the familiar case of the sample mean $T_n = \overline{X}_n$. We know that the variance of the sample mean depends on the distribution that generated the data, since $\mathbb{V}_F(\overline{X}_n) = \sigma^2/n$. We saw in Example 1.12 that the plug-in estimator for the variance is given by $\hat{\sigma}^2 = 1/n\sum_{i=1}^{n}(X_i - \overline{X}_n)^2$. Hence, our estimate for the variance of the sample mean is $\mathbb{V}_{\hat{\sigma}}(\overline{X}_n) = \hat{\sigma}^2/n$.   ▷

The case of the sample mean is the exception, however, not the rule. In most cases, the empirical variance cannot be directly represented as a function of the data. This is where the bootstrap comes into play.

1.7.2 Bootstrap Samples

A statistic T_n is a function of the data. Given an IID sample $X_1, \ldots, X_n \sim F$, we may compute the corresponding point estimate of the statistic $T_n = g(X_1, \ldots, X_n)$. In order to approximate the distribution for T_n, we may instead take many samples from our original distribution F, compute the value of our statistic for each sample, and use the result as a proxy for the actual distribution of T_n.

In general, however, we do not know the form of the distribution F that generated our data. Moreover, we only have access to a single sample from that distribution $X_1, \ldots, X_n \sim F$. This sample is our data set. Our goal is to use these data to draw inferences about our statistic T_n, beyond the simple point estimate $T_n = g(X_1, \ldots, X_n)$.

If we had access to the distribution function F, we could, of course, produce an arbitrary number of samples, compute the value of T_n for each sample, thereby approximating the distribution of T_n. To resolve this, we instead use the empirical distribution function $\hat{F}_n$ that is obtained from our single, real sample, i.e., our actual data set. But how can we draw a random sample from our empirical distribution?

Definition 1.33. *Given a set of* IID *data* $X_1, \ldots, X_n \sim F$, *a* bootstrap sample *of size* k *of the data is a set of* k *numbers* $X_1^*, \ldots, X_k^*$ *that are drawn randomly, with replacement, from the original data set* $X_1, \ldots, X_n$.

Since a bootstrap sample is drawn *with replacement* from the original data set, the bootstrap sample size k is not limited by the size of the data n. In practice, however, one typically considers a bootstrap sample to be the same size as the original data, i.e., with $k = n$.

Example 1.16. Consider the data set $\mathcal{D} = (R, G, B)$, representing the colors red, green, and blue. Our first draw from the set $\mathcal{D}$ will result in an R with a probability of $1/3$, G with a probability of $1/3$, and B with a probability of $1/3$, i.e.,

$$p(R) = p(G) = p(B) = 1/3.$$

Suppose we happen to draw a B. Since this draw occurs *with replacement*, it means that the probability distribution for our next draw is unchanged, i.e., we still have a $1/3$ probability of drawing a B. (Contrast this to a random draw *without replacement*: having first obtained a B, the probable outcome of our *second* draw would be $p(R) = p(G) = 1/2$, $p(B) = 0$, since we have already removed the B from the data set, and we are not "replacing" it.)

A bootstrap sample of size, say, $k = 10$ can be done in a single line of Python code:

```
np.random.choice(('R','G','B'), size=10, replace=True) .
```

This generated an output of

```
['B', 'G', 'B', 'B', 'R', 'G', 'G', 'B', 'B', 'R'].
```

(The output will be different each time, because it is generated randomly.)
▷

As one may have already suspected, the bootstrap sample is closely related to a sample from the empirical distribution.

Proposition 1.10. *A bootstrap sample* $X_1^*, \ldots, X_k^*$ *obtained from a set of data constitutes a random sample* $X_1^*, \ldots, X_k^* \sim \hat{F}_n$ *from the empirical distribution* $\hat{F}_n$ *generated by the data.*

Proof. Since the empirical distribution is equivalent to the distribution function of a set of equal-strength point masses, located at $X_1, \ldots, X_n$, a random number $X \sim \hat{F}_n$ is equivalent to selecting one of the data points $X_1, \ldots, X_n$ at random, i.e., with equal probability. Taking k IID samples from the empirical distribution $\hat{F}_n$ is therefore equivalent to drawing k values from the data set, with replacement, which is the definition of a bootstrap sample. □

1.7.3 Estimating the Variance of a Statistic

An immediate application of the bootstrap is estimating the variance of a statistic. The pseudocode is given in Algorithm 1.1.

Data: $X_1, \ldots, X_n \sim F$; a statistic $T_n = g(X_1, \ldots, X_n)$
Result: Estimate of the variance $v_{boot} \approx \mathbb{V}(T_n)$

1 **for** $b = 1$ **to** B **do**

2 $\quad$ Compute a bootstrap sample $X^*_{1,b}, \ldots, X^*_{n,b} \sim \hat{F}_n$;

3 $\quad$ Compute $T^*_{n,b} = g(X^*_{1,b}, \ldots, X^*_{n,b})$

4 **end**

5 Compute the sample mean $\overline{T}^*_n = \dfrac{1}{B} \sum\limits_{b=1}^{B} T^*_{n,b};$

6 Output $v_{boot} = \dfrac{1}{B} \sum\limits_{b=1}^{B} \left(T^*_{n,b} - \overline{T}^*_n \right)^2.$

Algorithm 1.1: Bootstrap Variance Estimation

Essentially, we perform many bootstrap samples from our data set, compute the value of the statistic T_n for each sample, storing the result, and use this collection of values as an approximate distribution for the statistic T_n. In particular, the variance of this vector of values yields an estimate for the variance of the statistic T_n.

In Python, functions are objects, and therefore may be passed in as arguments to other functions. Let's begin by defining a function for our statistic T_n. For example, if we were interested in computing the 90th quantile $Q_{0.9}$, we could code this using **numpy**'s built-in quantile function, as in Code Block 1.8.

```
1  def t(x):
2      # x is an array, set of data
3      return np.quantile(x, 0.9)
```

Code Block 1.8: Function definition of our statistic T_n

For a given set of data x, the function $t(x)$ outputs the point estimate of the quantile $Q_{0.9}$. We may pass this function, along with our observed data set, into a second function, **bootstrap**, defined in Code Block 1.9. The **bootstrap** function follows lines 1–4 of Algorithm 1.1. The bootstrap sample of the data x is computed by `np.random.choice(x, size=len(x))`. Passing the bootstrap sample into the function t produces a single simulated value for the statistic t. We do this many times and return the output as an array of values. To obtain the boostrap variance estimate, we simply compute the variance of this output array, as shown for the Boston housing

```
1  def bootstrap(x, t, B=10000):
2      # x is an array, set of data
3      # t = t(x) is a function: array --> scalar
4      # returns: array, sample from distribution for t
5      ts = np.zeros(B)
6      for b in range(B):
7          ts[b] = t( np.random.choice(x, size=len(x)) )
8      return ts
```

Code Block 1.9: Bootstrap sample of values for T_n

price data in Code Block 1.10.   The variance output is `1.368`, which is the

```
1  y = data['target']
2  ts = bootstrap(y, t)
3  print(ts.var())
4  plt.hist(ts, bins=30, normed=True)
```

Code Block 1.10: Boostrap variance for $Q_{0.90}$ for the Boston house price data

bootstrap estimate for $\mathbb{V}(Q_{0.90})$. (The standard error is the square-root of this, so $\hat{se}(Q_{0.90}) = 1.17$.) But since we retained *all* of our samples, we can further plot a histogram of the data, as shown in Figure 1.3. By eyeballing this histogram, we can conclude that we expect the quantile $Q_{0.90}$ to take a value between 32 and 37. And, indeed, we find that

$$\texttt{ts.mean() + 2 * ts.std() = 36.79}$$
$$\texttt{ts.mean() - 2 * ts.std() = 32.11'}$$

confirming our earlier statement that $\hat{T}_n \pm 2\,\hat{se}(T_n)$ is normally a good approximation for a confidence interval.

1.8 Transformations of Random Variables

In this section, we shall consider functions of random variables. In general, since a random variable X is a function from our sample space into the set of real numbers, i.e., $X : \Omega \to \mathbb{R}$, one may apply a function $g : \mathbb{R} \to \mathbb{R}$ to a random variable to obtain a new random variable $Y = g(X)$. Such mappings are referred to as *transformations* of random variables. Our primary application for this section is in defining several more advanced distributions later on, so the ready can skip without loss of continuity. For further details, we direct the reader to, e.g., Casella and Berger [2002].

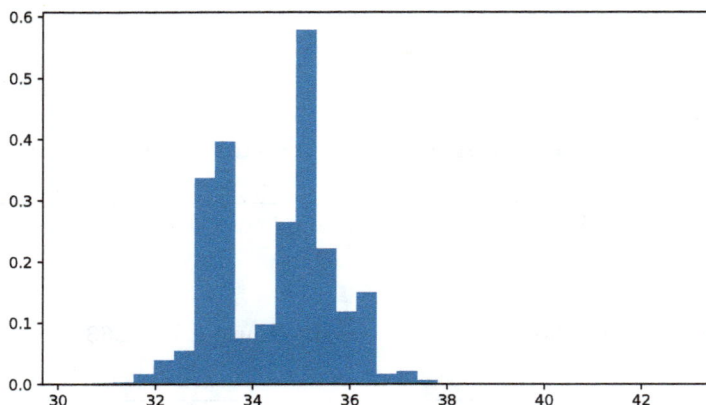

Fig. 1.3: Bootstrap sample of values for $Q_{0.90}$ for Boston housing data.

1.8.1 Univariate Transformations

If X is a random variable with CDF F_X, then any function $Y = g(X)$ is also a random variable. For any set $A \subset \mathbb{R}$, we have

$$\mathbb{P}(Y \in A) = \mathbb{P}(g(X) \in A) = \mathbb{P}(X \in g^{-1}(A)),$$

where we define the inverse mapping of the set A as $g^{-1}(A) = \{x \in \mathbb{R} : g(x) \in A\}$.

For convenience, let us define the following subsets of the real line:

$$\mathcal{X} = \{x \in \mathbb{R} : f_X(x) > 0\} \qquad \text{and} \qquad \mathcal{Y} = \{y \in \mathbb{R} : y = g(x) \text{ for some } x \in \mathcal{X}\}.$$

If X is a discrete random variable, then both $\mathcal{X}$ and $\mathcal{Y}$ are countable, and the probability mass function for Y is given by

$$f_Y(y) = \sum_{x \in g^{-1}(y)} f_X(x) \qquad \text{for } y \in \mathcal{Y}.$$

In other words, the probability that $Y = y$ is simply the sum of the probabilities that $X = x$, over every x such that $g(x) = y$.

The case of a continuous random variable is not so neatly packaged, as it required invertibility of the function g. Often, one must deal with piece-wise invertibility.

Theorem 1.7. *Let X be a continuous random variable with pdf $f_X(x)$, and consider the transformation $Y = g(X)$. Suppose there is a partition $A_0, A_1, \ldots, A_k$ of $\mathcal{X}$ such that $\mathbb{P}(X \in A_0) = 0$ and f_X is continuous on A_i.*

Furthermore, let us define the subsets $\mathcal{Y}_1, \ldots, \mathcal{Y}_k \subset \mathbb{R}$ *by the relations* $\mathcal{Y}_i = \{y : y = g(x),$ *for some* $x \in A_i\}$. *Finally, let us suppose that* g *is locally invertible on each* A_i, *i.e., suppose there exists functions* $g_1(x), \ldots, g_k(x)$, *defined on* $A_1, \ldots, A_k$, *respectively, such that*

1. $g(x) = g_i(x)$, *for* $x \in A_i$,
2. $g_i(x)$ *is monotone on* A_i,
3. $g_i^{-1}(y)$ *has a continuous derivative on* $\mathcal{Y}_i$, *for each* $i = 1, \ldots, k$.

Then the pdf of the random variable Y *is given by*

$$f_Y(y) = \sum_{i=1}^{k} f_X\left(g_i^{-1}(y)\right) \left| \frac{dg_i^{-1}(y)}{dy} \right| \mathbb{I}[y \in \mathcal{Y}_i]. \tag{1.34}$$

For a more detailed discussion of this theorem, along with supporting examples, see Casella and Berger [2002]. For intuition, however, let us consider the case in which we do not need to partition the domain $\mathcal{X}$.

Corollary 1.1. *Let* X, Y, *and* g *be as in Theorem 1.7, and further suppose that the function* g *is continuously differentiable on* $\mathcal{X}$ *with non-zero derivative. Then the pdf of the random variable* Y *is given by*

$$f_Y(y) = f_X\left(g_i^{-1}(y)\right) \left| \frac{dg^{-1}(y)}{dy} \right|. \tag{1.35}$$

1.8.2 Bivariate Transformations

Next, let us consider a transformation of two random variables X and Y. Suppose (X, Y) is a bivariate random vector which maps into (U, V) under the mapping $U = g_1(X, Y)$ and $V = g_2(X, Y)$. Given the joint distribution $f_{X,Y}(x, y)$ and the mapping functions $g : \mathbb{R}^2 \to \mathbb{R}^2$, how can we determine the joint distribution $f_{U,V}(u, v)$?

For the discrete case, the answer is quite intuitive, as seen in our next theorem.

Theorem 1.8. *Let* (X, Y) *be a discrete random vector with* PMF $f_{X,Y}(x, y)$, *and consider the vector mapping* $g : \mathbb{R}^2 \to \mathbb{R}^2$, *which defines a new discrete random vector* $(U, V) = g(X, Y)$. *Moreover, for any* $u, v \in \mathbb{R}$, *let us define the set*

$$g^{-1}(u, v) = \{(x, y) : g(x, y) = (u, v)\}.$$

Then the PMF *of the transformed random vector* (U, V) *is given by*

$$f_{U,V}(u, v) = \sum_{(x,y) \in g^{-1}(u,v)} f_{X,Y}(x, y). \tag{1.36}$$

Proof. Since the random vectors are discrete, they are defined on a countable subset of $\mathbb{R}^2$. By tracing the probability back to the original space, we may express the probability mass function as

$$
\begin{aligned}
f_{U,V}(u,v) &= \mathbb{P}((U,V) = (u,v)) \\
&= \mathbb{P}((X,Y) \in g^{-1}(u,v)) \\
&= \sum_{(x,y) \in g^{-1}(u,v)} f_{X,Y}(x,y).
\end{aligned}
$$

$\square$

The generalization to more than two dimensions is straightforward.

As we saw in the one-dimensional case, transforming probability *densities* always requires additional care as compared to their discrete analogue. Let us begin by defining the following subsets of $\mathbb{R}^2$. First, let $\mathcal{A} \subset \mathbb{R}^2$ be region of the (X,Y)-space with nonzero probability

$$
\mathcal{A} = \{(x,y) : f_{X,Y}(x,y) > 0\}.
$$

Similarly, let's define

$$
\mathcal{B} = g(\mathcal{A}) = \{(u,v) : (u,v) = g(x,y), \text{ for some } (x,y) \in \mathcal{A}\},
$$

so that $f_{U,V}(u,v) > 0$ if and only if $(u,v) \in \mathcal{B}$.

Theorem 1.9. *Let (X,Y) be a continuous random vector with pdf $f_{X,Y}(x,y)$, and consider the transformation $(U,V) = g(X,Y)$. Let $\mathcal{A},\mathcal{B} \subset \mathbb{R}^2$ be defined as above. Suppose there is a partition $\mathcal{A}_0, \mathcal{A}_1, \ldots, \mathcal{A}_k$ of $\mathcal{A}$ such that $\mathbb{P}((X,Y) \in \mathcal{A}_0) = 0$ and $f_{X,Y}$ is continuous on $\mathcal{A}_i$. Furthermore, let us define the subsets $\mathcal{B}_1, \ldots, \mathcal{B}_k \subset \mathbb{R}^2$ by the relations $\mathcal{B}_i = \{(u,v) : (u,v) = g(x,y), \text{ for some } (x,y) \in \mathcal{A}_i\}$. Finally, let us suppose that g is one-to-one and differentiable on each $\mathcal{A}_i$, i.e., suppose there exists functions $g_1(x,y), \ldots, g_k(x,y)$, defined on $\mathcal{A}_1, \ldots, \mathcal{A}_k$, respectively, such that*

1. *$g(x,y) = g_i(x,y)$, for $(x,y) \in \mathcal{A}_i$,*
2. *$g_i(x,y)$ is one-to-one on $\mathcal{A}_i$,*
3. *$g_i^{-1}(u,v)$ has a continuous derivative on $\mathcal{B}_i$, for each $i = 1, \ldots, k$.*

Then the pdf of the random vector (U,V) is given by

$$
f_{U,V}(u,v) = \sum_{i=1}^{k} f_{X,Y}\left(g_i^{-1}(u,v)\right) |J_i(u,v)| \, \mathbb{I}[(u,v) \in \mathcal{B}_i]. \tag{1.37}
$$

where the ith Jacobian is given by the determinant

$$
J_i(u,v) = \det \frac{\partial g_i^{-1}(u,v)}{\partial(u,v)} = \det \begin{bmatrix} \dfrac{\partial x}{\partial u} & \dfrac{\partial x}{\partial v} \\ \dfrac{\partial y}{\partial u} & \dfrac{\partial y}{\partial v} \end{bmatrix}.
$$

Of course, when $g : \mathcal{A} \to \mathcal{B}$ is one-to-one, there is no need to partition $\mathcal{A}$ into pieces. For an example, see the proof for Lemma 2.2.

1.8.3 Moment-generating Functions

Another useful tool that we shall encounter is sometimes useful in determining the distribution of a random variable.

Definition 1.34. *Let X be a random variable with a given CDF F. The moment-generating function (MGF) of X, denoted $M_X(t)$, is the function*

$$M_X(t) = \mathbb{E}[e^{tX}], \tag{1.38}$$

provided that such an expectation exists on some neighborhood of the origin.

The moment-generating function is a useful tool given the following.

Proposition 1.11. *Suppose the random variable X has a MGF $M_X(t)$, and let $M_X^{(n)}(t)$ represent the nth derivative of $M_X(t)$. Then*

$$\mathbb{E}[X^n] = M_X^{(n)}(0), \tag{1.39}$$

for $n \in \mathbb{Z}_+ = \{1, 2, \ldots\}$.

Proof. We note that the result follows as long as we can establish the slightly more general statement

$$M_X^{(n)}(t) = \mathbb{E}[X^n e^{tX}].$$

We proceed by induction. For the basis step, consider the case $n = 1$:

$$
\begin{aligned}
M_X'(t) &= \frac{d}{dt} \int_{-\infty}^{\infty} e^{tx} f_X(x)\, dx \\
&= \int_{-\infty}^{\infty} x e^{tx} f_X(x)\, dx \\
&= \mathbb{E}[X e^{tX}].
\end{aligned}
$$

Setting $t = 0$ yields the result for $n = 1$.

For the induction step, let us assume that $M_X^{(n)}(t) = \mathbb{E}[X^n e^{tX}]$ is true for some $n \geq 1$. This implies that

$$
\begin{aligned}
M_X^{(n+1)}(t) &= \frac{d}{dt} \int_{-\infty}^{\infty} x^n e^{tx} f_X(x)\, dx \\
&= \int_{-\infty}^{\infty} x^{n+1} e^{tx} f_X(x)\, dx \\
&= \mathbb{E}[X^{n+1} e^{tX}].
\end{aligned}
$$

This proves our proposition. □

Despite its ability to generate arbitrary moments $\mathbb{E}[X^n]$ for a random variable X, the MGF further possesses the following property.

Proposition 1.12. *Let X and Y be independent random variables with* MGFs M_X *and* M_Y, *resepctively. Then the* MGF *of the random variable* $Z = X + Y$ *is given by*

$$M_Z(t) = M_X(t)M_Y(t). \tag{1.40}$$

We leave the proof to the reader. We note, however, that Proposition 1.12 can sometimes lead to a simpler method for computing the PDF of the sum of two independent random variables. If we recognize the MGF of Z, we can sometimes easily draw conclusions on the distribution of the sum $X + Y$. This saves us the work of constructing a bivariate transformation of the form

$$U = X + Y$$
$$V = Y,$$

and then having to marginalize the result over V.

1.8.4 Location and Scale Families

Given any random variable with known distribution, it is natural to consider an associated family of random variables, known as the *location–scale family*, formed by horizontal translations and stretches of the underlying distribution. Specifically, we have the following.

Theorem 1.10. *Let Z be any continuous random variable with* PDF $f_Z(z)$. *Then the random variable X defined through the affine transformation*

$$X = \sigma Z + \mu,$$

for some $\mu \in \mathbb{R}$ and $\sigma \in \mathbb{R}_+$, has PDF *given by*

$$f_X(x) = \frac{1}{\sigma} f_Z\left(\frac{x - \mu}{\sigma}\right). \tag{1.41}$$

The theorem follows immediately from Corollary 1.1. We shall visit many examples in our next chapter. The transformation defined in Theorem 1.10 occurs often enough that we give special names to the parameters μ and σ, as follows.

Definition 1.35. *Given any* PDF $f(x)$ *and parameters $\mu \in \mathbb{R}$ and $\sigma > 0$, the family of distributions defined by Equation (1.41), i.e., the family of* PDFs *given by*

$$\frac{1}{\sigma} f\left(\frac{x - \mu}{\sigma}\right), \tag{1.42}$$

is called the location–scale family *with standard* PDF $f(x)$. *Moreover, we refer to μ as the* location parameter *and σ as the* scale parameter.

Note 1.10. The function defined by Equation (1.42) constitutes a PDF due to Theorem 1.10.   ▷

Note 1.11. For the special case $\sigma = 1$, we refer to the family $f(x - \mu)$ simply as a *location family.*

Similarly, for the special case $\mu = 0$, we refer to the family $(1/\sigma)f(x/\sigma)$ as a *scale family.*   ▷

Problems

1.1. Show that if the random variables X and Y are independent, then the conditional distributions reduce to

$$f(x|y) = f_X(x) \qquad \text{and} \qquad f(y|x) = f_Y(y).$$

1.2. Prove Proposition 1.5.

1.3. Find a counterexample to the assertion: uncorrelated random variables must be independent; i.e., find a pair of random variables X and Y that are *dependent*, but that nonetheless have covariance $\text{COV}(X, Y) = 0$.

1.4. Repeat Example 1.8 for the sample standard deviation S_n. In particular, for each sample size $n = 10^1, 10^2, \ldots, 10^5$, compute **n_samples** $= 10,000$ samples from the normal distribution $\mathcal{N}(\mu = 3, \sigma^2 = 4)$ with sample size n. Report the average sample standard deviation and the variance of the sample standard deviation across all **n_samples** samples. For $n = 10$ and $n = 100$, plot the histogram of sample standard deviations.

1.5. Let $X \sim \text{Unif}(a, b)$. Show that the variance of X is given by

$$\mathbb{V}(X) = \frac{(b - a)^2}{12}.$$

1.6. Prove Proposition 1.12.

1.7. The *Dvoretzky-Kiefer-Wolfowitz (DKW) inequality* states that, for any $\epsilon > 0$,

$$\mathbb{P}\left(\sup_x \left| F(x) - \hat{F}_n(x) \right| > \epsilon \right) \leq 2e^{-2n\epsilon^2}.$$

Use this to derive the $1 - \alpha$ confidence band:

$$\hat{F}_n(x) \pm \sqrt{\frac{1}{2n} \ln(2/\alpha)}$$

for the empirical distribution $\hat{F}_n(x)$.

2

A Menagerie of Distributions

The goal of this chapter is to provide a lexicon of basic distributions along with their definitions and properties. We will also indicate how one may evaluate and sample these distributions using built-in Python packages.

More details, illustrative examples, and discussion are available in many introductory statistics texts, such as Casella and Berger [2002], Hogg, *et al.* [2015], or Ross [2012].

2.1 Bernouilli and Binomial

2.1.1 Bernouilli Trials

A Bernouilli trial is a mathematical generalization of a coin-flip, as shown below.

Definition 2.1. *A* Bernouilli experiment *or* Bernouilli trial *is any random experiment over a two-point sample space* $\Omega = \{S, F\}$ *with fixed probability distribution* $\mathbb{P}(S) = p$. *A random variable* X *is said to be a* Bernouilli *random variable with probability of success* p, *denoted* $X \sim \text{Bern}(p)$, *if* $X : \Omega \to \mathbb{R}$ *is a random variable over* Ω *defined by* $X(S) = 1$ *and* $X(F) = 0$. *The points* S *and* F *are referred to as* success *and* failure, *respectively.*

An immediate consequence of this definition is the probability distribution over the random variable X.

Proposition 2.1. *Let* $X \sim \text{Bern}(p)$ *represent a Bernouilli random variable. Then the* PMF *for* X *is given by*

$$f(x) = p^x (1 - p)^{x-1}, \qquad (2.1)$$

for $x = 0, 1$.

Proof. Clearly, $\mathbb{P}(X = 1) = \mathbb{P}(S) = p$ and $\mathbb{P}(X = 0) = \mathbb{P}(F) = 1 - p$. Equation (2.1) is a compact way to express this fact, for if $x = 1$, then $f(1) = p$, and if $x = 0$, we have $f(0) = 1 - p$.   □

Proposition 2.2. *Let* $X \sim \text{Bern}(p)$ *represent a Bernouilli random variable. Then*

$$\mathbb{E}[X] = p \tag{2.2}$$
$$\mathbb{V}(X) = p(1 - p). \tag{2.3}$$

Proof. The expected value of X is given by

$$\mathbb{E}[X] = \sum_{x=0}^{1} x f(x) = 0 \cdot (1 - p) + 1 \cdot p = p.$$

The variance is given by

$$\mathbb{V}(X) = \sum_{x=0}^{1} (x - p)^2 f(x) = p^2(1 - p) + (1 - p)^2 p = p(1 - p).$$

□

Simulation in Python

The quickest way to simulate an outcome of a Bernouilli trial is to invoke `np.random.random()`, which generates a random number on the interval $[0, 1]$, and then check to see if it is less than the probability p. This is done in Code Block 2.1. If the random number generated by `np.random.random()` is less than p (which will occur with probability p), then the random variable x will equal one. Otherwise it will equal zero.

```
p = 0.3 # Probability of success.
x = int( np.random.random() < p )
```

Code Block 2.1: Simulation of a Bernouilli trial

2.1.2 Binomial Distribution

A Bernouilli trial rarely occurs in a vacuum. Typically, we are interested in a series of Bernouilli trials; e.g., we might flip a coin ten times. This gives rise to our second distribution family.

Definition 2.2. *Let* $X_1, \ldots, X_n \sim \text{Bern}(p)$ *be* IID *Bernouilli random variables. Then the random variable* $X = X_1 + \cdots + X_n$ *is a* binomial random variable, *which is denoted* $X \sim \text{Binom}(n, p)$.

Note 2.1. The binomial random variable can be interpreted as representing the total number of successes in a series of n Bernouilli trials, for which each trial has a probability of success p.   ▷

Proposition 2.3. *The* PMF *of a binomial random variable* $X \sim \text{Binom}(n, p)$ *is given by*

$$f(X = k) = \binom{n}{k} p^k (1 - p)^{n-k}, \tag{2.4}$$

where

$$\binom{n}{k} = \frac{n!}{k!(n-k)!},$$

spoken "n choose k", is the binomial coefficient.

Note 2.2. For convenience, we will sometime refer to the binomial distribution, as given by Equation (2.4), as $\text{Binom}(k; n, p)$.   ▷

Note 2.3. Normalization of the PMF given in Equation (2.4) follows directly from the binomial theorem:

$$(p + q)^n = \sum_{k=0}^{n} \binom{n}{k} p^k q^{n-k},$$

with $q = 1 - p$.   ▷

Proof. In order to determine the value of $f(k)$, the probability of observing exactly $X = k$ successes from a total of n Bernouilli trials, let us first consider the case in which those k successes occur for the first k Bernouilli trials, i.e., suppose we observe precisely:

$$\underbrace{S \cdots S}_{k \text{ times}} \cdot \underbrace{F \cdots F}_{n-k \text{ times}} .$$

The probability of this occurring is exactly the product of the individual probabilities, hence: $p^k (1 - p)^{n-k}$. However, this is only one possible way to arrive at a total of k successes. Any permutation of this result would also result in a total of k successes, simply in a different order. Since there are precisely $\binom{n}{k}$ ways of choosing k items from a set of n items, we conclude the total probability of achieving k successes is correctly given by Equation (2.4).   □

Proposition 2.4. *Let* $X \sim \text{Binom}(n, p)$ *be a binomial random variable. Then its expectation and variance are given by*

$$\mathbb{E}[X] = pn \tag{2.5}$$

$$\mathbb{V}(X) = np(1 - p). \tag{2.6}$$

Proof. This follows trivially from Propositions 1.1, 1.2, and 2.2: since

$$\mathbb{E}[X_i] = p \qquad \text{and} \qquad \mathbb{V}(X_i) = p(1-p),$$

it follows from Proposition 1.1 and 1.2 that

$$\mathbb{E}[X] = \sum_{i=1}^{n} p = np \qquad \text{and} \qquad \mathbb{V}(X) = \sum_{i=1}^{n} p(1-p) = np(1-p).$$

$\square$

Proposition 2.5. *Let* $X_1, \ldots, X_n \sim \text{Bern}(p)$ *be* IID *Bernouilli trials and consider the sample mean* $\overline{X} = \frac{1}{n} \sum_{i=1}^{n} X_i$. *Then*

$$\mathbb{E}[\overline{X}] = p \tag{2.7}$$

$$\mathbb{V}(\overline{X}) = \frac{p(1-p)}{n}. \tag{2.8}$$

Proof. From Proposition 1.1 and 1.2, it follows that

$$\mathbb{E}[\overline{X}] = \mathbb{E}\left[\frac{1}{n} \sum_{i=1}^{n} X_i\right] = \frac{1}{n} \sum_{i=1}^{n} \mathbb{E}[X_i] = p.$$

Similarly,

$$\mathbb{V}(\overline{X}) = \mathbb{V}\left(\frac{1}{n} \sum_{i=1}^{n} X_i\right) = \frac{1}{n^2} \sum_{i=1}^{n} \mathbb{V}(X_i) = \frac{1}{n^2} \cdot np(1-p) = \frac{p(1-p)}{n}.$$

$\square$

Note 2.4. Proposition 2.5 is simply an application of Theorem 1.2.   ▷

Simulation in Python

We can modify Code Block 2.1 to account for multiple Bernouilli trials by setting the `size` argument of `np.random.random`, thus generating a vector of random numbers, each of which may be compared in size to the probability p. This is done in Code Block 2.2.   The statement

```
1   p = 0.3 # Probability of success.
2   n = 10 # Number of trials.
3   x = np.sum( np.random.random(size=n) < p )
```

Code Block 2.2: Simulation of a series of Bernouilli trials

`np.random.random(size=n)` < `p` represents a Boolean vector of length n, i.e., by executing this command using $n = 5$, you might get, for example, something that looks like

$$\texttt{array([False, True, True, False, False]).}$$

Using the built-in `np.sum` function will then sum these values together, interpreting `True` as 1 and `False` as 0. The variable `x` therefore represents a single realization of the Binomial random variable $X \sim \mathrm{Binom}(n, p)$.

Of course, the `numpy` library is slightly more sophisticated than this, as it offers a built-in Binomial random variable generator. The same can be achieved by using the code shown in Code Block 2.3. Using the `size`

```
p = 0.3 # Probability of success
n = 10 # Number of trials
T = 1   # Number of i.i.d. draws from Binomial distribution
x = np.random.binomial(n, p, size=T)
```

Code Block 2.3: Binomial random variable

argument, we can further control the number of samples from the binomial distribution that we'd like to simulate. For example, executing line 4 using `T=5` could return a vector similar to

$$\texttt{array([3, 1, 4, 7, 1]) .}$$

Of course, if `T` is large, the relative proportion of any fixed value k should be approximately equal to the value given by Equation (2.4).

It turns out, the PMF given by Equation (2.4) is also a built-in function. To use it, we must import the `scipy` package. In Code Block 2.4, we pass in the entire array of possible values for the random variable $X = k$, for $k = 0, \ldots, 10$. If one is only interested in obtaining the probability of a particular value, say $k = 3$, one could simply pass $k = 3$ and obtain the value 0.266827932. This means that there will be a total of three out of ten successes approximately 26.68% of the time, if one repeats a large number of 10-trial experiments.

The plot of the PMF is shown in Figure 2.1. It was produced using `plt.bar(k, f)`.

Note 2.5. For a complete list of distributions available in `scipy`, please see the documentation available online at

https://docs.scipy.org/doc/scipy/reference/stats.html.

Each distribution has a built in CDF calculator, and a PMF or PDF calculator, depending on whether the random variable is discrete or continuous. In

```
1  p = 0.3 # Probability of success
2  n = 10 # Number of trials
3  k = np.arange(11) # Array from 0 to 10
4  f = scipy.stats.binom.pmf(k, n, p) # PMF
5  F = scipy.stats.binom.cdf(k, n, p) # CDF
```

Code Block 2.4: Distribution functions of a binomial random variable

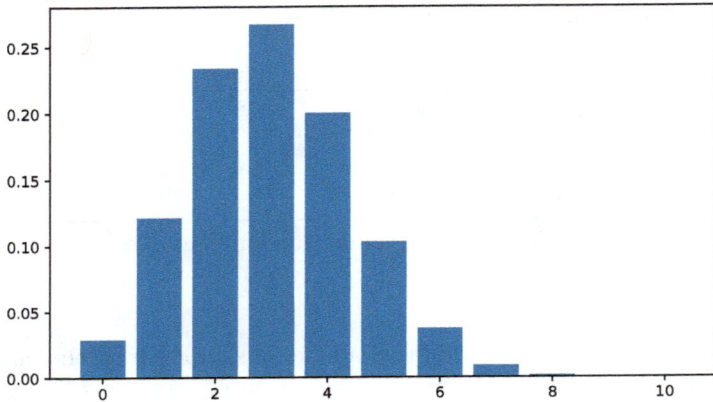

Fig. 2.1: PMF for $X \sim \text{Binom}(10, 0.30)$.

addition, each distribution has a built-in survival function (sf) and inverse survival function (isf) method, that operate similar to the above pmf and cdf methods, as well as basic statistics functions such as mean, median, var, and std. For example, executing scipy.stats.binom.var(n, p) returns a value of 2.1, which agrees with Equation (2.6).   ▷

2.2 Negative Binomial, Geometric, and Hypergeometric

In Section 2.1, we discussed Bernouilli trials and binomial random variables, which describe the probability distribution of the total number of successes out of a fixed number of repeated Bernouilli experiments. In this section, we discuss three closely related variations on that problem.

2.2.1 Negative Binomial Distribution

As we've seen, a binomial random variable represents the total number of successful outcomes from a *fixed* number of trials n. Thus, we decide *in advanced* how many experiments to make, and then set off to make that number. The negative binomial distribution is a variation on that theme.

Definition 2.3. *Let $r \in \mathbb{Z}_*$ be a fixed positive integer. A negative binomial random variable is a random variable X that represents the total number of trials needed before the rth success occurs. The probability distribution for X is said to have a* negative binomial distribution, *which we write as* $X \sim \mathrm{NBD}(r, p)$.

Note 2.6. A binomial random variable represents the total number of successes out of a fixed number of trials, whereas a negative binomial random variable represents the total number of trials required before a fixed number of successes is observed. ▷

Proposition 2.6 (Negative Binomial Distribution). *Let $X \sim \mathrm{NBD}(r, p)$ represent a negative binomial random variable. Then the* PMF *for X is given by*

$$f(x) = \binom{x-1}{r-1} p^r (1-p)^{x-r}, \tag{2.9}$$

for $x = r, r+1, \ldots$.

Proof. For a fixed x, Equation (2.9) represents the probability of the rth success occurring exactly on the xth trial. This implies that on the $(x-1)$th trial, there must have been only $r-1$ successes (which may have occurred on *any* of the preceding $x-1$ trials). The probability of observing $r-1$ successes (anywhere) in $x-1$ trials is given by the binomial distribution,

$$\mathrm{Binom}(r-1; x-1, p) = \binom{x-1}{r-1} p^{r-1} (1-p)^{x-1}.$$

The probability of observing a success on the xth trial is simply p. The probability of observing the rth success on the xth trial is therefore the product $p \cdot \mathrm{Binom}(r-1; x-1, p)$, which is equivalent to Equation (2.9). □

Proposition 2.7. *Let $X \sim \mathrm{NBD}(r, p)$ represent a negative binomial random variable. Then the mean and variance of X are given by*

$$\mathbb{E}[X] = \frac{r}{p} \tag{2.10}$$

$$\mathbb{V}(X) = \frac{r(1-p)}{p^2}. \tag{2.11}$$

The expected value of X given in Equation (2.10) should be exactly as we expected. For example, let us consider the case where $p = 0.10$ and $r = 10$, that is, each trial has a 10% probability of success, and we continue until we have observed exactly ten successes. It makes intuitive sense that, on average, we should require one hundred trials before observing the tenth success, which agrees exactly with Equation (2.10).

Proof. Following the elegant proof presented in Ross [2012], we can compute the expected value of the kth power of X as

$$\mathbb{E}[X^k] = \sum_{n=r}^{\infty} n^k \binom{n-1}{r-1} p^r (1-p)^{n-r}$$

$$= \frac{r}{p} \sum_{n=r}^{\infty} n^{k-1} \binom{n}{r} p^{r+1} (1-p)^{n-r}$$

$$= \frac{r}{p} \sum_{m=r+1}^{\infty} (m-1)^{k-1} \binom{m-1}{r} p^{r+1} (1-p)^{m-(r+1)}$$

$$= \frac{r}{p} \mathbb{E}\left[(Y-1)^{k-1}\right]$$

where the second line follows since $n\binom{n-1}{r-1} = r\binom{n}{r}$, the third line follows due to the reindexing $m = n + 1$, and where we recognize $Y \sim \text{NBD}(r + 1, p)$. Setting $k = 1$ yields Equation (2.10). Further, by setting $k = 2$, we have

$$\mathbb{E}[X^2] = \frac{r}{p}\mathbb{E}[Y-1] = \frac{r}{p}\left(\frac{r+1}{p} - 1\right),$$

from which we may, upon application of Proposition 1.3[1], obtain Equation (2.11).   □

Simulation in Python

Before delving into the built-in numpy and scipy methods, let us first build our own simulation from scratch. This is done in Code Block 2.5.

For this example, we threw in a couple of assertions (lines 10–12) to ensure that the input values take the appropriate form. Passing in $p = 5$ or $r = -6$ will throw an error.

The rest of the code is straightforward: we begin our sequence of experiments with n_success and n_trials set to zero. Then we simply add Bernouilli trials, one by one, until we reach a total of r successes. Line 17 simply increments n_trials by 1. (x+=1 is a Pythonic shorthand for x=x+1, though both are equivalent: they increment the value stored in variable x by one.) In Line 18, the phrase np.random.random() < p is evaluated as

---

[1] $\mathbb{V}(X) = \mathbb{E}[X^2] - \mathbb{E}[X]^2$

```python
def nbd_single_sample(r, p):
    # Returns a sample from NBD(r, p)
    # Returns number of trials before observing
    # rth success, with success probability p.
    # Inputs:
    #   r nonnegative int. number of success
    #   p float in (0, 1]
    # Output:
    #   n_trials (int) number of trials
    assert r >= 0
    assert isinstance(r, int)
    assert p > 0 and p <= 1

    n_success = 0
    n_trials = 0
    while n_success < r:
        n_trials += 1
        n_success += np.random.random() < p

    return n_trials
```

Code Block 2.5: Sample from NBD

being either **True** or **False**, depending on whether our random number (on the interval $(0, 1)$) is less than p. When adding **True** or **False** to an integer, Python instead adds the integer equivalents: 1 for **True** or 0 for **False**. Line 18 therefore simulates a single Bernouilli trial, with probability of success p, and adds the result to the running total **n_success**. Finally, notice that we use a while-loop, as opposed to a for-loop, as the total number of trials is not determined in advanced.

Recall that, for $X \sim \text{NBD}(10, 0.10)$, $\mathbb{E}[X] = 100$. The first three times I ran this script[2], by executing **nbd_single_sample(10, 0.10)** from the command line, I received the outputs 140, 108, 140. I ran it a few more times: 115, 126, 117, 136, 95. It seemed like there were a lot of outputs greater than 100, our expected average. But was this due to random chance? Or was there a mistake in the code?

To answer this question, let's consider a variation that instead returns an array of samples of a given sample size, as shown in Code Block 2.6[3]. This allows us to simulate many draws from $\text{NBD}(r, p)$ at once. From the command line, I executed **x = nbd_sample(10, 0.10, size=1000)**, and

---

[2] Since it is based on random numbers, the results will be different each time. Due to the random nature of the generator, the reader should not expect to receive the same outputs as printed herein.

[3] Normally, Code Blocks 2.5 and 2.6 would be combined into a single function. For the purpose of illustration, however, we decided to keep them separate.

```
1   def nbd_sample(r, p, size=1):
2       # Inputs:
3       #   r nonnegative int. number of success
4       #   p float in (0, 1]
5       # Output:
6       #   x (array) sample from nbd(r, p)
7
8       x = np.zeros(size)
9       for i in range(size):
10          x[i] = nbd_single_sample(r, p)
11
12      return x
```

Code Block 2.6: Multiple samples from NBD

then computed `x.mean()` and `x.var()`, obtaining outputs of 99.299 for the sample mean and 815.101599 for the sample variance. These are point estimators for μ and σ^2, and they closely agree with the exact, true values of $\mu = 100$, $\sigma^2 = 900$, as given by Equations (2.10) and (2.11).

If we were to run `x = nbd_sample(10, 0.10, size=1000)` many times, and compute the sample mean $\overline{X}$, using `x.mean()`, for each one, we would have an approximation of the sample distribution for $\overline{X}$. From Theorem 1.2, we have $\mathbb{E}[\overline{X}] = \mu = 100$ and $\mathbb{V}(\overline{X}) = \sigma^2/n = 900/1000 = 0.9$. Hence the standard error is given by $\text{se}(\overline{X}) = \sigma/\sqrt{n} \approx 0.9487$. Our original sample mean of 99.299 was therefore 0.74 standard errors below the expected value.

Numpy and Scipy commands

Unfortunately, there are several variations on the negative binomial distribution. A common variation is to consider $Y = X - r$, the number of failures before the rth success, see, for example, Casella and Berger [2002]. This is also the variation used in `numpy` and `scipy`'s libraries. These adjustments are shown in Code Block 2.7.

In particular, since `np.random.negative_binomial` returns the number of *failures* before the rth success, simply adding r (which will be added to each component of the output array) will yield the total number of *trials* before the rth success. Adjusting the `scipy.stats.nbinom` is a little more tricky, but can be accomplished by passing in an additional parameter `loc=r`, which effectively shifts the x-value by this number of units. In other words, if x represents the total number of trials, adding location `loc=r` will effectively shift the input to be $x - r$, which represents the number of failures, consistent with `scipy`'s definition of the negative binomial distribution.

A plot of the PMF for the negative binomial random variable $X \sim$ NBD$(10, 0.10)$ is shown in Figure 2.2.

```
1   r = 10    # Number of success.
2   p = 0.10 # Probability of success.
3   T = 1000 # Sample size.
4
5   x = np.random.negative_binomial(r, p, size=T) + r
6   scipy.stats.nbinom.pmf(100, r, p, loc=r) #Probability of exactly
       100 trials.
7   scipy.stats.nbinom.cdf(100, r, p, loc=r) #Probability less than or
       equal to 100 trials.
8   scipy.stats.nbinom.sf(100, r, p, loc=r) #Probability greater than
       100 trials (survival function).
9   scipy.stats.nbinom.mean(r, p, loc=r)   #E[X] = 100
10  scipy.stats.nbinom.var(r, p, loc=r)    #Var(X) = 900
11  scipy.stats.nbinom.median(r, p, loc=r) #Median = 97
```

Code Block 2.7: Numpy and Scipy NBDadjusted for our definition

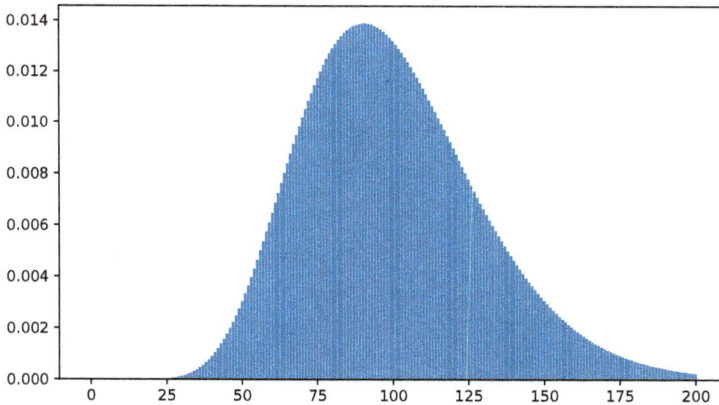

Fig. 2.2: PMF for $X \sim \text{NBD}(r = 10, p = 0.10)$.

2.2.2 Geometric Distribution

A geometric distribution is a special case of the negative binomial distribution, as defined below.

Definition 2.4. *A geometric random variable X with probability p, denoted $X \sim \text{Geom}(p)$, is a special case of the negative binomial random variable with $r = 1$.*

Thus, a geometric random variable represents the number of trials before observing a single success. It follows from Proposition 2.7, that the first success will occur, on average, on the $1/p$th trial. The full distribution is given below.

Proposition 2.8. *The* PMF *for a geometric random variable* $X \sim \text{Geom}(p)$ *is given by*

$$f(x) = p(1-p)^{x-1}, \qquad \qquad (2.12)$$

for $x = 1, 2, 3, \ldots$.

Proposition 2.9. *The geometric distribution is* memoryless, *since, for integers* $s > t$, *we have*

$$\mathbb{P}(X > s | X > t) = \mathbb{P}(X > s - t). \qquad (2.13)$$

The memoryless property of the geometric distribution means that it "forgets" its history. This is best illustrated by means of an example: suppose we have already observed ten failures, and hence we now know that the first success will occur for some value of $X > 10$. Given what we now know, the probability of the first success occurring on trial #12 is the same as what the probability of the first success occurring on trial #2 was, before we began the experiment. Another variation of this theme: at any point in time prior to observing the first success, it doesn't matter how many subsequent failures we have already observed, observing a success x *more* units in the future is equivalent to $f(x)$.

Proof. From the definition of conditional probability, we have

$$\mathbb{P}(X > s | X > t) = \frac{\mathbb{P}(X > s)}{\mathbb{P}(X > t)} = (1-p)^{s-t} = \mathbb{P}(X > s - t).$$

(Note that $\mathbb{P}(X > s, X > t) = \mathbb{P}(X > s)$, since $s > t$.) $\square$

Simulation in Python

The basic functionality to produce random samples from the geometric distribution is shown in Code Block 2.8.

```
1  p = 0.10 # 1\% Probability of success.
2  T = 1000 # Number of samples
3  x = np.random.geometric(p, size=T) # 1,000 samples from Geom(0.01)
4  n = np.arange(61) # Array of Total number of trials.
5  f = scipy.stats.geom.pmf(n, p) #PMF evaluated over n
6  F = scipy.stats.geom.cdf(n, p) #CDF
```

Code Block 2.8: Geometric distribution in Python

The PMF for a geometric random variable with $p = 0.10$ is shown in Figure 2.3.

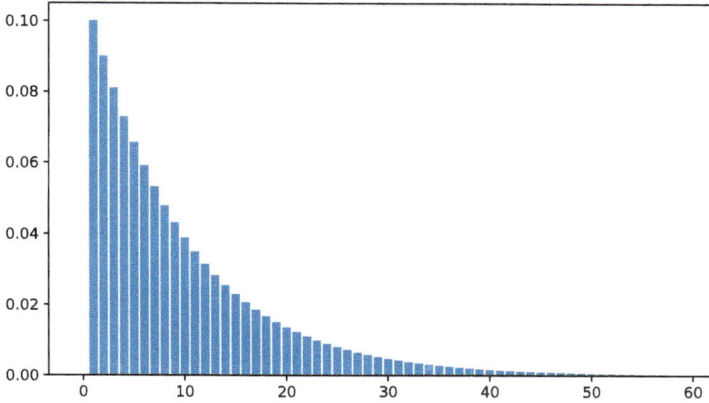

Fig. 2.3: PMF for $X \sim \text{Geom}(0.10)$.

2.2.3 Hypergeometric Distribution

Our final variation of the binomial distribution is the *hypergeometric distribution*. Like the binomial distribution, the hypergeometric distribution also represents the number of successes given a fixed total number of trials. The difference, however, is that the trials in the hypergeometric distribution are done *without replacement*, meaning that they are not independent.

Definition 2.5. *A random variable X is said to be a* hypergeometric *random variable with parameters N, M, and K, denoted $X \sim \text{HypGeom}(N, M, K)$, if it represents the number of red marbles in a random selection without replacement of K marbles from a bag initially consisting of M red marbles and $N - M$ white marbles.*

We can think of the hypergeometric random variable as the outcome of drawing a fixed number K marbles from a bag containing a total of N marbles, M of which are red. The K marbles are drawn *without replacement*, meaning that the outcome of each draw changes the probability of choosing a red marble on the following draw.

Proposition 2.10. *Let $X \sim \text{HypGeom}(N, M, K)$. Then the PMF for the random variable X is given by*

$$f(x) = \frac{\binom{M}{x}\binom{N - M}{K - x}}{\binom{N}{K}}, \tag{2.14}$$

for $x = 0, 1, 2, \ldots, K$. *We say that the random variable* X *has a* hypergeometric distribution.

Proof. The logic behind Equation (2.14) can be seen by counting. We start off with a bag of N marbles, M of which are red. From that bag, we draw K marbles, without replacement.

The total number of ways of selecting K marbles from a bag of N marbles is $\binom{N}{K}$. This represents the size of our sample space.

The probability mass $f(x)$ represents the probability that exactly $X = x$ of the K selected marbles are red. Out of the total number $\binom{N}{K}$ of possible outcomes, how many ways are there to generate an outcome with exactly x red marbles in the sample? First, we count the number of ways of selecting x red marbles out of the M red marbles in the bag; this is $\binom{M}{x}$. Second, we count the number of ways of selecting $K - x$ white marbles from the $N - M$ white marbles in the bag; this is $\binom{N-M}{K-x}$. Thus, the total number of possible ways of selecting x red marbles (and, therefore, $K - x$ white marbles) is the product $\binom{M}{x}\binom{N-M}{K-x}$. Dividing by the total number of possible outcomes (regardless of the count of red), we obtain our result.   □

Proposition 2.11. *Let* $X \sim \mathrm{HypGeom}(N, M, K)$. *Then the expected value and variance of* X *are given by*

$$\mathbb{E}[X] = \frac{KM}{N} \tag{2.15}$$

$$\mathbb{V}(X) = \frac{KM}{N}\frac{(N-M)(N-K)}{N(N-1)}. \tag{2.16}$$

We will omit the proof; though the interested reader can consult Casella and Berger [2002].

It is interesting to compare the mean and variance of the hypergeometric and binomial random variables. First, note that the ratio M/N represents the *initial* probability of drawing a red marble, i.e., the probability of drawing a red marble on the first draw. In this respect, Equation (2.15) is identical to the result in the case of a binomial random variable, i.e.,

$$\mathbb{E}[\mathrm{HypGeom}(N, M, K)] = \mathbb{E}[\mathrm{Binom}(n = K, p = M/N)].$$

The initial probability of drawing a white marble is of course $(N - M)/N$, which is a factor in Equation (2.16). If we were to suppose that the variance of the hypergeometric random variable was the same as the binomial random variable, we would end up with

$$np(1 - p) = \frac{KM}{N}\frac{(N-M)}{N},$$

which would be off by a factor of $(N - K)/(N - 1) < 1$. We conclude that the distribution of the hypergeometric random variable has an identical mean but smaller variance than the corresponding distribution of a binomial random variable, making the connection $p = M/N$. Therefore, the effect of the varying probability on each draw is to decrease the variance in the possible outcomes. This makes intuitive sense: each time we draw a red marble, it decreases the probability that the following marble will be red (because there is one fewer red marble in the bag). Similarly, when we draw a white marble, it increases the probability that the next marble is red. This acts as a sort of feedback mechanism that makes the total number of red marbles more likely to be closer to the expected value KM/N.

Simulation in Python

The parameters for the `numpy.random.hypergeometric` function are `ngood`, `nbad`, and `nsamples`, corresponding to M, $N-M$, and K, respectively. Usage is shown in line 5 of Code Block 2.9. The `scipy.stats.hypergeom` has

```
1   N = 100
2   M = 30
3   K = 10
4   T = 1000
5   samples = np.random.hypergeometric(M, N-M, K, size=T) # Generate T
        random samples
6   x = np.arange(11) # Array representing possible values for X
7   f = scipy.stats.hypergeom.pdf(x, N, M, K)
8   mu = scipy.stats.hypergeom.mean(N, M, K) # 3.0
9   sig2 = scipy.stats.hypergeom.var(N, M, K) # 1.9090
```

Code Block 2.9: Hypergeometric random variables in Python

the same variables presented above, i.e., $f(x) = \mathrm{HypGeom}(x; N, M, K)$[4]. A plot of the PMF for $\mathrm{HypGeom}(100, 30, 10)$ is shown in Figure 2.4.

Note 2.7. The output produced by the hypergeometric distribution, shown in Figure 2.4, looks very similar to the binomial distribution plotted in Figure 2.2. This makes intuitive sense, however, when we consider that we are only drawing 10 out of 100 marbles from the bag, and over the course of those ten draws, though the probability of choosing a red is changing, it is not changing greatly.

In fact, we can take this observation a step further and, holding the sample size K and the initial probability of a red marble $p = M/N$ fixed, take the limit to obtain

$$\lim_{N\to\infty} \mathrm{HypGeom}(x; N, pN, K) = \mathrm{Binom}(x; K, p).$$

---

[4] In the `scipy` docs, however, they use different variables to name these terms

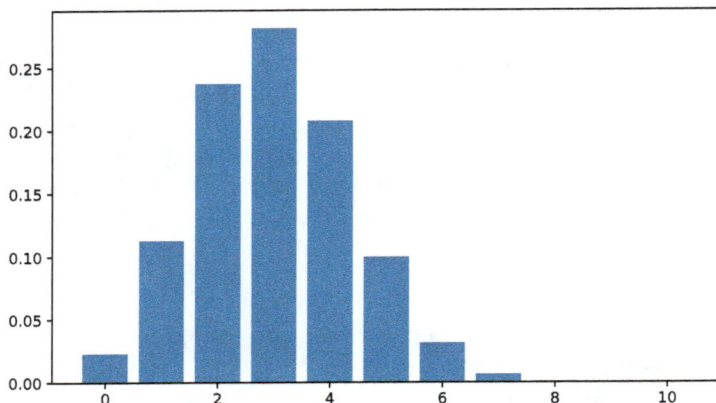

Fig. 2.4: PMF for $X \sim \text{HypGeom}(N = 100, M = 30, K = 10)$.

▷

2.3 Poisson and Exponential

In this section, we build a bridge from discrete to continuous distributions, introducing both the Poisson (discrete) and exponential (continuous) distributions. These two distributions are actually closely related, through a mechanism called the Poisson process.

2.3.1 Poisson Distribution

Definition 2.6. *The random variable X is said to have a* Poisson *distribution, denoted $X \sim \text{Poiss}(\lambda)$, if there is a number $\lambda > 0$ such that the* PMF *for X is given by*

$$f(x) = \frac{e^{-\lambda}\lambda^x}{x!}, \qquad (2.17)$$

for $x = 0, 1, 2, \ldots$.

The Poisson distribution is actually an interesting approximation of the binomial distribution, as shown in Proposition 2.12.

Proposition 2.12. *Let $X \sim \text{Binom}(n, p)$ be a binomial random variable. As n increases, with $\lambda = np$ held fixed, the random variable X approaches the Poisson random variable with parameter λ, i.e.,*

$$\lim_{n \to \infty} \text{Binom}(x; n, p = \lambda/n) = \text{Poiss}(x; \lambda). \qquad (2.18)$$

Proof. Let $X \sim \text{Binom}(n, p)$ and $\lambda = np$. Then the PMF for X is given by

$$\text{Binom}(x; n, p = \lambda/n) = \binom{n}{x} p^x (1 - p)^{n-x}$$

$$= \frac{n!}{x!(n-x)!} \left(\frac{\lambda}{n}\right)^x \left(1 - \frac{\lambda}{n}\right)^{n-x}$$

$$= \frac{n(n-1)\cdots(n-x+1)}{n^x} \frac{\lambda^x}{x!} \frac{(1-\lambda/n)^n}{(1-\lambda/n)^x}$$

Now, from calculus, we have

$$\lim_{n \to \infty} (1 - \lambda/n)^x = 1,$$

$$\lim_{n \to \infty} (1 - \lambda/n)^n = e^{-\lambda}, \text{ and}$$

$$\lim_{n \to \infty} \frac{n(n-1)\cdots(n-x+1)}{n^x} = 1.$$

The result follows. □

Proposition 2.12 shows that for large n and moderate $\lambda = np$, the number of successes may be approximated by a Poisson random variable.

The Poisson distribution is also used to model phenomena in which we are waiting for events to occur. In this context, the number of events occurring in a fixed interval of time can be modeled using the Poisson distribution. This relies on an underlying assumption that, for small time intervals, the probability of arrival is proportional to the length of the interval. We will examine this in more detail in Section 2.3.3.

Proposition 2.13. *Let $X \sim \text{Poiss}(\lambda)$. Then the mean and variance of X are given by*

$$\mathbb{E}[X] = \mathbb{V}(X) = \lambda. \tag{2.19}$$

For a proof, see Casella and Berger [2002]. For some insights into why this may not be surprising, consider the substitution $p = \lambda/n$ into Equations (2.5) and (2.6):

$$\mathbb{E}[X] = np = \lambda, \quad \mathbb{V}(X) = np(1-p) = \lambda\left(1 - \frac{\lambda}{n}\right).$$

We thus obtain Equation (2.19) in the limit as $n \to \infty$.

Finally, we note that the Poisson random variable is additive, in the following sense.

Proposition 2.14. *Let $X \sim \text{Poiss}(\lambda_1)$ and $Y \sim \text{Poiss}(\lambda_2)$ be independent Poisson random variables. Then the random variable $X + Y$ is the Poisson random variable $X + Y \sim \text{Poiss}(\lambda_1 + \lambda_2)$.*

We leave the proof for Exercise 2.3.

Simulation in Python

Simulating random numbers from a Poisson distribution in `numpy` and evaluating the PMF and CDF and other statistical quantities in `scipy` are done in the obvious way, as shown in Code Block 2.10.

```python
lam = 5 # lambda has a reserved meaning in Python
T = 1000
samples = np.random.poisson(lam, size=T) # T random samples from
    Poisson
x = np.arange(16)
f = scipy.stats.poisson.pmf(x, lam)
F = scipy.stats.poisson.cdf(x, lam)
```

Code Block 2.10: Simulation of a Poisson random variable

A plot of a Poisson random variable is shown in Figure 2.4.

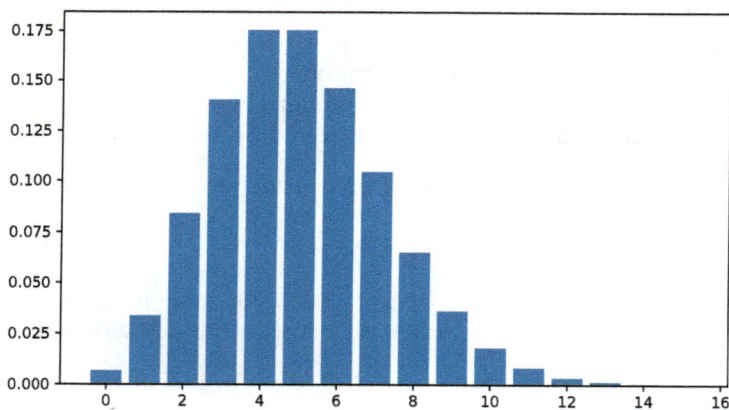

Fig. 2.5: PMF for $X \sim \text{Poiss}(5)$.

2.3.2 Exponential Distribution

We next turn to our first example of a continuous random variable: the exponential.

Definition 2.7. *A nonnegative, continuous random variable* X *is said to have an* exponential distribution *with scale parameter* β, *denoted* $X \sim \mathrm{Exp}(\beta)$, *if its* PDF *is of the form*

$$f(x) = \frac{1}{\beta} e^{-x/\beta}, \tag{2.20}$$

where $0 < \beta < \infty$, *for* $x \in [0, \infty)$. *Alternatively, the exponential distribution is sometimes parameterized by the* rate parameter $\lambda = 1/\beta$.

Note 2.8. Since the exponential random variable is nonnegative, its normalization is given by

$$\int_0^\infty \frac{1}{\beta} e^{-x/\beta} \, dx = 1.$$

▷

Proposition 2.15. *Let* $X \sim \mathrm{Exp}(\beta)$. *Then its mean and variance are given by*

$$\mathbb{E}[X] = \beta, \tag{2.21}$$
$$\mathbb{V}(X) = \beta^2. \tag{2.22}$$

Proof. It is straightforward to show that

$$\mathbb{E}[X] = \int_0^\infty \frac{x}{\beta} e^{-x/\beta} \, dx = \beta \qquad \text{and} \qquad \mathbb{E}[X^2] = \int_0^\infty \frac{x^2}{\beta} e^{-x/\beta} \, dx = 2\beta^2.$$

These can be combined, using Proposition 1.3, to yield the variance given by Equation (2.22). □

Proposition 2.16. *Let* $X \sim \mathrm{Exp}(\beta)$. *Then the* CDF *for* X *is given by*

$$F(x) = 1 - e^{-x/\beta}. \tag{2.23}$$

Similarly, its survival function *is given by*

$$S(x) = 1 - F(x) = e^{-x/\beta}. \tag{2.24}$$

Proposition 2.17. *The exponential distribution is* memoryless, *in the sense of Equation (2.13), i.e., for* $s > t \geq 0$, *we have*

$$\mathbb{P}(X > s | X > t) = \mathbb{P}(X > s - t).$$

Proof. From the definition of conditional probability, we have

$$\mathbb{P}(X > s | X > t) = \frac{\mathbb{P}(X > s)}{\mathbb{P}(X > t)} = e^{-(s-t)/\beta} = \mathbb{P}(X > s - t).$$

□

```
1  beta = 10
2  T = 1000
3  samples = np.random.exponential(scale=beta, size=T)
4  x = np.linspace(0, 30, num=100) # A linear array of values between
      0 and 30
5  f = scipy.stats.expon.pdf(x, scale=beta)
6  F = scipy.stats.expon.cdf(x, scale=beta)
```

Code Block 2.11: Simulation of the exponential random variable in Python

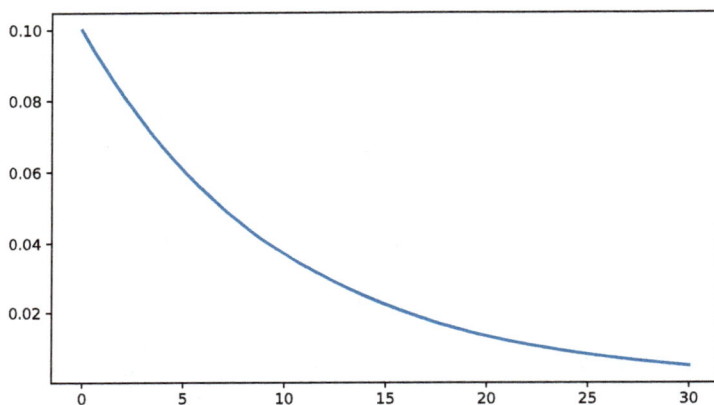

Fig. 2.6: PMF for $X \sim \text{Exp}(10)$.

Simulation in Python

Python code for generating random numbers from an exponential distribution and evaluating the PDF and CDF is shown in Code Block 2.11. A plot of the PDF is shown in Figure 2.6.

2.3.3 Poisson Process

The Poisson and exponential distributions are closely related to a mechanism often used in modeling a stream of events occurring at random intervals.

Definition 2.8. *For each $t \geq 0$, let N_t denote the number of occurrences of a particular event on the interval $[0, t]$. The nonnegative random variables N_t are said to constitute a* Poisson process *with parameter $\lambda > 0$ if they satisfy the following axioms*

1. *The process begins at $t = 0$; i.e., $N_0 = 0$;*
2. *The numbers of events in nonoverlapping intervals are independent; i.e., $N_{t_2} - N_{t_1}$ and $N_{s_2} - N_{s_1}$ are independent whenever $[t_1, t_2] \cap [s_1, s_2] = \emptyset$;*
3. *The distribution of the number of events in a given interval depends only on its length; i.e., for any $s, t > 0$, the random variables N_s and $N_{t+s} - N_t$ are identically distributed;*
4. *The probability of an event in a sufficiently short interval is proportional to its length; i.e., $\mathbb{P}(N_h = 1) = \lambda h + o(h)^5$; and*
5. *The probability of simulatneous events in a sufficiently short interval is essentially zero; i.e., $\mathbb{P}(N_h > 1) = o(h)$.*

Note 2.9. Due to axiom 2, axioms 4 and 5 apply to any sufficiently short interval, i.e., $\mathbb{P}(N_{t+h} - N_t = 1) = \lambda h + o(h)$, and $\mathbb{P}(N_{t+h} - N_t > 1) = o(h)$.
▷

Theorem 2.1. *Let N_t define a Poisson process with parameter λ. Then N_t is a Poisson random variable with parameter λt; i.e.,*

$$N_t \sim \text{Poiss}(\lambda t). \tag{2.25}$$

Proof. Fix $t > 0$. The random variable N_t represents the number of events occurring on the interval $I = [0, t]$. Let us divide the interval I into n equally spaced subintervals of length $h = t/n$, so that $I_i = [x_{i-1}, x_i]$ for $i = 1, \ldots, n$, where $x_i = ih$. If n is large enough, we may combine axioms 3 and 4 to show that the probability of an event occurring on the ith subinterval is given by

$$p = \mathbb{P}(N_{x_i} - N_{x_{i-1}} = 1) \approx \lambda h.$$

Similarly, by axioms 3 and 5, the probability of two or more events occurring on any subinterval is essentially zero. Since our subintervals are disjoint, axiom 2 guarantees that our n subintervals constitute a sequence of n IID Bernouilli trials, with probability $p = \lambda h = \lambda t/n$. Therefore, for sufficiently large n, the random variable N_t may be approximated by

$$N_t \approx \text{Binom}(n, \lambda t/n),$$

that approximation becoming exact as $n \to \infty$. However, we saw from Proposition 2.12 that the right hand side becomes $\text{Poiss}(\lambda t)$ as $n \to \infty$. This proves the result.   □

In order to show the relationship between the Poisson process and the exponential distribution, we first need to define interarrival times.

Definition 2.9. *Let N_t represent a Poisson process with parameter λ. Then the nth interarrival time, denoted T_n, represents the time of the first event, if $n = 1$, or the time between the $(n-1)$th event and the nth event, if $n \geq 2$.*

---
[5] *Little-Oh Notation*: We write $f(h) = o(h)$ if $f(h)/h \to 0$ as $h \to 0$.

Note 2.10. If we let
$$t_n = \inf\{t \geq 0 : N_t \geq n\}$$
represent the time of the nth event (notice that $t_0 = 0$), then the nth interarrival time T_n is given by $T_n = t_n - t_{n-1}$.   ▷

Theorem 2.2. *Let N_t define a Poisson process with parameter λ, and let $\{T_n\}$ represent the sequence of interarrival times. Then the T_ns are IID exponential random variables with scale parameter $\beta = 1/\lambda$; i.e.,*

$$T_n \sim \exp(\beta = 1/\lambda). \tag{2.26}$$

Proof. Let us first consider T_1. For a given, fixed value $t > 0$, the probability that T_1 occurs after t is given by

$$\mathbb{P}(T_1 > t) = \mathbb{P}(N_t = 0) = \text{Poiss}(0; \lambda t) = e^{-\lambda t},$$

since $N_t \sim \text{Poiss}(\lambda t)$[6]. However, we recognize this from Equation (2.24) as the survival function for an exponential random variable with scale parameter $\beta = 1/\lambda$. Thus, $T_1 \sim \text{Exp}(1/\lambda)$.

Next, let us consider T_n for $n \geq 2$, and suppose that the $(n-1)$th event occurred at time $s = t_{n-1}$. Consider now any $t > s$. The probability of the nth interarrival time being greater than $t - s$ is given by

$$\mathbb{P}(T_n > t|t_{n-1} = s) = \mathbb{P}(N_t - N_s = 0) = \mathbb{P}(N_{t-s} = 0) = e^{-\lambda(t-s)}.$$

Thus, the Poisson process is memoryless, and $T_n \sim \text{Exp}(1/\lambda)$.   □

Simulation in Python

We can use the theory presented in this section to build a simple simulator for a Poisson process, as shown in Code Block 2.12. Note that by appending the event time to the list of events (line 9) inside of the while-loop, we are automatically ensuring that the event times are less than the input parameter t.

In order to simulate a single sample from $N_t \sim \text{Poiss}(\lambda t)$, we can simply take the length of the output, e.g., `len(poiss_proc(10, lambda_rate=3))`.

The output of the `poiss_proc` can be plotted visually, as shown in Code Block 2.13. Note that line 3 leverages the `emp_dist` function, defined previously in Code Block 1.5. The output of `emp_dist` is multiplied by the number of samples, in order to *un-normalize* the CDF. Using `num=2000` in numpy's `linspace` method ensures a resolution of 0.05 for successive event times. The resulting output is shown in Figure 2.7. Four additional runs of `poiss_proc(100, lambda_rate=0.10)` are shown in Figure 2.8.

---
[6] Recall that $0! = 1$

```
1   def poiss_proc(t, lambda_rate=1):
2       # Inputs:
3       #    t (float) -- length of time to run simulation.
4       #    lambda_rate -- the rate parameter of the Poisson process.
5       # Output:
6       #    events -- a list of event times.
7       assert lambda_rate > 0
8
9       # Set current_time to time of first event.
10      current_time = np.random.exponential(scale=1/lambda_rate)
11      events = [] # List of Events
12
13      while current_time < t:
14          events.append(current_time) # Add event time to list
15          delta_t = np.random.exponential(scale=1/lambda_rate)
16          current_time += delta_t # Time of next event
17
18      events = np.array(events) # Convert list to np array.
19
20      return events
```

Code Block 2.12: Simulation of a Poisson process

```
1   x = np.linspace(0, 100, num=2000)
2   samples = poiss_proc(100, lambda_rate=0.10)
3   y = emp_dist(x, samples) * len(samples)
4   plt.plot(x, y)
```

Code Block 2.13: Generating plot of Poisson process

2.4 The Normal Distribution

In this section, we introduce the normal distribution, discuss its properties, and conclude with a discussion of one of the key results of statistics: the central limit theorem.

2.4.1 Normal Distribution

The normal distribution (or *Gaussian* distribution) is the distribution which gives rise to the familiar "bell-shaped curve" of classical statistics. It also plays a key part in sampling distributions via the central limit theorem, which we'll discuss later in the section.

Definition 2.10. *The continuous random variable X is a* normal random variable with mean μ and variance σ^2, *denoted $X \sim \mathrm{N}(\mu, \sigma^2)$, if its* PDF

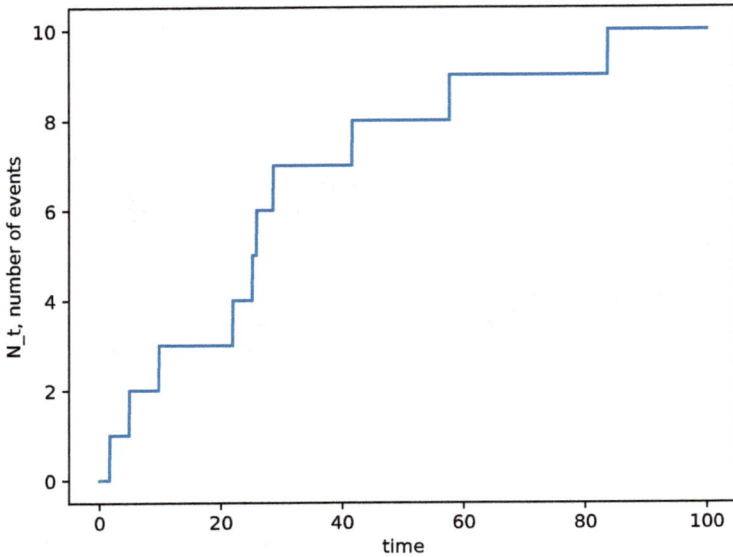

Fig. 2.7: Poisson process N_t, with rate $\lambda = 0.10$.

satisfies the normal distribution, *given by*

$$f(x) = \frac{1}{\sigma\sqrt{2\pi}} e^{-(x-\mu)^2/(2\sigma^2)}, \tag{2.27}$$

for $-\infty < x < \infty$. *The special case of* $Z \sim \mathrm{N}(0,1)$ *is referred to as a* standard normal random variable.

Note 2.11. Given the normal random variable $X \sim \mathrm{N}(\mu, \sigma^2)$, the transformed random variable $Z = (X - \mu)/\sigma$ constitutes a *standard* normal random variable; i.e., $Z \sim \mathrm{N}(0,1)$. See Lemma 2.1 for a proof.   ▷

Proposition 2.18. *The normal distribution, as given by Equation (2.27), is properly normalized; i.e., the total probability under the curve is 1.*

Proof. To prove normalization, let us define the total probability as

$$I = \int_{-\infty}^{\infty} \frac{1}{\sigma\sqrt{2\pi}} e^{-(x-\mu)^2/(2\sigma^2)} \, dx.$$

By introducing the change of variable $z = (x - \mu)/\sigma$, we can rewrite I as

$$I = \int_{-\infty}^{\infty} \frac{1}{\sqrt{2\pi}} e^{-z^2/2} \, dz.$$

(a)

(b)

(c)

(d)

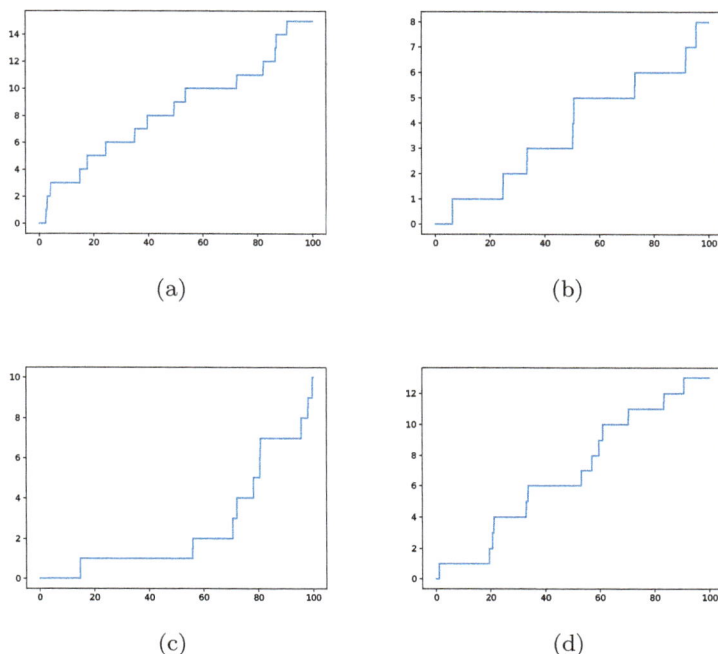

Fig. 2.8: Four simulations of a Poisson process.

Next, let us consider the quantity I^2, as given by

$$I^2 = \frac{1}{2\pi} \left(\int_{-\infty}^{\infty} e^{-x^2/2}\, dx \right) \left(\int_{-\infty}^{\infty} e^{-y^2/2}\, dy \right)$$

$$= \frac{1}{2\pi} \int_{-\infty}^{\infty} \int_{-\infty}^{\infty} e^{-(x^2+y^2)/2}\, dx\, dy$$

$$= \frac{1}{2\pi} \int_{0}^{2\pi} \int_{0}^{\infty} e^{-r^2/2} r\, dr\, d\theta,$$

where the last equality follows by converting to polar coordinates. The final expression is easily evaluated to yield $I^2 = 1$. Since $I > 0$, it follows that $I = 1$ and the proposition is proved. □

Proposition 2.19. *Let $X \sim \mathrm{N}(\mu, \sigma^2)$. Then the mean and variance of X are (shockingly) given by*

$$\mathbb{E}[X] = \mu \qquad and \qquad \mathbb{V}(X) = \sigma^2. \tag{2.28}$$

Proof. It is a straightforward matter to show that

$$\mathbb{E}[X] = \int_{-\infty}^{\infty} \frac{x}{\sigma\sqrt{2\pi}} e^{-(x-\mu)^2/(2\sigma^2)}\, dx = \mu$$

and

$$\mathbb{E}[X^2] = \int_{-\infty}^{\infty} \frac{x^2}{\sigma\sqrt{2\pi}} e^{-(x-\mu)^2/(2\sigma^2)}\, dx = \mu^2 + \sigma^2.$$

The result follows.   □

Definition 2.11. *Let $Z \sim N(0,1)$ be the standard normal random variable. We shall denote the standard-normal* CDF *and survival function using the variable $\Phi(z)$ and $\Psi(z)$, respectively; i.e., we may write*

$$\Phi(z) = \mathbb{P}(Z \le z) = \int_{-\infty}^{z} \frac{1}{\sigma\sqrt{2\pi}} e^{-x^2/2}\, dx, \qquad (2.29)$$

$$\Psi(z) = \mathbb{P}(Z > z) = \int_{z}^{\infty} \frac{1}{\sigma\sqrt{2\pi}} e^{-x^2/2}\, dx. \qquad (2.30)$$

Further, for $\alpha \in (0,1)$, we shall use z_α to represent the point for which

$$\Psi(z_\alpha) = \mathbb{P}(Z > z_\alpha) = \alpha. \qquad (2.31)$$

In other words, $z_\alpha = \Psi^{-1}(\alpha)$.

Though the CDF, survival function, and inverse survival function cannot be evaluated analytically, numerical evaluations are no challenge to the computer. Before turning to simulation, however, there is one additional feature worth pointing out.

Simulation in Python

The numpy and scipy code for handling normal random variables is shown in Code Block 2.14. First, note that np.random.randn represents the standard normal random variable, so samples must be amplified by the standard deviation σ and shifted by the mean μ. The result of line 4 is an array of samples from $N(\mu, \sigma^2)$.

The scipy library handles the mean and variance using the location and scale parameters, as shown on lines 6–11. Lines 6–8 generate the PMF, the CDF, and the survival function for the distribution $N(\mu, \sigma^2)$, evaluated at x, which may be a number or an array of numbers.

The built-in inverse survival function (line 9) generates the point x_0 at which 5% of the total probability still remains, i.e., $C(x_0) = 0.95$. This is confirmed in lines 10 and 11.

The normal random variable $X \sim N(100, 225)$, as generated by Code Block 2.14, is plotted in Figure 2.9. The area to the right of x0=124.6728 is shaded, representing a total probability of 0.05.

```
1   mu = 100
2   sigma = 15
3   n_samples = 1000
4   samples = np.random.randn(n_samples) * sigma + mu
5   x = np.linspace(50, 150, num=200)
6   f = scipy.stats.norm.pdf(x, loc=mu, scale=sigma) # PDF
7   F = scipy.stats.norm.cdf(x, loc=mu, scale=sigma) # CDF
8   S = scipy.stats.norm.sf(x, loc=mu, scale=sigma) # Survival Function
9   x0 = scipy.stats.norm.isf(0.05, loc=mu, scale=sigma) # Inv. Survival
10  scipy.stats.norm.sf(x0, loc=mu, scale=sigma) # 0.05
11  scipy.stats.norm.cdf(x0, loc=mu, scale=sigma) # 0.95
```

Code Block 2.14: Normal random variable

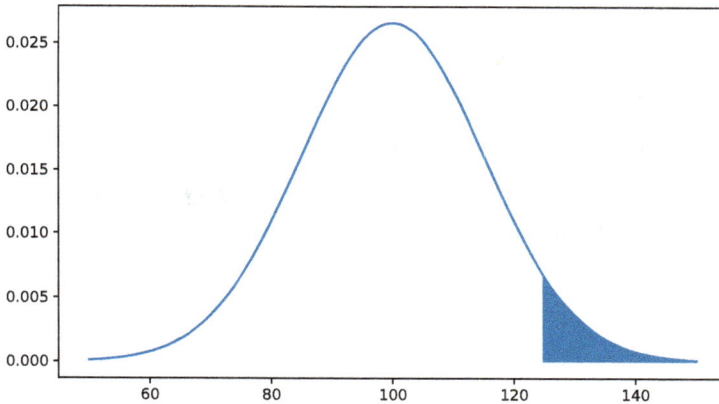

Fig. 2.9: Normal random variable $X \sim \mathrm{N}(100, 225)$.

2.4.2 Sums of Normal Random Variables

Normal random variables have certain additive properties that are quite useful.

Lemma 2.1. *Let* $X \sim \mathrm{N}(\mu, \sigma^2)$ *be a normal random variable. Then the random variable* $Y = aX + b$, *for* $a \neq 0$, *is also a random variable, and* $Y \sim N(a\mu + b, a^2\sigma^2)$.

Proof. Consider the transformation $y = g(x) = ax + b$. The mapping $g : \mathbb{R} \to \mathbb{R}$ is one-to-one, so there is no need to partition the space. From Equation (1.35) (Corollary 1.1), we have

$$f_Y(y) = f_X((y-b)/a)/a$$
$$= \frac{1}{a\sigma\sqrt{2\pi}} e^{-((y-b)/a-\mu)^2/(2\sigma^2)}$$
$$= \frac{1}{a\sigma\sqrt{2\pi}} e^{-(y-(a\mu+b))^2/(2a^2\sigma^2)},$$

which we recognize as the PDF of a random variable $Y \sim \mathrm{N}(a\mu + b, a^2\sigma^2)$. □

This lemma immediately validates Note 2.11, which is important on its own right, so we restate the result next as a corollary to the preceding lemma.

Corollary 2.1. *Let* $X \sim \mathrm{N}(\mu, \sigma^2)$ *be a normal random variable. Then the transformed random variable*

$$Z = \frac{X - \mu}{\sigma} \tag{2.32}$$

is a standard normal random variable, i.e., $Z \sim \mathrm{N}(0, 1)$.

Proof. This follows from Lemma 2.1 by setting $a = 1/\sigma$ and $b = -\mu/\sigma$. □

Lemma 2.2. *Let* $X, Y \sim \mathrm{N}(0, 1)$ *be two independent and identically distributed standard normal random variables. Then the random variable* $U = X + Y$ *is a normal random variable, and* $U \sim \mathrm{N}(0, 2)$.

Proof. Due to the independence of X and Y, the joint distribution is given by

$$f_{X,Y}(x, y) = \frac{1}{2\pi} e^{-x^2/2} e^{-y^2/2} = \frac{1}{2\pi} e^{-(x^2+y^2)/2}.$$

To proceed, we will be applying Theorem 1.9. Let us define the mapping $(u, v) = g(x, y)$ by the relations

$$u = x + y \qquad \text{and} \qquad v = x - y,$$

which has the inverse mapping

$$x = \frac{u+v}{2} \qquad \text{and} \qquad y = \frac{u-v}{2}.$$

The mapping $g : \mathbb{R}^2 \to \mathbb{R}^2$ is one-to-one, so we do not need to partition the domain. The Jacobian of this transformation is given by

$$J(u, v) = \det \frac{\partial g^{-1}(u, v)}{\partial(u, v)} = \det \begin{bmatrix} \dfrac{\partial x}{\partial u} & \dfrac{\partial x}{\partial v} \\ \dfrac{\partial y}{\partial u} & \dfrac{\partial y}{\partial v} \end{bmatrix} = \det \begin{bmatrix} 1/2 & 1/2 \\ 1/2 & -1/2 \end{bmatrix} = -1/2.$$

From Equation (1.37), we have

$$f_{U,V}(u, v) = f_{X,Y}\left(\frac{u+v}{2}, \frac{u-v}{2}\right)\left|\frac{-1}{2}\right| = \frac{1}{4\pi}e^{-(u^2+v^2)/4}.$$

This can be factored as

$$f_{U,V}(u, v) = \left[\frac{1}{2\sqrt{\pi}}e^{-u^2/4}\right]\left[\frac{1}{2\sqrt{\pi}}e^{-v^2/4}\right],$$

which we recognize as the product of two normal distributions with mean $\mu = 0$ and variance $\sigma^2 = 2$. The distribution for U can be obtained by marginalizing over the random variable V, from which we obtain

$$f_U(u) = \int_{-\infty}^{\infty} f_{U,V}(u, v)\, dv = \frac{1}{2\sqrt{\pi}}e^{-u^2/4}.$$

This proves that $U = X + Y \sim N(0, 2)$. (Consequently, it also proves that $V = X - Y \sim N(0, 2)$.) $\qquad\square$

Lemma 2.2 was a warm-up for our next lemma.

Lemma 2.3. *Let $X \sim N(0, \sigma^2)$ and $Y \sim N(0, \tau^2)$ be two independent and identically distributed standard normal random variables. Then the random variable $Z = X + Y$ is a normal random variable, and $Z \sim N(0, \sigma^2 + \tau^2)$.*

Proof. The joint distribution is given by

$$f_{X,Y}(x, u) = \frac{1}{2\pi\sigma\tau}e^{-(x^2/\sigma^2 + y^2/\tau^2)/2}.$$

Consider the transformation

$$u = \frac{x+y}{\sqrt{\sigma^2 + \tau^2}} \quad \text{and} \quad v = \frac{\tau^2 x - \sigma^2 y}{\sigma\tau\sqrt{\sigma^2 + \tau^2}}.$$

The Jacobian of the transformation (exercise) is $|J(u, v)| = \sigma\tau$. Further, note that

$$u^2 + v^2 = \frac{x^2}{\sigma^2} + \frac{y^2}{\tau^2}.$$

It follows that the joint distribution for the random vector (U, V) is given by

$$f_{U,V}(u, v) = \frac{1}{2\pi}e^{-(u^2+v^2)/2} = \left[\frac{1}{\sqrt{2\pi}}e^{-u^2/2}\right]\left[\frac{1}{\sqrt{2\pi}}e^{-v^2/2}\right].$$

By marginalizing over V, we conclude that $U \sim N(0, 1)$. From Lemma 2.1, we further conclude that the random variable $Z = X + Y \sim N(0, \sigma^2 + \tau^2)$, since $Z = \sqrt{\sigma^2 + \tau^2}\,U$. $\qquad\square$

Lemma 2.4. *Let $X \sim N(\mu, \sigma^2)$ and $Y \sim N(\nu, \tau^2)$ be two independent normal random variables. Then the random variable $Z = X + Y$ is normal, and $Z \sim N(\mu + \nu, \sigma^2 + \tau^2)$.*

Proof. First, consider $U = X - \mu$ and $V = Y - \nu$. By Lemma 2.1, we have $U \sim N(0, \sigma^2)$ and $V \sim N(0, \tau^2)$.

Next, consider $W = U + V$. By Lemma 2.3, it follows that $W \sim N(0, \sigma^2 + \tau^2)$. (In terms of our original variables X and Y, $W = X + Y - (\mu + \nu)$.)

Finally, consider $Z = W + (\mu + \nu)$. By Lemma 2.1, $Z \sim N(\mu + \nu, \sigma^2 + \tau^2)$. However, in terms of our original variables, $Z = X + Y$. This completes the result. $\qquad\square$

Theorem 2.3. *Suppose $X_1, \ldots, X_n$ are mutually independent normal random variables with $X_i \sim N(\mu_i, \sigma_i^2)$, for $i = 1, \ldots, n$. Then the random variable*

$$X = \sum_{i=1}^{n} c_i X_i$$

is also a normal random variable; moreover

$$X \sim N\left(\sum_{i=1}^{n} c_i \mu_i, \sum_{i=1}^{n} c_i^2 \sigma_i^2 \right). \tag{2.33}$$

Proof. First, let us define $Y_1, \ldots, Y_n$ by the relations $Y_i = c_i X_i$. From Lemma 2.1, $Y_i \sim N(c_i \mu_i, c_i^2 \sigma_i^2)$. Applying Lemma 2.4 repeatedly then produces the result. $\qquad\square$

Note 2.12. There are many ways to go about proving Theorem 2.3. A very simple and elegant proof involves *moment-generating functions*, as shown in Casella and Berger [2002]. It can also be proved using convolution integrals, as done in Ross [2012]. $\qquad\triangleright$

Corollary 2.2. *Let $X_1, \ldots, X_n \sim N(\mu, \sigma^2)$ be IID samples, then the sample mean $\overline{X} = (1/n) \sum_{i=1}^{n} X_i$ has the normal distribution $\overline{X} \sim N(\mu, \sigma^2/n)$.*

Proof. This follows from Theorem 2.3 using $c_i = 1/n$, $\mu_i = \mu$, and $\sigma_i = \sigma$. $\square$

An illustration of this Corollary can be found in Figure 1.1.

2.4.3 The Central Limit Theorem

The result from Corollary 2.2 is not only exactly true for IID samples of a normal random variable, but also approximately true for IID samples from many different distributions.

Theorem 2.4 (Central Limit Theorem). *Let* $X_1, \ldots, X_n \sim F$ *be* IID *samples from a distribution with finite mean* μ *and finite variance* σ^2. *Let* $\overline{X}_n = (1/n) \sum_{i=1}^{n} X_i$ *represent the sample mean. Then the sequence of distributions*

$$W_n = \frac{\overline{X}_n - \mu}{\sigma/\sqrt{n}} \tag{2.34}$$

approaches the standard normal distribution $N(0, 1)$ *in the limit as* $n \to \infty$[7].

Note 2.13. In practice, we may approximate W using the standard normal for $n > 30$. ▷

The central limit theorem adds to Theorem 1.2, which gave us the expected value and variance of the sample mean as $\mathbb{E}[\overline{X}_n] = \mu$ and $\mathbb{V}(\overline{X}_n) = \sigma^2/n$. Theorem 2.4 tells us that, in addition, whenever we have large samples, the sampling distribution of $\overline{X}_n$ is approximately a normal with mean μ and variance σ^2/n.

Simulation in Python

We can see Theorem 2.4 in action by constructing a simulations in Python. Let's choose the distribution that is most unlike the normal distribution: a good candidate is the geometric distribution, as shown in Figure 2.3. In particular, let's draw samples from Geom(0.10), which has $\mathbb{E}[X] = p = 10$ and $\mathbb{V}(X) = (1 - p)/p^2 = 90$. The code is given in Code Block 2.15.

```
1   n_samples = 1000
2   sample_size = 10
3   p = 0.10
4   mu = 1 / p
5   sigma = np.sqrt( (1 - p) / p**2 )
6   x_bar = np.zeros(n_samples) # Sample Means
7   for i in range(n_samples):
8       x = np.random.geometric(p, size=sample_size)
9       x_bar[i] = x.mean() # Sample Mean.
10
11  plt.hist(x_bar, bins=50, normed=True)
12  x = np.linspace(0, 22, num=100)
13  f = scipy.stats.norm.pdf(x, loc=1/p,
        scale=sigma/np.sqrt(sample_size))
14  plt.plot(x, f, color='b', linewidth=2)
```

Code Block 2.15: Sampling distribution for $\overline{X}_{10}$ for $X \sim \text{Geom}(0.10)$

---

[7] I.e., the sequence $\{W_n\}$ converges to $Z \sim N(0, 1)$ *in distribution* as $n \to \infty$

We collect a total of 1,000 samples, each of sample size 10, and compute the sample mean $\overline{X}_{10}$ for each one. The resulting histogram for these 1,000 sample means is shown in Figure 2.10, plotted concurrently with $N(10, 9)$.

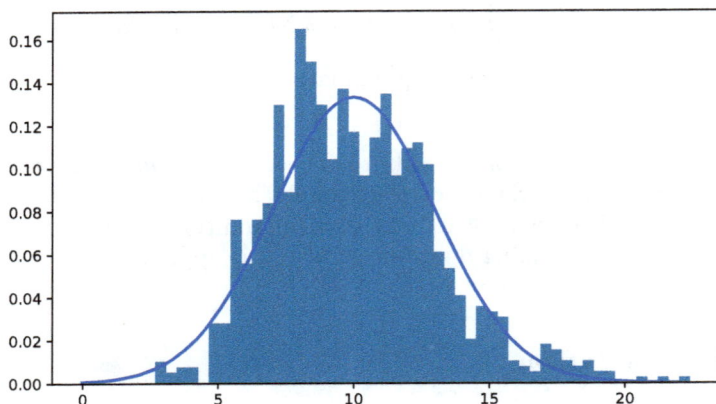

Fig. 2.10: Distribution of sample mean $\overline{X}_{10}$, with $N(10, 9)$.

The result is certainly not bad. But let's see what happens if we choose a larger sample size. Let's increase the sample size to 30. The histogram for $\overline{X}_{30}$ and the PDF for the approximating normal distribution $N(10, 3)$ are shown in Figure 2.11.

As the sample size n increases, the distribution of the sample mean $\overline{X}_n$ will more closely resemble that of the normal distribution $N(\mu, \sigma^2/n)$. Note also the change of scales between Figures 2.10 and 2.11: the variance in the sample mean decreases as the sample size increases.

2.5 Chi-Squared, T, and F

In this section, we introduce several important distributions for hypothesis testing and analysis of variance.

2.5.1 The Chi-Squared Distribution

Definition 2.12. *Let* $Z_1, \ldots, Z_p \sim N(0, 1)$ *represent* IID *standard normal random variables. Then a random variable* X *is said to be a* chi-squared *random variable with* p *degrees of freedom, denoted* $X \sim \chi_p^2$, *if it can be expressed as*

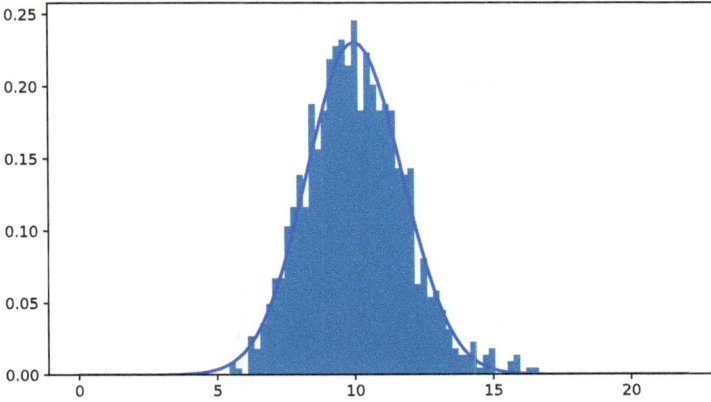

Fig. 2.11: Distribution of sample mean $\overline{X}_{30}$, with $N(10, 3)$.

$$X = \sum_{i=1}^{p} Z_i^2. \tag{2.35}$$

Proposition 2.20. *If $X \sim \chi_p^2$ and $Y \sim \chi_q^2$ are two independent chi-squared distributions, with degrees of freedom p and q, respectively, then the sum is a chi-squared random variable with $p + q$ degrees of freedom; i.e., $X + Y \sim \chi_{p+q}^2$.*

Proof. This follows immediately from Definition 2.12.   □

We next seek to uncover an analytical form for the PDF of a chi-squared random variable. We will proceed using mathematical induction. We state our basis step in the following lemma.

Lemma 2.5. *Let $X \sim \chi_1^2$ represent a chi-squared random variable with 1 degrees of freedom. Then the PDF for X is given by*

$$f(x) = \frac{1}{\sqrt{2\pi x}} e^{-x/2}, \tag{2.36}$$

for $0 < x < \infty$.

Proof. First consider a standard normal random variable $Z \sim N(0, 1)$, with PDF

$$f_Z(z) = \frac{1}{\sqrt{2\pi}} e^{-z^2/2}.$$

Let us consider the transformation $x = z^2$, and proceed by following the recipe outlined in Theorem 1.7. The mapping $x = g(z) = z^2$ is not one-to-one, so we will partition the domain into three subsets

$$\mathcal{A}_0 = \{0\}, \qquad \mathcal{A}_1 = (-\infty, 0), \qquad \text{and} \qquad \mathcal{A}_2 = (0, \infty).$$

Note that, consistent with Theorem 1.7, $\mathbb{P}(X \in \mathcal{A}_0) = 0$. Also, $\mathcal{X}_i = \{x \in \mathbb{R} : x = g(z) \text{ for some } z \in \mathcal{A}_i\} = (0, \infty)$ for $i = 1, 2$. Moreover, $g_1^{-1}(x) = -\sqrt{x}$ and $g_2^{-1}(x) = \sqrt{x}$, from which it follows

$$\left| \frac{dg_1^{-1}(x)}{dx} \right| = \left| \frac{dg_2^{-1}(x)}{dx} \right| = \frac{1}{2\sqrt{x}}.$$

Applying Equation (1.34) to this context, we have

$$f_X(x) = \sum_{i=1}^{2} f_Z(g_i^{-1}(x)) \cdot \frac{1}{2\sqrt{x}} = \frac{1}{\sqrt{2\pi x}} e^{-x/2},$$

which proves the result. (Note that the $1/2$ disappears as there are two identical terms in the sum.) □

Lemma 2.6. *Let* $X \sim \chi_2^2$ *represent a chi-squared random variable with 2 degrees of freedom. Then the* PDF *for* X *is given by*

$$f(x) = \frac{1}{2} e^{-x/2}, \tag{2.37}$$

for $0 < x < \infty$.

Proof. Let $X, Y \sim N(0, 1)$ be standard normal random variables, with joint distribution

$$f_{X,Y}(x, y) = \frac{1}{2\pi} e^{-(x^2 + y^2)/2}.$$

Next, let us consider the transformation

$$u = x^2 + y^2 \qquad \text{and} \qquad v = x/y.$$

Note, the random variable $U = X^2 + Y^2$ is a χ_2^2 random variable, so it suffices to show that the distribution for U is given by Equation (2.37).

To proceed, we shall follow the recipe outlined in Theorem 1.9. Since the mapping $g : \mathbb{R}^2 \to (0, \infty) \times \mathbb{R}$ defined above is not one-to-one, we must partition the plane into three regions:

$$\mathcal{A}_0 = \{(x, y) \in \mathbb{R}^2 : y = 0\}$$
$$\mathcal{A}_1 = \{(x, y) \in \mathbb{R}^2 : y > 0\}$$
$$\mathcal{A}_2 = \{(x, y) \in \mathbb{R}^2 : y < 0\}.$$

Note that $\mathbb{P}((x, y) \in \mathcal{A}_0) = 0$. The Jacobian of the transformation can be readily be shown to equal $|J(u, v)| = 1/(2(1 + v^2))$. Applying Equation (1.37), we may write

$$f_{U,V}(u,v) = \sum_{i=1}^{2} f_{X,Y}(g_i^{-1}(u,v))\frac{1}{2(1+v^2)} = \frac{1}{2\pi(1+v^2)}e^{-u/2}.$$

Recall from calculus that

$$\int_{-\infty}^{\infty} \frac{dv}{1+v^2} = \pi.$$

Therefore, by margianlizing over the random variable V, we obtain Equation (2.37), and the result is proven. □

Before presenting the PDF of a general chi-squared random variable, let us first define the *gamma function*, which is a useful generalization of the factorial.

Definition 2.13. *The* gamma function $\Gamma(z)$ *is the function defined by the integral*

$$\Gamma(z) = \int_0^{\infty} x^{z-1}e^{-x}\,dx. \tag{2.38}$$

By integrating by parts, it is not difficult to show that

$$\Gamma(z+1) = z\Gamma(z). \tag{2.39}$$

Since $\Gamma(1) = 1$, it follows that $\Gamma(n) = (n-1)!$. The gamma function, however, is defined for any positive real number (as well as negative non-integers). A particularly useful value to know is $\Gamma(1/2) = \sqrt{\pi}$. Using the recurrence relation given by Equation (2.39), one can easily determine values of $\Gamma(n/2)$, for any positive integer n.

Theorem 2.5. *Let* $X \sim \chi_p^2$ *represent a chi-squared random variable with* p *degrees of freedom. Then the* PDF *for* X *is given by*

$$f(x) = \frac{1}{\Gamma(p/2)2^{p/2}}x^{p/2-1}e^{-x/2}, \tag{2.40}$$

for $0 < x < \infty$. *This distribution is known as the* chi-squared distribution *with* p *degrees of freedom.*

Proof. We shall proceed by using a modified form of mathematical induction: for our induction step, we will increment the parameter p by *two* instead of by one. We will therefore require two basis steps: for $p = 0$ and for $p = 1$. The basis steps, however, have already been proved in Lemmas 2.5 and 2.6, as one can easily verify that Equations (2.36) and (2.37) are a special cases of Equation (2.40) for $p = 1, 2$, respectively.

For the *induction step*, we must prove the logical statement: if Equation (2.40) is true for a given value of p, then it must also be true for $p + 2$.

To begin, let us assume that Equation (2.40) is valid for a given value of $p \geq 1$. Let $X \sim \chi_p^2$ be a chi-squared random variable with p degrees of freedom, and let $Y \sim \chi_2^2$ be a chi-squared random variable with 2 degrees of freedom. From Equation (2.37) and our assumed distribution for X, given by Equation (2.40), we see that the joint distribution is given by

$$f_{X,Y}(x,y) = \frac{x^{p/2-1}}{\Gamma(p/2)2^{p/2+1}} e^{-(x+y)/2}.$$

Now consider the transformation $g : (0,\infty) \times (0,\infty) \to (0,\infty) \times (0,\pi/2)$ defined by

$$u = x + y \quad \text{and} \quad v = \tan^{-1}\left(\frac{y}{x}\right).$$

Note that due to Proposition 2.20, the random variable $U = X + Y$ is a chi-squared random variable with $p + 2$ degrees of freedom; i.e., $U \sim \chi_{p+2}^2$. This relation is invertible, with inverse transformation given by

$$x = \frac{u}{1 + \tan(v)} \quad \text{and} \quad y = \frac{u\tan(v)}{1 + \tan(v)}.$$

The Jacobian is therefore

$$J(u,v) = \det \begin{bmatrix} \dfrac{1}{1+\tan(v)} & \dfrac{-u\sec^2(v)}{(1+\tan(v))^2} \\ \dfrac{\tan(v)}{1+\tan(v)} & \dfrac{u\sec^2(v)}{(1+\tan(v))^2} \end{bmatrix} = \frac{u\sec^2(v)}{(1+\tan(v))^2}.$$

The joint distribution for the random vector (U,V) is therefore given by

$$\begin{aligned} f_{U,V}(u,v) &= \frac{1}{\Gamma(p/2)2^{p/2+1}} \left(\frac{u}{1+\tan(v)}\right)^{p/2-1} e^{-u/2} \cdot \frac{u\sec^2(v)}{(1+\tan(v))^2} \\ &= \frac{u^{p/2}e^{-u/2}}{\Gamma(p/2)2^{p/2+1}} \frac{\sec^2(v)}{(1+\tan(v))^{p/2+1}} \end{aligned}$$

However, one can easily show that

$$\int_0^{\pi/2} \frac{\sec^2(v)}{(1+\tan(v))^{p/2+1}} \, dv = \frac{2}{p}.$$

Therefore, the marginal distribution for the random variable U is given by

$$f_U(u) = \int_0^{\pi/2} f_{U,V}(u,v)\, dv = \frac{u^{p/2}e^{-u/2}}{(p/2)\cdot\Gamma(p/2)2^{p/2+1}} = \frac{u^{p/2}e^{-u/2}}{\Gamma(p/2+1)2^{p/2+1}}.$$

However, this expression is equivalent to Equation (2.40) with the replacement $p \to p + 2$.

In conclusion, we have shown each of the following conditions:

- Equation (2.40) is true for $p = 1$ (Lemma 2.5);
- Equation (2.40) is true for $p = 2$ (Lemma 2.6); and
- whenever Equation (2.40) is true for p, then it must be true for $p + 2$.

We conclude that Equation (2.40) holds for all positive integers p. $\qquad\square$

Sampling a Normal Distribution

We saw previously, in Corollary 2.2, that when sampling from a normal distribution, the sample mean is also a normal random variable. Next, we show the relation between the sample variance with the chi-squared distribution.

Theorem 2.6. *Let $X_1, \ldots, X_n \sim N(\mu, \sigma^2)$ be IID samples from a normal distribution. Then the sample mean $\overline{X}_n = (1/n)\sum_{i=1}^{n} X_i$ and the sample variance $S_n^2 = 1/(n-1)\sum_{i=1}^{n}(X_i - \overline{X}_n)^2$ are independent and, moreover,*

$$\frac{(n-1)S^2}{\sigma^2} \sim \chi_{n-1}^2 \qquad (2.41)$$

is a chi-squared random variable with $n-1$ degrees of freedom.

Proof. First, to show that $\overline{X}_n$ is independent of S_n^2, we can rewrite S_n^2 as

$$S_n^2 = \frac{1}{n-1}\left(\left[\sum_{i=2}^{n}(X_i - \overline{X}_n)\right]^2 + \sum_{i=2}^{n}(X_i - \overline{X}_n)^2\right),$$

which is a function of only $(X_2 - \overline{X}_n, \ldots, X_n - \overline{X}_n)$. From there, one can show that each of these $n-1$ random variables are independent of $\overline{X}_n$, which completes the proof of the first statement. See Casella and Berger [2002] for details.

Next, in order to prove Equation (2.41), let us consider the random variable

$$W = \sum_{i=1}^{n}\left(\frac{X_i - \mu}{\sigma}\right)^2.$$

Now, due to Corollary 2.1 and Definition 2.12, it follows that $W \sim \chi_n^2$ is a chi-squared random variable with n degrees of freedom. Further, we have

$$W = \sum_{i=1}^{n}\left[\frac{(X_i - \overline{X}_n) + (\overline{X}_n - \mu)}{\sigma}\right]^2 = \sum_{i=1}^{n}\left(\frac{X_i - \overline{X}_n}{\sigma}\right)^2 + \frac{(\overline{X}_n - \mu)^2}{\sigma^2/n},$$

since the cross terms vanish:

$$2\sum_{i=1}^{n}\frac{(\overline{X}_n - \mu)(X_i - \overline{X}_n)}{\sigma^2} = \frac{2(\overline{X}_n - \mu)}{\sigma^2}\sum_{i=1}^{n}(X_i - \overline{X}_n) = 0.$$

However, we already know from Corollary 2.2 that the sample mean $\overline{X}_n \sim N(\mu, \sigma^2/n)$, and therefore the random variable

$$Z = \frac{\overline{X}_n - \mu}{\sigma/\sqrt{n}} \sim N(0, 1)$$

is a standard normal random variable, independent from S_n^2.

Combining the above, we have

$$W = \frac{(n-1)}{\sigma^2}S_n^2 + Z^2,$$

where $W \sim \chi_n^2$ and $Z^2 \sim \chi_1^2$. We conclude that $(n-1)S_n^2/\sigma^2 \sim \chi_{n-1}^2$. $\square$

Simulation in Python

The notation for invoking the chi-squared random variable in numpy and scipy is given in Code Block 2.16. Several chi-squared distributions are plotted in Figure 2.12.

```
1  dof = 5
2  sample_size = 100
3  samples = np.random.chisquare(dof, size=sample_size)
4  x = np.linspace(0, 20)
5  f = scipy.stats.chi2.pdf(x, dof)
```

Code Block 2.16: Chi-squared random variable in Python

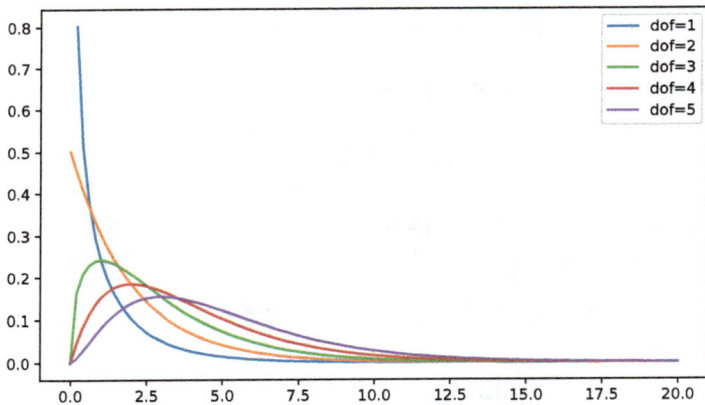

Fig. 2.12: Chi-squared distribution with $p = 1, 2, 3, 4, 5$ degrees of freedom.

2.5.2 Student's T Distribution

If we have a random sample $X_1, \ldots, X_n \sim N(\mu, \sigma^2)$, we have seen that the quantity

$$Z = \frac{\overline{X}_n - \mu}{\sigma/\sqrt{n}}$$

is a standard normal random variable, i.e., $Z \sim N(0, 1)$, where $\overline{X}_n = (1/n) \sum_{i=1}^{n} X_i$ is the sample mean. But what if we do not know the true variance? Could we use the sample variance as a proxy? Let us substitute S_n for σ and then make the following clever rearrangement:

$$\frac{\overline{X}_n - \mu}{S_n/\sqrt{n}} = \frac{(\overline{X}_n - \mu)/(\sigma/\sqrt{n})}{\sqrt{S_n^2/\sigma^2}}.$$

the numerator $(\overline{X}_n - \mu)/(\sigma/\sqrt{n}) \sim N(0,1)$ is a standard normal random variable, and the denominator is $\sqrt{\chi_{n-1}^2/(n-1)}$, independent of the numerator. This motivates the following definition.

Definition 2.14. *Let $X \sim N(0,1)$ and $Y \sim \chi_p^2$ represent independent random variables: X a standard normal and Y a chi-squared with p degrees of freedom. Then the random variable $T = X/\sqrt{Y/p}$ is said to be a t random variable with p degrees of freedom, denoted $T \sim t_p$.*

Proposition 2.21. *Let $X_1, \ldots, X_n \sim N(\mu, \sigma^2)$ be IID normal random variables, and let $\overline{X}_n = (1/n)\sum_{i=1}^n X_i$ be the sample mean and $S_n^2 = 1/(n-1)\sum_{i=1}^n (X_i - \overline{X}_n)^2$ the sample standard deviation. Then the random variable*

$$T = \frac{\overline{X} - \mu}{S_n/\sqrt{n}} \tag{2.42}$$

is a t-random variable with $n-1$ degrees of freedom; i.e., $T \sim t_{n-1}$.

Proof. As previously shown, the random variable T can be rearranged as

$$T = \frac{(\overline{X}_n - \mu)/(\sigma/\sqrt{n})}{\sqrt{S_n^2/\sigma^2}} = \frac{Z}{\sqrt{X/(n-1)}},$$

where $Z = (\overline{X}_n - \mu)/(\sigma/\sqrt{n}) \sim N(0,1)$ and $X = (n-1)S_n^2/\sigma^2 \sim \chi_{n-1}^2$. The result follows from Definition 2.14. $\qquad\square$

Proposition 2.22. *Let $T \sim t_p$ be a t-random variable with p degrees of freedom. Then its PDF is given by*

$$f_T(t) = \frac{\Gamma((p+1)/2)}{\Gamma(p/2)\sqrt{p\pi}}(1 + t^2/p)^{-(p+1)/2}, \tag{2.43}$$

for $-\infty < t < \infty$. This is known as the Student's t distribution with p degrees of freedom.

Proof. Let $X \sim N(0,1)$ and $Y \sim \chi_p^2$, as in Definition 2.14. Then the joint distribution is given by

$$f_{X,Y}(x,y) = \frac{1}{\sqrt{2\pi}}e^{-x^2/2}\frac{1}{\Gamma(p/2)2^{p/2}}y^{p/2-1}e^{-y/2},$$

for $-\infty < x < \infty$ and $0 < y < \infty$. Next, consider the transformation

$$t = \frac{x}{\sqrt{y/p}} \qquad \text{and} \qquad u = y.$$

It is straightforward to show that the Jacobian is given by $J(t, u) = \sqrt{u/p}$. Therefore, the joint distribution of (U, V) is given by

$$f_{U,V}(u, v) = f_{X,Y}\left(t\sqrt{u/p}, u\right)\sqrt{u/p} = \frac{u^{p/2-1/2}e^{-u/2(1+t^2/p)}}{\sqrt{2p\pi}\,\Gamma(p/2)2^{p/2}}$$

It can be shown that by integrating with respect to u from zero to infinity that one obtains the result. (See Casella and Berger [2002].)   □

Simulation in Python

The numpy and scipy calls for the Student's t distribution are shown in Code Block 2.17. Student's t distributions with several different degrees of freedom are shown in Figure 2.13. As the number of degrees of freedom increases, the Student's t distribution approaches a standard normal distribution.

```
1  dof = 5
2  sample_size=100
3  samples = np.random.standard_t(dof, size=sample_size)
4  x = np.linspace(-5, 5, num=100)
5  f = scipy.stats.t.pdf(x, dof)
```

Code Block 2.17: Student's t distribution in Python

2.5.3 Snedecor's F Distribution

We next introduce a distribution that is useful when comparing the variances of samples from two normal distributions.

Definition 2.15. *Let $U \sim \chi_p^2$ and $V \sim \chi_q^2$ be independent chi-squared distributions with p and q degrees of freedom. Then the random variable F defined by*

$$F = \frac{U/p}{V/q} \tag{2.44}$$

is said to be a Snedecor's F random variable with p and q degrees of freedom, denoted $F \sim F_{p,q}$.

Note 2.14. If $F \sim F_{p,q}$, it follows trivially that $1/F \sim F_{q,p}$.   ▷

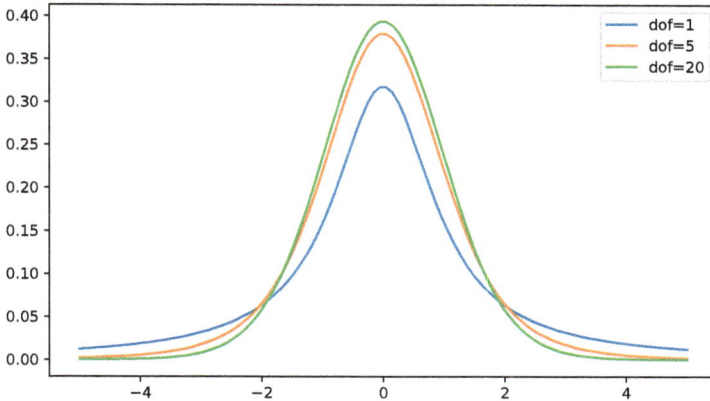

Fig. 2.13: Student's t distribution with $p = 1, 5, 20$ degrees of freedom.

The Snedecor's F distribution can be used to understand the relationship between the ratio of sample variances to the ratio of population (true) variances, as shown in the following proposition.

Proposition 2.23. *Let $X_1, \dots, X_n \sim N(\mu_X, \sigma_X^2)$ and $Y_1, \dots, Y_m \sim N(\mu_Y, \sigma_Y^2)$ constitute random samples from independent normal distributions, and let S_X^2 and S_Y^2 be the sample variances of the respective samples. Then the random variable*

$$F = \frac{S_X^2/S_Y^2}{\sigma_X^2/\sigma_Y^2} \sim F_{n-1,m-1} \tag{2.45}$$

has a Snedecor's F distribution with $n - 1$ and $m - 1$ degrees of freedom.

Proof. We may rearrange Equation (2.45) as

$$F = \frac{S_X^2/S_Y^2}{\sigma_X^2/\sigma_Y^2} = \frac{S_X^2/\sigma_X^2}{S_Y^2/\sigma_Y^2}.$$

However, from Theorem 2.6, we have that

$$U = \frac{(n-1)S_X^2}{\sigma_X^2} \sim \chi_{n-1}^2 \quad \text{and} \quad V = \frac{(m-1)S_Y^2}{\sigma_Y^2} \sim \chi_{m-1}^2,$$

thus

$$F = \frac{U/(n-1)}{V/(m-1)},$$

and the result follows from Definition 2.15. □

The PDF for the Snedecor's F distribution is given in the following proposition, which we state without proof.

Proposition 2.24. *Let $F \sim F_{p,q}$ be a Snedecor's F distribution with p and q degrees of freedom. Then the pdf for F is called the* Snedecor's F *distribution and is given by*

$$f_F(x) = \frac{\Gamma\left(\frac{p+q}{2}\right)}{\Gamma(p/2)\Gamma(q/2)} \left(\frac{p}{q}\right)^{p/2} x^{p/2-2} \left[1 + (p/q)x\right]^{-(p+q)/2}, \qquad (2.46)$$

for $0 < x < \infty$.

The F distribution is also closely related to the Student's t distribution.

Proposition 2.25. *Let $X \sim t_q$ be a Student's t random variable with q degrees of freedom. Then $X^2 \sim F_{1,q}$.*

Proof. Since $X \sim t_q$, we can write

$$X = \frac{Z}{\sqrt{Y/p}},$$

where $Z \sim N(0,1)$ and $Y \sim \chi_p^2$. Hence, it follows that

$$X^2 = \frac{Z^2}{Y/p},$$

and recalling that $Z^2 \sim \chi_1^2$, the result follows.   □

Simulation in Python

The numpy and scipy commands for the Snedecor's F distribution are given in Code Block 2.18. The PDFs of the Snedecor's F-distribution for several select values of p and q are shown in Figure 2.14.

```
p = 7
q = 11
n_samples = 30
samples = np.random.f(p, q, size=n_samples)
x = np.linspace(0, 5, num=100)
f = scipy.stats.f.pdf(x, p, q)
```

Code Block 2.18: F distribution in Python

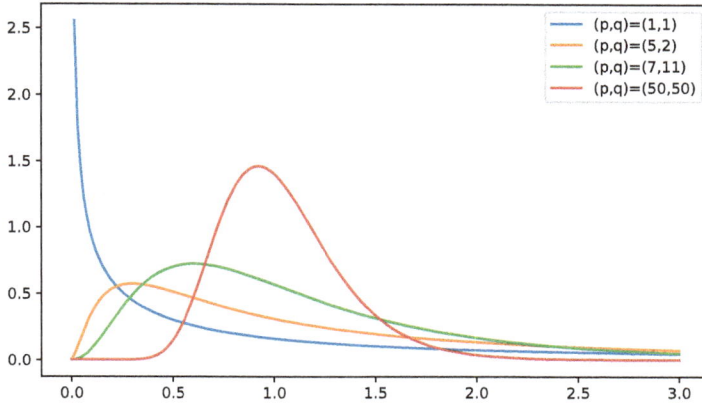

Fig. 2.14: Snedecor's F-distribution with (1,1), (2,5), (7,11), and (50,50) degrees of freedom.

2.6 Beta and Gamma

The binomial distribution is parameterized by the probability of success p. But what if we do not know the value for p? Is it possible to specify a distribution for the parameter p? If so, what would such a distribution look like?

Questions like these are resolved using Bayesian statistics. In this context, there are two probability distributions that are quite useful: the beta and the gamma. They are used for creating *distributions for unknown parameters*. We will explore how this is done later; the purpose of this section is simply to introduce the beta and gamma distributions.

2.6.1 Beta Distribution

The beta distribution is often used a distribution for probabilities, as its domain is the open interval $(0, 1)$. Before defining the beta distribution, however, we will first define the beta function.

Definition 2.16. *The* beta function *is defined to be the following functions of two variables α and β*

$$B(\alpha, \beta) = \int_0^1 x^{\alpha-1}(1-x)^{\beta-1}\,dx. \tag{2.47}$$

The beta function is closely related to the gamma function (Definition 2.13), as shown in the following proposition.

Proposition 2.26. *The beta function can be computed using the gamma function via the identity*

$$B(\alpha, \beta) = \frac{\Gamma(\alpha)\Gamma(\beta)}{\Gamma(\alpha + \beta)}. \tag{2.48}$$

In practice, we use Equation (2.48) when evaluating the beta function, as opposed to dealing directly with Equation (2.47). Now that we have introduced the beta function, we are ready to define the beta distribution.

Definition 2.17. *The continuous random variable X defined on the interval $(0, 1)$ is said to be a beta random variable with parameters α and β, denoted $X \sim \text{Beta}(\alpha, \beta)$, if its PDF is given by the beta distribution*

$$f(x) = \frac{1}{B(\alpha, \beta)} x^{\alpha-1} (1 - x)^{\beta-1}, \tag{2.49}$$

for $0 < x < 1$, and for positive parameters $\alpha > 0$ and $\beta > 0$.

In order to determine the mean and variance, we will rely on the following lemma.

Lemma 2.7. *Let $X \sim \text{Beta}(\alpha, \beta)$. Then the expected value of X^n is given by*

$$\mathbb{E}[X^n] = \frac{\Gamma(\alpha + n)\Gamma(\alpha + \beta)}{\Gamma(\alpha)\Gamma(\alpha + \beta + n)}.$$

Proof. To compute the expected value of X^n, we take the integral

$$\mathbb{E}[X^n] = \frac{1}{B(\alpha, \beta)} \int_0^1 x^n x^{\alpha-1} (1 - x)^{\beta-1} \, dx$$

$$= \frac{1}{B(\alpha, \beta)} \int_0^1 x^{\alpha+n-1} (1 - x)^{\beta-1} \, dx.$$

However, from Equation (2.47), we recognize this integral as $B(\alpha + n, \beta)$, hence

$$\mathbb{E}[X^n] = \frac{B(\alpha + n, \beta)}{B(\alpha, \beta)}.$$

A simple application of Equation (2.48) produces the result.   □

Theorem 2.7. *Let $X \sim \text{Beta}(\alpha, \beta)$ be a beta random variable. Then the mean and variance of X are given by*

$$\mathbb{E}[X] = \frac{\alpha}{\alpha + \beta} \tag{2.50}$$

$$\mathbb{V}(X) = \frac{\alpha\beta}{(\alpha + \beta)^2 (\alpha + \beta + 1)} \tag{2.51}$$

Proof. From Lemma 2.7, we have

$$\mathbb{E}[X] = \frac{\Gamma(\alpha+1)\Gamma(\alpha+\beta)}{\Gamma(\alpha)\Gamma(\alpha+\beta+1)} = \frac{\alpha}{\alpha+\beta},$$

where the last equality holds due to Equation (2.39), since $\Gamma(z+1)/\Gamma(z) = z$. A second application of this identity yields

$$\Gamma(z+2) = (z+1)\Gamma(z+1) = (z+1)z\Gamma(z).$$

Using this, we obtain

$$\mathbb{E}[X^2] = \frac{\Gamma(\alpha+2)\Gamma(\alpha+\beta)}{\Gamma(\alpha)\Gamma(\alpha+\beta+2)} = \frac{(\alpha+1)\alpha}{(\alpha+\beta+1)(\alpha+\beta)}.$$

Applying Equation (1.4) to the above two results and simplifying produces the result. □

Simulation in Python

The numpy and scipy methods for simulating a beta distribution are given in Code Block 2.19. The PDF for several different beta random variables is shown in Figure 2.15.

```
1  alpha = 5
2  beta = 10
3  n_samples = 30
4  samples = np.random.beta(alpha, beta, size=n_samples)
5  x = np.linspace(0, 1, num=100)
6  f = scipy.stats.beta.pdf(x, alpha, beta)
```

Code Block 2.19: Beta random variable in Python

2.6.2 Gamma Distribution

Definition 2.18. *The continuous random variable X defined on the interval $(0, \infty)$ is said to be a* gamma *random variable with parameters α and β, denoted $X \sim \text{Gamma}(\alpha, \beta)$, if its PDF is given by the* gamma distribution

$$f(x) = \frac{1}{\Gamma(\alpha)\beta^\alpha} x^{\alpha-1} e^{-x/\beta}, \tag{2.52}$$

for $0 < x < \infty$, and for positive parameters $\alpha > 0$ and $\beta > 0$.

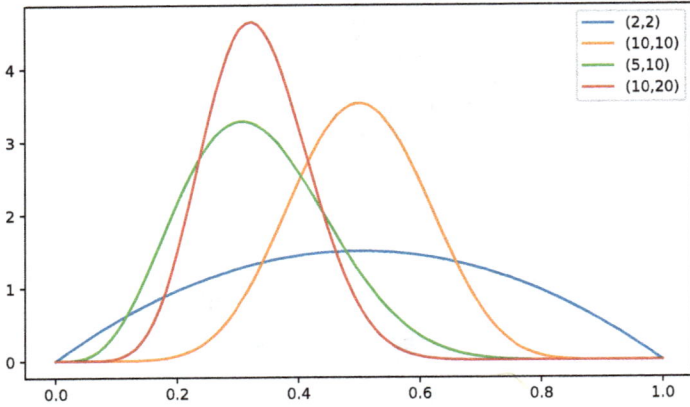

Fig. 2.15: PDF for several beta random variables.

Note 2.15. The parameters of a gamma random variable $X \sim \text{Gamma}(\alpha, \beta)$ are sometimes referred to as the shape parameter (α) and scale parameter (β). It is important to point out that, like the exponential distribution, the gamma distribution can be parameterized either by the scale parameter or the rate parameter $\lambda = 1/\beta$. (To make matters worse, some authors use β as the rate parameter as well; sometimes k and θ are used as the shape and scale parameters.)   ▷

Note 2.16. For the special case of $\alpha = 1$, Equation (2.52) is equivalent to an exponential distribution with scale parameter β; i.e., the exponential distribution is a special case of the gamma distribution: $\text{Exp}(\beta) = \text{Gamma}(1, \beta)$.
▷

Proposition 2.27. *Let* $X \sim \text{Gamma}(\alpha, \beta)$ *represent a gamma random variable with shape* α *and scale* β*. Then the mean and variance are given by*

$$\mathbb{E}[X] = \alpha\beta \tag{2.53}$$
$$\mathbb{V}(X) = \alpha\beta^2. \tag{2.54}$$

Proof. The mean is given by

$$\mathbb{E}[X] = \frac{1}{\Gamma(\alpha)\beta^\alpha} \int_0^\infty x^\alpha e^{-x/\beta} \, dx$$
$$= \frac{1}{\Gamma(\alpha)\beta^\alpha} \cdot \Gamma(\alpha+1)\beta^{\alpha+1},$$

and the result follows by application of Equation (2.39). Note: the second line follows by recognizing $x^\alpha e^{-x/\beta}$ as the kernel of a $\Gamma(\alpha + 1, \beta)$ distribution, since Equation (2.52) must integrate to unity.

The proof for Equation (2.54) follows along similar lines. $\qquad\square$

Proposition 2.28. *The* MGF *for a Gamma random varaible* $X \in \mathrm{Gamma}(\alpha, \beta)$ *is given by*

$$M_X(t) = (1 - \beta t)^\alpha,$$

for $t < 1/\beta$.

Proof. The MGF for a Gamma random variable can be expressed as

$$M_X(t) = \frac{1}{\Gamma(\alpha)\beta^\alpha} \int_0^\infty x^{\alpha-1} e^{-x/(\frac{\beta}{1-\beta t})} \, dx.$$

(See Exercise 2.4.) However, we recognize the integrand as the kernel of a Gamma distribution with $a = \alpha$ and $b = \frac{\beta}{1-\beta t}$. Hence

$$\int_0^\infty x^{\alpha-1} e^{-x/(\frac{\beta}{1-\beta t})} \, dx = \int_0^\infty x^{a-1} e^{-x/b} \, dx = \Gamma(a)b^a = \Gamma(\alpha) \left(\frac{\beta}{1 - \beta t} \right)^\alpha.$$

This completes the result. $\qquad\square$

Proposition 2.29. *Given two independent Gamma random variables with the same scale parameter* $X_1 \sim \mathrm{Gamma}(\alpha_1, \beta)$ *and* $X_2 \sim \mathrm{Gamma}(\alpha_2, \beta)$, *the sum* $X_1 + X_2$ *is the Gamma random variable* $X_1 + X_2 \sim \mathrm{Gamma}(\alpha_1 + \alpha_2, \beta)$.

Proof. From Proposition 1.12, the MGF of $Y = X_1 + X_2$ is

$$M_{X_1+X_2}(t) = M_{X_1}(t)M_{X_2}(t) = (1 - \beta)^{\alpha_1+\alpha_2},$$

where the last equality follows from Proposition 2.28. However, we recognize the result as the MGF for a $\mathrm{Gamma}(\alpha_1+\alpha_2, \beta)$ distribution, which completes the result. $\qquad\square$

Finally, we note a connection between the Gamma distribution and the chi-squared distribution.

Proposition 2.30. *The chi-squared distribution is a special case of the Gamma distribution:*

$$\chi_k^2 = \mathrm{Gamma}(k/2, 2).$$

Simulation in Python

Due to the varying definitions of the gamma random variable (whether it is parameterized by rate or scale), one must take an extra step of precaution to ensure that one is passing the correct arguments into the `numpy` and `scipy` methods. This is shown in Code Block 2.20. Several members of the family of gamma random variables are shown in Figure 2.16.

```
1  alpha = 5 #shape
2  beta = 2 #scale
3  n_samples = 30
4  samples = np.random.gamma(alpha, scale=beta, size=n_samples)
5  x = np.linspace(0, 30, num=200)
6  f = scipy.stats.gamma.pdf(x, alpha, scale=beta)
```

Code Block 2.20: Gamma random variable in Python

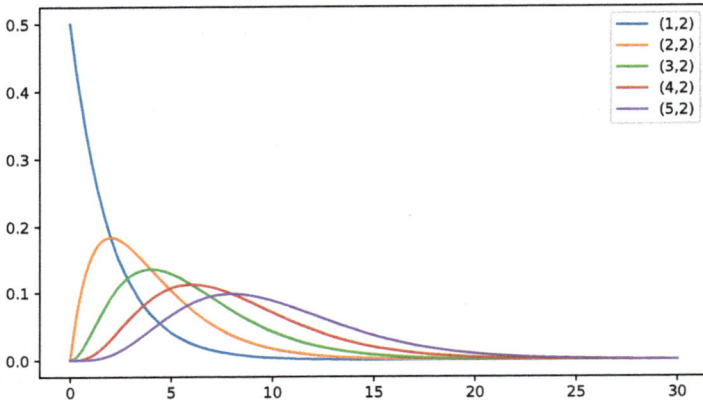

Fig. 2.16: PDF for several gamma random variables.

2.6.3 Gamma's Relation to the Poisson Process

We have already seen in Theorem 2.2 that the interarrival times of a Poisson process with parameter λ follow an exponential distribution with scale parameter $\beta = 1/\lambda$. Since the exponential distribution is a special case of the gamma distribution, it is natural to ask whether the gamma distribution is also related to the Poisson process. Our next theorem provides the answer.

Theorem 2.8. *Let N_t represent a Poisson process with rate parameter λ. Let X_k represent the arrival time of the kth event; i.e., $X_k = \inf\{t > 0 : N_t \geq k\}$. Then the distribution for X_k is given by*

$$X_k \sim \text{Gamma}(k, 1/\lambda); \tag{2.55}$$

i.e., the arrival time of the kth event follows a gamma random variable with shape parameter $\alpha = k$ and scale parameter $\beta = 1/\lambda$.

Proof. We proceed by induction. The basis step is trivial, as pointed out in Note 2.16. Specifically, it follows from Equation (2.52) that a Gamma$(1, \beta)$ random variable is equivalent to a Exp(β) random variable. Since the random variable X_1 is the first arrival time, it follows from Theorem 2.2 that $X_1 \sim$ Exp$(1/\lambda)$, which is equivalent to $X_1 \sim$ Gamma$(1, 1/\lambda)$.

For the induction step, let us assume that Equation (2.55) holds for some fixed $k \geq 1$. Theorem 2.2 also tells us that the interarrival time T_{k+1} for the $(k+1)$th event follows the same exponential distribution, i.e., $T_{k+1} \sim$ Exp$(1/\lambda)$. It follows that, assuming Equation (2.55) holds for k, the joint distribution is given by

$$f_{X_k, T_{k+1}}(x, t) = \frac{\lambda^k}{\Gamma(k)} x^{k-1} e^{-\lambda x} \cdot \lambda e^{-\lambda t} = \frac{\lambda^{k+1}}{\Gamma(k)} x^{k-1} e^{-\lambda(x+t)}.$$

We will follow the transformation from the proof of Theorem 2.5, namely,

$$u = x + t \qquad \text{and} \qquad v = \tan^{-1}\left(\frac{t}{x}\right),$$

for $0 < u < \infty$ and $0 < v < \pi/2$. We saw previously that $x = u/(1 + \tan(v))$ and the Jacobian is given by

$$J(u, v) = \frac{u \sec^2(v)}{(1 + \tan(v))^2}.$$

The joint distribution for the variables $U = X_k + T_{k+1}$ and $V = \tan^{-1}(T_{k+1}/X_k)$ is therefore given by

$$f_{U,V}(u, v) = \frac{\lambda^{k+1}}{\Gamma(k)} \frac{u^{k-1}}{(1 + \tan(v))^{k-1}} e^{-\lambda u} \frac{u \sec^2(v)}{(1 + \tan(v))^2}$$

$$= \frac{\lambda^{k+1}}{\Gamma(k)} \frac{\sec^2(v)}{(1 + \tan(v))^{k+1}} u^k e^{-\lambda u}.$$

Noting that

$$\int_0^{\pi/2} \frac{\sec^2(v)}{(1 + \tan(v))^{k+1}} \, dv = \frac{1}{k},$$

we conclude that the marginal distribution for U is

$$f_U(u) = \frac{\lambda^{k+1}}{k\Gamma(k)} u^k e^{-\lambda u}.$$

But $U = X_k + T_{k+1} = X_{k+1}$, the arrival time of the kth event. Since $k\Gamma(k) = \Gamma(k+1)$, we recognize the preceding equation as the gamma distribution Gamma$(k+1, 1/\lambda)$.

We have therefore proven that

1. Equation (2.55) is true for $k = 1$, and
2. if Equation (2.55) is true for any $k \geq 1$, it follows that it must also be true for $k + 1$.

We conclude that the theorem holds for all positive integers k. □

2.6.4 Beta-Binomial

Earlier in this section, we introduced the beta distribution by alluding to the fact that it can serve as a useful distribution for an unknown probability. Let's take the first step in this direction and derive our first *hierarchical* distribution: the beta-binomial distribution.

Definition 2.19. *Let $P \sim \text{Beta}(\alpha, \beta)$ be a beta random variable and let $X|(P = p) \sim \text{Binom}(n, p)$, the value of X conditioned on the realization $P = p$, be a binomial random variable. Then we say that the discrete random variable X is a* beta-binomial *random variable with parameters n, α, and β, which we denote as $X \sim \text{BetaBin}(n, \alpha, \beta)$.*

Note 2.17. For brevity, moving forward we shall use the shorthand $X|P \sim \text{Binom}(n, P)$ to represent the more verbose $X|(P = p) \sim \text{Binom}(n, p)$. In either case, however, the statement is interpreted as meaning that the conditional distribution of X, given the realization $P = p$, is a binomial distribution with parameters n and p.   ▷

The beta-binomial random variable is an example of a more general structure known as a hierarchical model, as defined in the following.

Definition 2.20. *A* hierarchical model *for a random variable X is any statistical model for X in which one or more of its parameters is itself a random variable; i.e., a hierarchical model is any model of the form*

$$X|Y \sim \mathcal{D}(\theta; Y)$$
$$Y \sim \mathcal{E}(\phi);$$

where $\mathcal{D}$ and $\mathcal{E}$ are given distributions with parameters $\mathcal{D}(\theta, y)$ and $\mathcal{E}(\phi)$. The parameters θ, ϕ, and y may represent scalar or vector parameters. The parameter(s) ϕ are sometimes referred to as hyperparameters *for X. The distribution for a random variable X modeled by a hierarchical model is sometimes referred to as a* mixture distribution *or* compound distribution.

Proposition 2.31. *Let $X \sim \text{BetaBin}(n, \alpha, \beta)$. Then the mean and variance of X are given by*

$$\mathbb{E}[X] = \frac{n\alpha}{\alpha + \beta} \tag{2.56}$$

$$\mathbb{V}(X) = \frac{\alpha\beta n(\alpha + \beta + n)}{(\alpha + \beta)^2(\alpha + \beta + 1)}. \tag{2.57}$$

Proof. From Equation (1.19), we have

$$\mathbb{E}[X] = \mathbb{E}[\mathbb{E}[X|P]] = \mathbb{E}[nP] = \frac{n\alpha}{\alpha + \beta}.$$

We have further seen that the expected value of the square of a binomial random variable is given by

$$\mathbb{E}[X^2|P] = np(1-p) + n^2p^2 = np + p^2(n^2 - n)$$

and that the expected value of the square of a beta random variable is given by

$$\mathbb{E}[P^2] = \frac{\alpha(\alpha+1)}{(\alpha+\beta+1)(\alpha+\beta)}.$$

Thus, we can write

$$\mathbb{E}[X^2] = \mathbb{E}[\mathbb{E}[X^2|P]] = \mathbb{E}[nP + P^2(n^2 - n)] = \frac{n\alpha}{\alpha+\beta} + \frac{(n^2-n)\alpha(1+\alpha)}{(\alpha+\beta+1)(\alpha+\beta)}.$$

Finally, the variance, as given by Equation (2.57), can be derived using $\mathbb{V}(X) = \mathbb{E}[X^2] - \mathbb{E}[X]^2$. □

It turns out that Equation (1.19) from Theorem 1.1 has a variance counterpart, which we state in our next theorem. For a proof, see Casella and Berger [2002].

Theorem 2.9. *Given random variables X, Y,*

$$\mathbb{V}(X) = \mathbb{E}[\mathbb{V}(X|Y)] + \mathbb{V}(\mathbb{E}[X|Y]), \tag{2.58}$$

provided that the expectations exist.

Proposition 2.32. *Let $X \sim \text{BetaBin}(n, \alpha, \beta)$. Then the PMF for X is known as the* beta-binomial distribution *and is given by*

$$f(x) = \binom{n}{x} \frac{B(\alpha+x, \beta+n-x)}{B(\alpha, \beta)}, \tag{2.59}$$

for $x = 0, 1, 2, \ldots, n$.

Proof. First, we have $X|P \sim \text{Binom}(n, p)$, and so therefore

$$f_{X|P}(x|p) = \binom{n}{p} p^x (1-p)^{n-x}.$$

Further, $P \sim \text{Beta}(\alpha, \beta)$, and thus

$$f_P(p) = \frac{1}{B(\alpha, \beta)} p^{\alpha-1}(1-p)^{\beta-1}.$$

In order to obtain the probability mass function for f_X, we can marginalize over P, weighting each value of $P = p$ with its respective probability. Thus,

$$f_X(x) = \int_0^1 f_{X|P}(x|p) f_P(p) \, dp$$

$$= \binom{n}{x} \frac{1}{B(\alpha, \beta)} \int_0^1 p^{\alpha+x-1}(1-p)^{\beta+n-x-1} \, dp.$$

Recognizing the integrand as the kernel of the Beta($\alpha + x, \beta + n - x$) distribution proves the result. □

Simulation in Python

numpy does not contain a built-in beta-binomial random number generator; this, however, can easily be remedied, as shown in Code Block 2.21. Note

```
1   n = 10
2   alpha = 5
3   beta = 10
4   n_samples = 30
5   def betabinomSampler(n, alpha, beta, size=None):
6       p = np.random.beta(alpha, beta, size=size)
7       x = np.random.binomial(n, p)
8       return x
9   samples = betabinomSampler(n, alpha, beta, size=n_samples)
10  x = np.arange(11)
11  f = scipy.stats.betabinom.pmf(x, n, alpha, beta)
```

Code Block 2.21: Simulation of a beta-binomial random variable

that the default size of betabinomSampler is size=None, not size=1. This way, if no argument is passed in for size, the output will be a scalar, not an np.array (as would be the case if using size=1). The PMF for a BetaBin$(10, 5, 10)$ random variable is shown in Figure 2.17.

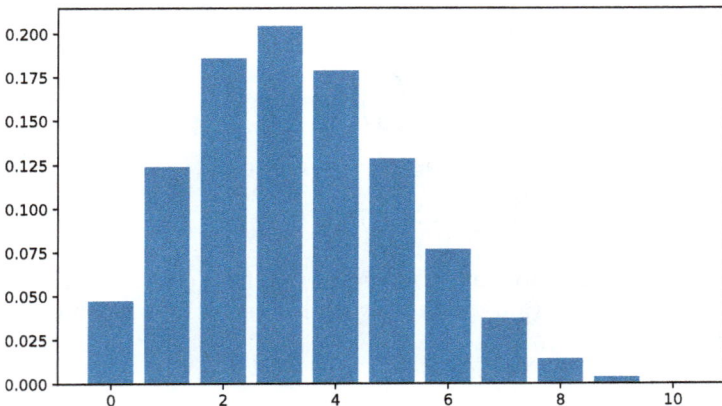

Fig. 2.17: PMF for BetaBin$(10, 5, 10)$.

It is of course natural to ask how the beta-binomial compares with a simple binomial distribution. Since the expected value of $P \sim \text{Beta}(5, 10)$ is $1/3$, we may add the PMF for $\text{Binom}(10, 0.33)$ to our plot, as shown in Figure 2.18. We observe that the beta-binomial distribution peaks at the same spot, but it is more spread out, as compared with the corresponding binomial distribution. This makes intuitive sense: since the probability in the beta-binomial hierarchical model is allowed to vary around $p = 0.33$, we expect to see more samples further away from the expected value, as is indeed the case.

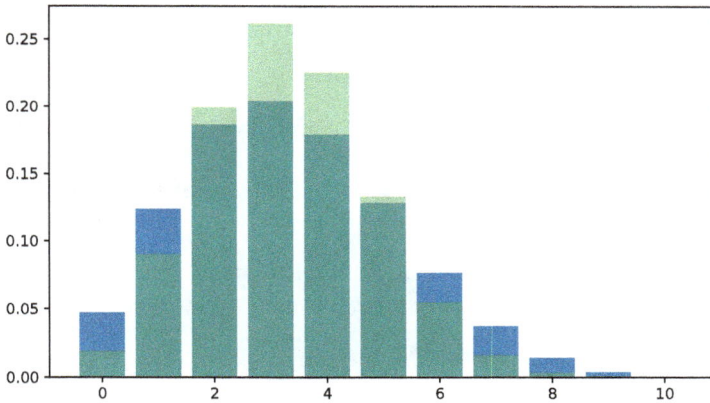

Fig. 2.18: Concurrent plot of PMF for $\text{BetaBin}(10, 5, 10)$ (blue) and $\text{Binom}(10, 0.33)$ (green).

2.7 Multivariate Distributions

We conclude the chapter with two important generalizations of distributions to higher dimensions.

2.7.1 The Multivariate Normal Distribution

Definition 2.21. *The random vector X is a k-dimensional multivariate normal random variable with parameters μ and Σ, denoted $X \sim N(\mu, \Sigma)$, if its PDF is of the form of a multivariate normal distribution; i.e., if*

$$f(x; \mu, \Sigma) = \frac{1}{(2\pi)^{k/2}|\Sigma|^{1/2}} e^{-(x-\mu)^T \Sigma^{-1}(x-\mu)/2}, \tag{2.60}$$

for some constant vector $\mu \in \mathbb{R}^k$, and a constant, symmetric, positive definite matrix $\Sigma \in \mathbb{R}^{k \times k}$, with determinant $|\Sigma|$.

Proposition 2.33. *Let $X \sim N(\mu, \Sigma)$ for some $\mu \in \mathbb{R}^k$ and symmetric, positive definite $\Sigma \in \mathbb{R}^{k \times x}$. Then the mean and variance–covariance matrix of the random vector X are given by*

$$\mathbb{E}[X] = \mu, \tag{2.61}$$

$$\mathbb{V}(X) = \Sigma. \tag{2.62}$$

The following is a generalization of Corollary 2.1.

Proposition 2.34. *Let $X \sim N(\mu, \Sigma)$ be a k-dimensional multivariate normal random vector. Then the transformed random vector*

$$Z = \Sigma^{-1/2}(X - \mu) \tag{2.63}$$

is a k-dimensional standard normal random vector; i.e., $Z \sim N(0_k, I_k)$, where $0_k \in \mathbb{R}^k$ is the zero vector and $I_k \in \mathbb{R}^{k \times k}$ is the identity matrix. Conversely, given $Z \sim N(0_k, I_k)$, the transformed random vector $X = \Sigma^{1/2} Z + \mu \sim N(\mu, \Sigma)$.

Proposition 2.35. *Let $X \sim N(\mu, \Sigma)$ represent a k-dimensional normal random vector. Then the components of X are independent, i.e., $X_i \perp\!\!\!\perp X_j$ for $i \neq j$, if and only if the covariance matrix Σ is diagonal, i.e., if and only if $\Sigma = \text{diag}(\sigma_1^2, \ldots, \sigma_k^2)$, for positive constants $\sigma_1^2, \ldots, \sigma_k^2$. Moreover, given independence, each component is a normal random variable with $X_i \sim N(\mu_i, \sigma_i^2)$.*

Proof. First, let us assume that the components of X are independent. This implies that they are uncorrelated, by Proposition 1.6, and thus $\text{COV}(X_i, X_j) = 0$ whenever $i \neq j$, and $\text{COV}(X_i, X_i) = \sigma_i^2$. Since $\Sigma_{ij} = \text{COV}(X_i, X_j)$, by the definition of a variance–covariance matrix, it follows that independence implies that Σ must be a diagonal matrix.

Second, let us assume that the variance–covariance matrix Σ is diagonal, which implies that the components of X are uncorrelated (but not necessarily independent). To prove independence, we must show that the PDF given by Equation (2.60) can be factored. To start, recall from linear algebra that the determinant of a diagonal matrix $\Sigma = \text{diag}(\sigma_1^2, \ldots, \sigma_k^2)$ is the product of its diagonal entries; i.e., $|\Sigma| = \sigma_1^2 \cdots \sigma_k^2$. Furthermore, we have that $\Sigma^{-1} = \text{diag}(\sigma_1^{-2}, \ldots, \sigma_k^{-2})$, and thus

$$(x - \mu)^T \Sigma^{-1}(x - \mu) = \sum_{i=1}^{k}(x_i - \mu)^2/\sigma_i^2.$$

Combining these results, it follows that, for the case of a diagonal covariance matrix, the PDF given by Equation (2.60) can be expressed

$$f_{X_1,\ldots,X_n}(x;\mu,\Sigma) = \prod_{i=1}^{k} \frac{1}{\sigma_i\sqrt{2\pi}} e^{-(x_i-\mu_i)^2/\sigma_i^2} = \prod_{i=1}^{k} f_{X_i}(x_i;\mu_i,\sigma_i^2).$$

Moreover, we recognize these factors as independent normal distributions, hence $X_i \sim N(\mu_i,\sigma_i^2)$. This completes the proof. $\qquad\square$

To understand the multivariate normal distribution a step further, a consequence of our next result will show that the contours of the multivariate normal PDF are ellipsoids.

Theorem 2.10. *Let $X \sim N(\mu,\Sigma)$ represent a k-dimensional multivariate normal distribution, then there exists a linear transformation of the form $Y = V^T(X-\mu)$, for some matrix $V \in \mathbb{R}^{k\times k}$, such that the components of Y are independent normal random variables. Moreover, the matrix V is the orthogonal matrix whose columns are the eigenvectors of Σ.*

Proof. Since the covariance matrix Σ is symmetric and positive definite, it can be factored as

$$\Sigma = VDV^T,$$

where $D = \text{diag}(\sigma_1^2,\ldots,\sigma_k^2)$ is the diagonal matrix of positive eigenvalues $\sigma_1^2,\ldots,\sigma_k^2$, and $V = [v_1,\ldots,v_k]$ is the (orthogonal) matrix of associated eigenvectors $v_1,\ldots,v_k$. Moreover, the inverse matrix is obtained simply by inverting the eigenvalues; i.e.,

$$\Sigma^{-1} = VD^{-1}V^T.$$

Now consider the transformation $y = V^T(x-\mu)$. The exponent in Equation (2.60) is thus equivalent to

$$(x-\mu)^T\Sigma^{-1}(x-\mu) = (x-\mu)^TVD^{-1}V^T(x-\mu) = y^TD^{-1}y,$$

excluding the factor of $-1/2$ for brevity. Moreover, since the matrix V of eigenvectors is orthogonal, it follows that the determinant

$$|\Sigma| = |V|\,|D|\,|V^T| = |D|,$$

since $|V^T| = |V|^{-1}$ for orthogonal matrices. It follows that the random vector $Y \sim N(0_k, D)$. Since D is diagonal, the components of Y are independent, by Proposition 2.35. Moreover, it follows that $Y_i \sim N(0,\sigma_i^2)$. $\square$

Note 2.18. We refer to the columns of matrix V as the *principal axes* of the variance–covariance matrix Σ. Moreover, the eigenvalues σ_i^2 are referred to as the *principal variances* of Σ. $\qquad\qquad\triangleright$

In the case of a univariate normal random variable, we saw the usefulness of the z-score as a way for determining "how far" a given value is from the mean. The generalization of the z-score to higher dimensions is given in the following definition.

Definition 2.22. *Given a symmetric, positive-definite matrix $\Sigma \in \mathbb{R}^{k \times k}$, the* Mahalanobis distance *between two vectors $x, y \in \mathbb{R}^k$ is given by*

$$d_\Sigma(x, y) = \sqrt{(x - y)^T \Sigma^{-1}(x - y)}. \tag{2.64}$$

In particular, if X is a random vector with mean $\mathbb{E}[X] = \mu \in \mathbb{R}^k$ and variance–covariance matrix $\mathbb{V}(X) = \Sigma \in \mathbb{R}^{k \times k}$, we say that the Mahalanobis distance m *of an observed value $x \in \mathbb{R}^k$ is the number*

$$m(x) = d_\Sigma(x, \mu). \tag{2.65}$$

Proposition 2.36. *Let $X \sim \mathrm{N}(\mu, \Sigma)$ be a k-dimensional normal random vector. The contours of the Mahalanobis distance $m(x)$ are concentric ellipsoids centered at μ.*

Proof. Consider again the transformation $Y = V^T(X - \mu)$, as given in Theorem 2.10, and let $\Sigma = VDV^T$ be the principal component decomposition of Σ. The level set $d_\Sigma(x, \mu) = m$ is equivalent to

$$m^2 = (x - \mu)^T V D^{-1} V^T (x - \mu) = y^T D^{-1} y.$$

Since D^{-1} is diagonal, this is equivalent to

$$\frac{y_1^2}{\sigma_1^2} + \cdots + \frac{y_k^2}{\sigma_k^2} = z^2, \tag{2.66}$$

the equation of an ellipsoid in $\mathbb{R}^k$. We conclude that the level sets $d_\Sigma(x, \mu) = m$ are ellipsoids, centered at μ, with principal axes $v_1, \ldots, v_k$ and principal widths $y_i = \pm m \sigma_i$. $\qquad \square$

Corollary 2.3. *Let $X \sim \mathrm{N}(\mu, \Sigma)$ be a k-dimensional normal random vector. Then the random variable $M^2 = d_\Sigma(X, \mu)$, corresponding to the squared Mahalanobis distance of X, is a chi-squared random variable with k degrees of freedom; i.e., $M^2 \sim \chi_k^2$.*

Proof. Because the components Y_i of the random vector Y, as defined in the proofs of Theorem 2.10, are normal random variables with $Y_i \sim \mathrm{N}(0, \sigma_i^2)$, it follows that the scaled random variables $Y_i/\sigma_i \sim \mathrm{N}(0, 1)$ are standard normal random variables. Equation (2.66) therefore implies that M^2 is a chi-squared random variable with k degrees of freedom, following Definition 2.12. $\qquad \square$

The probability of a multivariate normal random vector X having a Mahalanobis distance $d_\Sigma(X, \mu)$ less than m is given for the case of $k = 1, 2, 3, 4$ in Table 2.1[8]. Notice, in particular, that the row $k = 1$ recovers

---

[8] These probabilities can be computed in Python using `scipy.stats.chi2.cdf(m**2, k)`, for dimension k and Mahalanobis distance m.

the basic facts about a standard normal random variable: 68% probability of being within one standard deviation from the mean; 95% probability of being within two standard deviations from the mean; and 99.7% probability of being within three standard deviations from the mean. Notice that as the dimension increases, the probability of being within a fixed Mahalanobis distance to the mean diminishes.

	$m = 1$	$m = 2$	$m = 3$
$k = 1$	0.6827	0.9545	0.9973
$k = 2$	0.3935	0.8647	0.9889
$k = 3$	0.1987	0.7385	0.9707
$k = 4$	0.0902	0.5940	0.9389

Table 2.1: Cumulative probabilities at various Mahalanobis distances and dimensions.

This has dire implications for computing a 95% confidence ellipsoid for high-dimensional normal distributions. The Mahalanobis distance that contains 95% of the probability, i.e., the solution to $\mathbb{P}(d_\Sigma(X, \mu) < m) = 0.95$, is shown in Table 2.2. These data can be generated using the built-in inverse survival function, `np.sqrt(scipy.stats.chi2.isf(0.05, k))`. We see that the "two standard deviations from the mean" rule begins to fail as we move up in dimension.

$k = 1$	$k = 2$	$k = 3$	$k = 4$	$k = 5$	$k = 6$	$k = 7$	$k = 8$	$k = 9$	$k = 10$
1.9600	2.4477	2.7955	3.0802	3.3272	3.5485	3.7506	3.9379	4.1133	4.2787

Table 2.2: Mahalanobis distance containing 95% probability.

Finally, we conclude with a generalization of Theorem 2.4.

Theorem 2.11 (Multivariate Central Limit Theorem). *Suppose the k-dimensional random vectors $X_1, \ldots, X_n \sim F$ constitute an IID sample from a distribution F with finite mean $\mu \in \mathbb{R}^k$ and variance $\Sigma \in \mathbb{R}^{k \times k}$. Let $\overline{X}_n = (1/n) \sum_{i=1}^n X_i \in \mathbb{R}^k$ represent the sample mean. Then the sequence of random vectors*

$$Z_n = \sqrt{n} \Sigma^{-1/2} (\overline{X}_n - \mu) \tag{2.67}$$

approaches the k-dimensional standard normal random vector $Z \sim N(0_k, I_k)$ in the limit as $n \to \infty$; i.e., $Z_n \to Z$ in distribution as $n \to \infty$.

Note 2.19. An alternative way of stating Theorem 2.11 is

$$\sqrt{n}(\overline{X}_n - \mu) \to N(0_k, \Sigma), \tag{2.68}$$

as $n \to \infty$. This follows from Proposition 2.34.   ▷

Python

The Mahalanobis distance, as given by Definition 2.22, can easily be set to code, as shown in Code Block 2.22.

```python
def mahalanobis(x, y, Sigma):
    return np.sqrt( (x-y).T @ np.linalg.inv(Sigma) @ (x-y))
```

Code Block 2.22: Mahalanobis Distance

Next, we turn to the case of level sets of the Mahalanobis distance for the special case of dimension $k = 2$, which will be common in examples. Such a routine is provided in Code Block 2.23. (We will use this code later, in Example 5.6.)

```python
def ellipse(mu, Sigma, z=1, num=100):
    t = np.linspace(0, 2*np.pi, num=num)

    D, V = np.linalg.eig(Sigma)
    # D: Array of Eigenvalues
    # V: (Orthogonal) Matrix of Eigenvectors

    f1 = z * np.sqrt(D[0])
    f2 = z * np.sqrt(D[1])
    # Scale Factors along principal axes.

    Y = np.array([ f1 * np.cos(t), f2 * np.sin(t)])
    # 2 x 100 matrix

    X = V @ Y + mu.reshape((2,1)) * np.ones(num)
    # Transform back to original axes.

    return X[0, :], X[1, :]
```

Code Block 2.23: Code to generate ellipse.

Line 4 returns the diagonal matrix of eigenvalues and the matrix of eigenvectors for Σ. Lines 8–9 computes the width of the ellipse along its principal axes, as given by Equation (2.66). We can parameterize this ellipse relative to its principal axes coordinates using

$$y(t) = \begin{bmatrix} z\sigma_1 \cos(t) \\ z\sigma_2 \sin(t) \end{bmatrix},$$

for $0 \le t \le 2\pi$. This is done on line 12. Finally, we transform back into the original coordinates using $x = Vy + \mu$ (line 15).

Moreover, we can take random samples and compute the PDF and CDF of a multivariate normal using the built-in numpy and scipy packages, as shown in Code Block 2.24. It should be noted that the CDF of a multivariate distribution is defined in the sense that $F([x_0, y_0]) = \mathbb{P}(X < x_0$ and $Y < y_0)$.

One way to generate a contour plot of the PDF is shown in lines 18–21. The output is shown in Figure 2.19.

```
mu = np.array([2, 3])
Sigma = np.array([[5/2, -3/2], [-3/2, 5/2]])
# D = array([4., 1.])
# V = array([[ 0.70710678, 0.70710678],
#            [-0.70710678, 0.70710678]])

n_samples=100
X = np.random.multivariate_normal(mu, Sigma, size=n_samples)
# 100 x 2
x0 = np.array([2,3])
scipy.stats.multivariate_normal.pdf([2,3], mean=mu, cov=Sigma)
# 0.07957747154594767
scipy.stats.multivariate_normal.cdf([2,3], mean=mu, cov=Sigma)
# P(X < 2 and Y < 3)
# 0.14758361765043332

## Contour Plot of PDF
x, y = np.mgrid[0:5:.01, 0:5:.01]
pos = np.dstack((x, y))
f = scipy.stats.multivariate_normal.pdf(pos, mean=mu, cov=Sigma)
plt.contourf(x, y, f)
```

Code Block 2.24: Random samples from multivariate normal

2.7.2 The Multinomial Distribution

We next turn to our final distribution of the chapter, completing the circle by returning to our Bernoulli roots. For additional detail, see Härdle and Simar [2019]. To motivate our definition, let us first consider a simple example.

Example 2.1. Suppose that a jar contains marbles of an assortment of k different colors, where the ith color is represented with proportion $p_i \in [0, 1]$, such that $\sum_{i=1}^{k} p_i = 1$. A total of n marbles are selected at random

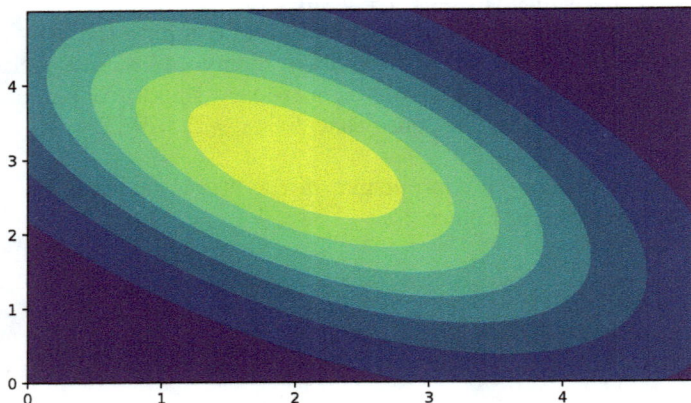

Fig. 2.19: Contour plot of multivariate normal PDF produced by Code Block 2.24.

from the jar; i.e., we perform n independent draws *with replacement*[9]. Let $X = (X_1, \ldots, X_k)$ be the vector of counts following our n draws; i.e., let X_i represent the number of marbles of color i obtained from the n draws, such that $\sum_{i=1}^{n} X_i = n$. The vector X is said to constitute a *multinomial random variable*.   ▷

We will build up the theory to tackle such problems step-by-step. First, however, we present some useful notation.

Definition 2.23. *The* standard $(k-1)$-simplex Δ^{k-1}, *or* probability simplex, *is the subset of* $\mathbb{R}_*^k$ *defined by*

$$\Delta^{k-1} = \left\{ x \in \mathbb{R}_*^k : \sum_{i=1}^{k} x_i = 1 \right\},$$

where $\mathbb{R}_* = [0, \infty)$ *is the set of nonnegative real numbers.*

The set Δ^{k-1} is useful in describing a vector of probabilities. Specifically, a vector $p = (p_1, \ldots, p_k)$ that satisfies $p_i \geq 0$ and $\sum_{i=1}^{k} p_i = 1$ and alternatively be specified by the simple statement that $p \in \Delta^{k-1}$. Another way of viewing the set Δ^{k-1} is the *convex hull* of the set of standard basic vectors $\mathbb{U}^k \subset \mathbb{R}^k$, defined by

---

[9] *With replacement* means that the probabilities of selecting a particular color does not change as we draw more marbles; i.e., it is as though we return each marble to the jar before drawing the next.

$$\mathbb{U}^k = \left\{ x \in \mathbb{B}^k : \sum_{i=1}^{k} x_i = 1 \right\} = \{e_1, \ldots, e_k\},$$

where e_i is the ith standard basis vector $e_i = (\delta_{i1}, \ldots, \delta_{ik})$.

Next, let us describe a single experiment that has k mutually exclusive and exhaustive outcomes $A_1, \ldots, A_k$, with probabilities $p_1, \ldots, p_k$. This is a generalization of the Bernoulli random variable, for which there were precisely two such outcomes. The random vector to describe such a scenario is therefore of the form $X = (X_1, \ldots, X_k)$.

Definition 2.24. *A random vector X is said to be a* multivariate Bernoulli *random variable with probability $p \in \Delta^{k-1}$, denoted $X \sim$ MultiBern(p), if X is of the form $X \in \mathbb{U}^k = \{e_1, \ldots, e_k\}$ and has PMF given by $f_X(e_i) = p_i$, i.e.,*

$$f_X(x_1, \ldots, x_k) = p_1^{x_1} \cdots p_k^{x_k}. \tag{2.69}$$

In particular, note that a multivariate Bernoulli trial has precisely k possible outcomes. The mean and variance–covariance matrix of a multivariate Bernoulli random vector is given as follows.

Proposition 2.37. *Let $X \sim$ MultiBern(p) be a multivariate Bernoulli random vector with $p \in \Delta^{k-1}$. Then $\mathbb{E}[X] = p$ and the variance and covariance of the components of X are given by*

$$\Sigma_{ij} = \mathrm{COV}(X_i, X_j) = \begin{cases} p_i(1 - p_i) & \text{for } i = j \\ -p_i p_j & \text{for } i \neq j \end{cases} \tag{2.70}$$

Proof. The marginal distribution of X_i is $X_i \sim$ Bern(p_i), and hence $\mathbb{V}(X_i) = p_i(1 - p_i)$.

For the second part, consider X_i and X_j, with $i \neq j$. From Proposition 1.5, we have

$$\mathrm{COV}(X_i, X_j) = \mathbb{E}[X_i X_j] - \mathbb{E}[X_i]\mathbb{E}[X_j].$$

Now, $\mathbb{E}[X_i] = p_i$ and $\mathbb{E}[X_j] = p_j$. Moreover, $\mathbb{E}[X_i X_j] = 0$, since $X_i X_j = 0$ for all $X \in \mathbb{U}^k$; i.e., X_i and X_j cannot *both* equal one. The result follows. $\square$

Note 2.20. The matrix Σ from Proposition 2.37 is singular due to a relation of linear dependence among its columns created by the constraint $\sum_{i=1}^{k} p_i = 1$; moreover rank(Σ) = $k - 1$. This is easiest to see for the case of $k = 2$, since

$$\det \begin{bmatrix} p_1(1 - p_1) & -p_1 p_2 \\ -p_1 p_2 & p_2(1 - p_2) \end{bmatrix} = p_1 p_2(1 - p_1 - p_2) = 0,$$

since $p_1 + p_2 = 1$. $\triangleright$

In order to describe the scenario in Example 2.1, we can therefore conduct a sequence of multivariate Bernoulli trials, similar to the construction of the binomial random variable.

Definition 2.25. *Let* $B_1, \ldots, B_n \sim \text{MultiBern}(p)$ *represent a sequence of* k-*dimensional Bernoulli trials, for some* $p \in \Delta^{k-1}$. *The random vector* $X = \sum_{i=1}^{n} B_i \in \mathbb{Z}_*^k$ *of nonnegative integers, satisfying* $\sum_{i=1}^{k} X_i = n$, *is a* multinomial random vector, *denoted* $X \sim \text{Multi}(n, p)$.

Note 2.21. The sum $X = \sum_{i=1}^{n} B_i$ that defines the vector X is a vector sum, whereas the sum $\sum_{i=1}^{k} X_i = n$, which constrains the vector X, represents a sum of its components. Moreover, the latter sum is redundant and stated only for clarity: Since $\sum_{j=1}^{k} B_{ij} = 1$ for any $B_i \in \text{MultiBern}(p)$ (since B_i is a standard unit vector), it follows automatically that $\sum_{i=1}^{n} \sum_{j=1}^{n} B_{ij} = n$.
▷

In order to construct the PMF for a multinomial random vector, we will require an understanding of the multinomial coefficients.

Definition 2.26. *Given* $x \in \mathbb{Z}_*^k$, *with* $\sum_{i=1}^{k} x_i = n$, *the* multinomial coefficient *is defined by*

$$\binom{n}{x_1, \ldots, x_k} = \frac{n!}{x_1! \cdots x_k!}. \tag{2.71}$$

An important interpretation of the multinomial coefficient is given in our next proposition. This interpretation is crucial in deriving the PMF for a multinomial random vector.

Proposition 2.38. *The number of ways of distributing* n *objects into* k *distinct bins, such that* x_i *objects are assigned to the* ith *bin, is given by the multinomial coefficient defined by Equation* (2.71).

Proof. We proceed bin-by-bin. The number of ways of selecting x_1 objects for the first bin is simply

$$\binom{n}{x_1} = \frac{n!}{x_1!(n - x_1)!}.$$

Once this selection has been made, we have $(n - x_1)$ objects remaining that are eligible to be distributed into the second bin. The number of ways to select x_2 of these remaining objects is

$$\binom{n - x_1}{x_2} = \frac{(n - x_1)!}{x_2!(n - x_1 - x_2)!}.$$

Therefore, the number of ways of selecting x_1 objects for the first bin *and* x_2 objects for the second bin, leaving $(n - x_1 - x_2)$ remaining objects that have yet to be selected, is given by

$$\binom{n}{x_1}\binom{n - x_1}{x_2} = \frac{n!}{x_1! x_2!(n - x_1 - x_2)!}.$$

Proceeding in this fashion, we find that the number of ways to assign x_1 objects to the first bin, $\ldots$, x_{k-1} objects to the $(k-1)$st bin, is given by

$$\binom{n}{x_1} \cdots \binom{n - x_1 - \cdots - x_{k-2}}{x_{k-1}} = \frac{n!}{x_1! \cdots x_{k-1}!(n - x_1 - \cdots - x_{k-1})!}.$$

However, $n - x_1 - \cdots - x_{k-1} = x_k$, and so the above expression reduces to Equation (2.71). Finally, we note that there is precisely one way to select x_k objects for the final bin out of the remaining x_k objects. This completes the result. $\square$

Proposition 2.39. *The* PMF *of a multinomial random vector* $X \sim \text{Multi}(n, p)$, *for some* $p \in \Delta^{k-1}$, *is given by the* multinomial distribution

$$f(x_1, \ldots, x_k) = \frac{n!}{x_1! \cdots x_k!} p_1^{x_1} \cdots p_k^{x_k} = n! \prod_{i=1}^{k} \frac{p_i^{x_i}}{x_i!}, \qquad (2.72)$$

such that $\sum_{i=1}^{k} X_i = n$.

Proof. Let $B_1, \ldots, B_n \sim \text{MultiBern}(p)$ represent a sequence of n IID multivariate Bernoulli trials, with $X = \sum_{i=1}^{n} B_i$, as given by Definition 2.25. We must show that the probability of outcome $X = (x_1, \ldots, x_k)$ is given by Equation (2.72).

However, stating that the random vector $X = (x_1, \ldots, x_k)$ is equivalent to requiring that

- precisely x_1 of the Bernoulli trials resulted in e_1;
- precisely x_2 of the Bernoulli trials resulted in e_2; $\ldots$ and
- precisely x_k of the Bernoulli trials resulted in e_k.

Now, the probability of observing any one particular sequence $\{B_1, \ldots, B_n\}$ satisfying the above conditions is given by

$$p_1^{x_1} \cdots p_k^{x_k}.$$

However, from Proposition 2.38, we know that there are precisely

$$\binom{n}{x_1, \ldots, x_k} = \frac{n!}{x_1! \cdots x_k!}$$

different ways in which the n Bernoulli trials can produce a result consistent with the above requirements. The result follows. $\square$

The fact that the distribution Equation (2.72) is normalized is an immediate consequence of the multinomial theorem, which is stated as follows.

Theorem 2.12 (Multinomial Theorem). *For positive integers k and n, let $\mathcal{A} = \{x \in \mathbb{Z}_*^k : \sum_{i=1}^k x_i = n\}$. Then for any real numbers $p_1, \ldots, p_k$,*

$$(p_1 + \cdots + p_k)^n = \sum_{x \in \mathcal{A}} \frac{n!}{x_1! \cdots x_k!} p_1^{x_1} \cdots p_k^{x_k}.$$

Normalization follows since the probabilities in Definition 2.23 sum to unity; i.e., $\sum_{i=1}^k p_i = 1$.

Another important property of a multinomial random vector X is that the marginal distribution of any component is a binomial distribution. We state this result in the following.

Proposition 2.40. *Let $X \sim \mathrm{Multi}(n, p)$. Then the marginal distribution of each component X_i is given by $X_i \sim \mathrm{Binom}(n, p_i)$, for $i = 1, \ldots, k$.*

Proof. We follow the proof presented in Casella and Berger [2002]. Without loss of generality, let us compute the marginal distribution for X_k. Given a particular value $X_k = x_k \in [0, n]$, we must sum the PMF given in Equation (2.72) over all of the permissible values of $(x_1, \ldots, x_{k-1})$, which is given by the set $\mathcal{B} = \{(x_1, \ldots, x_{k-1}) : \sum_{i=1}^{k-1} x_i = m - x_k\}$. Hence

$$f_{X_k}(x_k) = \sum_{\mathcal{B}} \frac{n!}{x_1! \cdots x_k!} p_1^{x_1} \cdots p_k^{x_k} \frac{(n - x_k)!}{(n - x_k)!} \frac{(1 - p_k)^{n - x_k}}{(1 - p_k)^{n - x_k}}$$

$$= \frac{n!}{x_k!(n - x_k)!} p_k^{x_k} (1 - p_k)^{n - x_k} \sum_{\mathcal{B}} \frac{(n - x_k)!}{x_1! \cdots x_{k-1}!} \frac{p_1^{x_1} \cdots p_{k-1}^{x_{k-1}}}{(1 - p_k)^{n - x_k}}$$

$$= \mathrm{Binom}(x_k; n, p_k) \sum_{\mathcal{B}} \frac{(n - x_k)!}{x_1! \cdots x_{k-1}!} \left(\frac{p_1}{1 - p_k}\right)^{x_1} \cdots \left(\frac{p_{k-1}}{1 - p_k}\right)^{x_{k-1}}$$

The last line follows since $n - x_k = x_1 + \cdots + x_{k-1}$. Now, the term in the summation is equivalent to the PMF for a $\mathrm{Multi}(n - x_k, q)$ random vector, with reduced probability vector $q_i = p_i/(1 - p_k)$, for $i = 1, \ldots, k - 1$. This expression sums to unity, by Theorem 2.12. The result follows. □

Finally, since we define the multinomial random vector as a sum of IID multivariate Bernoulli trials, the variance–covariance of a multinomial random vectort immediately follows from Proposition 2.37. We state the result as follows.

Proposition 2.41. *Let $X \sim \mathrm{Multi}(n, p)$, where $p \in \Delta^{k-1}$. Then $\mathbb{E}[X] = np$ and the variance-covariance matrix is given by $n\Sigma$, where Σ is the variance-covariance matrix of a Bernoulli random vector with probability $p \in \Delta^{k-1}$, as given by Proposition 2.37.*

Example usage of a multinomial random variable in Python is shown in Code Block 2.25. Note that `scipy.stats.multinomial` has a limited number of methods: `pmf` and a few others we have not discussed (`logpmf`, `rvs` (random samples), `entropy`, and `cov`).

```
1  n = 10
2  p = [0.2, 0.3, 0.5]
3  n_samples = 1000
4  samples = np.random.multinomial(n, p, size=n_samples)
5  # Returns n_samples x len(p) matrix: print( samples.shape ) ->
       (1000, 3)
6  # First row: samples[0, :] --> array([3, 0, 7])
7
8  scipy.stats.multinomial.pmf([2, 3, 5], n, p) # 0.08505 # Does not
       have CDF, SF
9  scipy.stats.multinomial.cov(n, p)
10 # array([[ 1.6, -0.6, -1. ],
11 #        [-0.6,  2.1, -1.5],
12 #        [-1. , -1.5,  2.5]])
```

Code Block 2.25: Multinomial distribution in Python

Reduced Multinomial Random Vector

Let X be a k-dimensional multinomial random vector $X \in \text{Multi}(n, p)$, where $p \in \Delta^{k-1}$. We have seen that the components of X are *dependent*, since the final component can always be determined by the preceding components through the relationship

$$X_k = n - X_1 - \cdots - X_{k-1}.$$

Moreover, we've seen that this relation of linear dependence leads to a singularity in the corresponding variance–covariance matrix $\Sigma = \mathbb{V}(X)$. As a practical implication, Σ is therefore noninvertible, rendering it difficult to apply the central limit theorem to such a case. To remedy this, we define the following *reduced* multinomial random vector, as follows.

Definition 2.27. *A random vector $X \in \mathbb{R}^{k-1}$ is said to be a* reduced multinomial random vector, *denoted $X \in \text{RedMulti}(n, p)$, for a given $n \in \mathbb{Z}_+$ and $p \in \Delta^{k-1}$, if its distribution is of the form*

$$f(x_1, \ldots, x_{k-1}) = \frac{n!}{x_1! \cdots x_{k-1}!(n - x_1 - \cdots - x_{k-1})!} p_1^{x_1} \cdots p_{k-1}^{x_{k-1}} p_k^{n-x_1-\cdots-x_{k-1}}$$

$$= \frac{n! \, p_k^n}{(n - x_1 - \cdots - x_{k-1})!} \prod_{i=1}^{k-1} \frac{1}{x_i!} \left(\frac{p_i}{p_k} \right)^{x_i},$$

on the domain $0 \leq \sum_{i=1}^{k-1} X_i \leq n$; as long as $p_k \neq 0$.

In particular, for the special case $n = 1$, we call the random vector X a reduced multivariate Bernoulli random vector, *denoted $X \sim \text{RedMultiBern}(p)$, with corresponding distribution*

$$f(x_1, \ldots, x_{k-1}) = p_k \prod_{i=1}^{k-1} \left(\frac{p_i}{p_k}\right)^{x_i},$$

on the domain $X \in \{0_{k-1}, e_1, \ldots, e_{k-1}\} \subset \mathbb{R}^{k-1}$.[10]

Note 2.22. The reduced multinomial distribution is obtained by substituting $X_k = n - \sum_{i=1}^{k-1} X_i$ in Equation (2.72). We are thus eliminating a redundant variable X_k. (We may continue to reference $p_k = 1 - \sum_{i=1}^{k-1} p_i$.)
▷

Note 2.23. In the construction of Definition 2.27, we require that $p_k \neq 0$. This is typically not a problem, as it is customary to have $p_i > 0$, for $i = 1, \ldots, k$. (Otherwise, what is the point of including the corresponding outcome.) Should $p_k = 0$,[11] the situation is easily resolved by permuting the indices.
▷

Note 2.24. For the case $k = 2$, the multinomial distribution $\text{Multi}(n, p)$, for $p \in \Delta^1$, is a function of *two* random variables: the number of successes *and* the number of failures. The *reduced* multinomial distribution is the binomial distribution, which is a function of *only* the number of successes.
▷

Proposition 2.42. *Let* $X \in \text{RedMultiBern}(p)$ *be a* $(k-1)$-*dimensional reduced multivariate Bernoulli random vector, with* $p \in \Delta^{k-1}$. *Then its mean and variance–covariance matrix* $\Sigma = \mathbb{V}(X)$ *are given by*

$$\mathbb{E}[X_i] = p_i \tag{2.73}$$

$$\Sigma_{ij} = p_i \delta_{ij} - p_i p_j, \tag{2.74}$$

respectively, where $\delta_{ij} = \mathbb{I}[i = j]$ *is the Kronecker-delta function, and where* $i, j = 1, \ldots, k-1$.

Note 2.25. The expressions given by Equations (2.73) and (2.74) are equivalent to the results of Proposition 2.37. In particular, Equation (2.74) is a restatement of Equation (2.70). The difference is that the matrix given by Equation (2.70) is the full variance–covariance matrix, which lives in $\Sigma \in \mathbb{R}^{k \times k}$ and is singular, whereas the matrix specified in Equation (2.74) is the *reduced* variance–covariance matrix, which lives in $\Sigma \in \mathbb{R}^{(k-1) \times (k-1)}$ and is nonsingular.
▷

---

[10] This is equivalent to $X_i \in \{0, 1\}$, such that $\sum_{i=1}^{k-1} X_i \leq 1$.

[11] For instance, suppose that a small island population exists in which every member's favorite color is either red, blue, green, or yellow. Experimenters arrive and conduct a survey in which inhabitants can respond with red, blue, green, yellow, or chartreuse. There is an exactly 0% probability that any inhabitant has the favorite color of chartreuse, and therefore $p_5 = 0$. To determine the reduced multinomial distribution, the experimenters must either remove X_5 from consideration altogether, as it serves no purpose, or permute the indices (say, to chartreuse, red, blue, green, yellow) and then proceed as usual.

We leave the proof to the reader; see Exercise 2.7.

Since the variance–covariance matrix of a reduced multinomial random vector is nonsingular, we can compute its inverse. This is provided in the following.

Proposition 2.43. *Let $X \in \text{RedMultiBern}(p)$ be a $(k-1)$-dimensional reduced multivariate Bernoulli random vector, with $p \in \Delta^{k-1}$. Then its variance–covariance matrix $\Sigma = \mathbb{V}(X)$, as given by Equation (2.74), is invertible. Moreover, its inverse is given by*

$$\Sigma_{ij}^{-1} = p_i^{-1} \delta_{ij} + p_k^{-1}, \tag{2.75}$$

for $i, j = 1, \ldots, k-1$.

Proposition 2.43 is a special case of the result proved in Withers and Nadarajah [2014]. It is also given in Mood [1950].

Note 2.26. More generally, any matrix $\Sigma \in \mathbb{R}^{s \times s}$ of the form

$$\Sigma = \text{diag}(p) - pp^T, \tag{2.76}$$

where pp^T is the outer product $(pp^T)_{ij} = p_i p_j$, has the matrix inverse

$$\Sigma^{-1} = \text{diag}(p)^{-1} + (1 - \nu)^{-1} J_s, \tag{2.77}$$

where $J_s \in \mathbb{R}^{s \times s}$ is a matrix of all ones and where $\nu = \sum_{i=1}^{s} p_i$, whenever $\nu \neq 1$. ▷

Note 2.27. Let us compare Proposition 2.37 and 2.42 in light of Note 2.26. Let $p = (p_1, \ldots, p_k) \in \Delta^{k-1} \subset \mathbb{R}^k$ and let $\tilde{p} = (p_1, \ldots, p_{k-1}) \in \mathbb{R}^{k-1}$ represent the first $k-1$ components of p. Note that $||p||_1 = \sum_{i=1}^{k} p_i = 1$, whereas $||\tilde{p}||_1 = \sum_{i=1}^{k-1} \tilde{p}_i = 1 - p_k < 1$.[12]

The variance–covariance matrix for the original multivariate Bernoulli random vector $X \sim \text{MultiBern}(p)$ given by Equation (2.70) is equivalent to

$$\Sigma = \mathbb{V}(X) = \text{diag}(p) - pp^T \in \mathbb{R}^{k \times k}.$$

Similarly, the variance–covariance matrix of the *reduced* random vector $\tilde{X} \sim \text{RedMultiBern}(p)$ given by Equation (2.74) is equivalent to

$$\tilde{\Sigma} = \mathbb{V}(\tilde{X}) = \text{diag}(\tilde{p}) - \tilde{p}\tilde{p}^T \in \mathbb{R}^{(k-1) \times (k-1)}. \tag{2.78}$$

Note that $\tilde{\Sigma}$ consists of the first $(k-1)$ rows and $(k-1)$ columns of Σ. Both of these are in the form of Equation (2.76), with one *crucial* distinction:

---

[12] Here, the quantity $||x||_1 = \sum_{i=1}^{n} |x_i|$ represents the ℓ_1 norm, or *Manhattan norm*, of the vector $x \in \mathbb{R}^n$. In this context, the absolute values in the sum are unnecessary, since $p_i \geq 0$, for $i = 1, \ldots, k$.

$\nu = ||p||_1 = 1$, whereas $\tilde{\nu} = ||\tilde{p}||_1 = 1 - p_k$. Thus, the former is not invertible, whereas the inverse of the reduced variance–covariance matrix is given by Equation (2.77), which gives us

$$\tilde{\Sigma}^{-1} = \mathrm{diag}(\tilde{p})^{-1} + p_k^{-1} J_{k-1}, \qquad (2.79)$$

which is equivalent to Equation (2.75).[13]   ▷

Problems

2.1. In the proof of Proposition 2.7, show that

$$n\binom{n-1}{r-1} = r\binom{n}{r}.$$

2.2. Prove that for any two independent continuous random variables X and Y with PDFs $f_X(x)$ and $f_Y(y)$, respectively, the PDF of the random variable $Z = X + Y$ is given by the following *convolution integral*

$$f_Z(z) = f_X * f_Y(z) \triangleq \int_{-\infty}^{\infty} f_X(w) f_Y(z-w)\, dw.$$

Hint: consider the bivariate transformation

$$Z = X + Y$$
$$W = X.$$

2.3. Prove Proposition 2.14.

2.4. Complete the first step in the proof of Proposition 2.28: Let $X \sim$ Gamma(α, β). Show that the MGF for X is given by

$$M_X(t) = \frac{1}{\Gamma(\alpha)\beta^\alpha} \int_0^\infty x^{\alpha-1} e^{-x/(\frac{\beta}{1-\beta t})}\, dx.$$

2.5. Let $X \sim$ Gamma(α, β) be a gamma random variable with PDF $f_X(x)$. Show that for $\sigma > 0$, the PDF $(1/\sigma) f_X(x/\sigma)$ is the PDF of a Gamma$(\alpha, \sigma\beta)$ random variable.

2.6. Variance of Beta–Binomial Random Variable
(a) Prove Theorem 2.9.
(b) Use Equation (2.58) to provide an alternate proof of Equation (2.57).

2.7. Prove Proposition 2.42.

---

[13] Recall that the inverse of a diagonal matrix is obtained by inverting each of its diagonal elements.

2.8. Let $X_1, \ldots, X_n$ be independent samples, where $X_i \sim \mathrm{N}(\mu, \sigma^2/w_i)$, for a known set of weights $w_i > 0$. Let $\omega = \sum_{i=1}^{n} w_i$ be the total weight, and define

$$\overline{X} = \frac{1}{\omega} \sum_{i=1}^{n} w_i X_i \qquad \text{and} \qquad S^2 = \frac{1}{n-1} \sum_{i=1}^{n} w_i \left(X_i - \overline{X}\right)^2.$$

Show that $\overline{X} \sim \mathrm{N}(\mu, \sigma^2/\omega)$ and $(n-1)S^2/\sigma^2 \sim \chi^2_{n-1}$.

3

Getting Testy

The goal of this chapter is to apply the knowledge of various distributions that we built up in Chapter 2 to our first task of statistical inference: classical hypothesis testing and confidence intervals. Excellent supplemental reading, discussion, and examples can be found in Casella and Berger [2002], Hogg, *et al.* [2015], and Shao [2003].

3.1 Hypothesis Testing

The goal of this section is to introduce basic terminology surrounding classical hypothesis testing before diving into specific examples.

3.1.1 The Null Hypothesis

In general, a *hypothesis* is a statement about one or more population parameters, and a *hypothesis test* is a procedure for inferring the truth or validity of a hypothesis given a set of data.

There is a twist, however, in that hypothesis testing does not seek to prove what is true, but rather to disprove what is not true. Therefore, instead of trying to prove a hypothesis, we will be devising methods for disproving its complement, otherwise known as the *null hypothesis*. The null hypothesis earns its name as it is typically a statement about what is *not* there: in the language of experimentation, a null hypothesis is usually of the form that our experiment had no effect. Formally, we have the following.

Definition 3.1. *A hypothesis test concerning some population parameter $\theta \in \Theta$, where Θ is the parameter space, consists of a null hypothesis of the form*

$$H_0 : \theta \in \Theta_0$$

and a decision rule that determines, as a function of a sample from the population, whether we should reject the null hypothesis.

Definition 3.2. *The complement of a null hypothesis is called the* alternative hypothesis, *denoted* $H_A : \theta \in \Theta_0^c$. *Whenever we do not reject the null hypothesis, we say that we* accept the null hypothesis.

Definition 3.3. *A* two-sided test *is any hypothesis test with null hypothesis of the form* $H_0 : \theta = \theta_0$. *A* one-sided test *is any hypothesis test with null hypothesis of the form* $H_0 : \theta \leq \theta_0$ *or* $H_0 : \theta \geq \theta_0$.

Note 3.1. We were very purposeful in how we set up Definitions 3.1 and 3.2: a hypothesis test is a decision on whether or not to reject the null hypothesis. "Acceptance" of a null hypothesis is simply the absence of its rejection; it is an unfortunate term which should not be misinterpreted as implying the truth of the null hypothesis. For example, if the sample size is $n = 1$, i.e., our data set consists of a single datum, it should be virtually impossible to reject the null hypothesis, simply due to having an insufficient amount data to support such a claim. In such a case, we are said to *accept the null hypothesis.* But this simply means that we could not rule it out. Thus, the act of accepting the null hypothesis really means accepting the null hypothesis *for now*; i.e., we allow it to live to die another day[1].   ▷

To be slightly more careful with the meaning of a *decision rule* in Definition 3.1, we have the following.

Definition 3.4. *Let* $\mathbb{X}$ *be the sample space for a distribution. A* decision rule *is a statistic* $r : \mathbb{X} \to \mathbb{A}$ *that determines an action* $a \in \mathbb{A}$ *as a function of the data* $\mathcal{X} \in \mathbb{X}$, *where* $\mathbb{A}$ *is the set of permissible actions, i.e.,* action space.

In particular, the action space for any decision rule in a hypothesis test is always binary; i.e., $\mathbb{A} = \mathbb{B} \triangleq \{0, 1\}$.

Typically, however, the decision rule in a hypothesis test consists of a *test statistic* $T : \mathbb{X} \to \mathbb{R}$ along with set of values $R \subset \mathbb{R}$ for which we would reject the null hypothesis. Formally, we have the following.

Definition 3.5. *Let* $T : \mathbb{X} \to \mathbb{R}$ *be a statistic and* $R \subset \mathbb{R}$, *and define and the function* $\mathbb{I}_R : \mathbb{R} \to \mathbb{B}$ *as the indicator function* $\mathbb{I}_R(x) = \mathbb{I}[x \in R]$. *Whenever a decision rule* r *for a hypothesis test consists of the composition* $r = \mathbb{I}_R \circ T$, *then the statistic* T *is referred to as a* test statistic *and the region* R *is referred to as the* rejection region *corresponding to* T.

When a rejection region is in standard form $R_c = [c, \infty)$, *the value of* c *is referred to as the* critical value *of the test. (In some cases, a simple transformation such as* $T \to -T$ *or* $T \to |T|$ *is required to achieve standard form.)*

Note 3.2. Some authors use *rejection region* to refer to the subset $\mathcal{R} \subset \mathbb{X}$ of the sample space defined by $\mathcal{R} = \{\mathcal{X} \in \mathbb{X} : r(\mathcal{X}) = 1\}$.   ▷

---

[1] Be weary of statisticians who accept you as their friend.

Example 3.1. Suppose we have a friend who is a magician who persistently challenges us to annoying coin tosses. Our magician friend always takes *heads.* We are beginning to suspect that he has a trick coin that is more likely to come up *heads* than it is *tails.* (The dastard!)

We devise the following hypothesis test: our null hypothesis is that the coin is fair or better, i.e., $\mathbb{P}(\text{heads}) \leq 0.50$. Our decision rule is to reject the null hypothesis if the observed rate of heads is greater than or equal to 60%. In particular, our test statistic is simply the sample mean $T(\mathcal{X}) = \overline{X}_n$, and our rejection region is the interval $R = [0.60, 1]$. Thus, we reject the null hypothesis (and call our friend a *cheat!*) whenever we observe more than 60% heads.

This, of course, is not a very *good* hypothesis test. In particular, if our magician friend makes a single coin flip and it comes up heads, we have already rejected the null hypothesis and concluded that the coin is not fair. It is, however, a very good illustration of the concepts defined thus far into the chapter.   ▷

3.1.2 Types of Errors

Next, we explore the types of errors that we might encounter when performing a hypothesis test. Indeed, any hypothesis test has four possible outcomes: our test must either accept or reject the null hypothesis, and the null hypothesis must be (actually) true or false. These are illustrated in Table 3.1.

| | | Truth | |
		H_A	H_0
Decision	Reject H_0	Correct Decision	Type I Error
	Accept H_0	Type II Error	Correct Decision

Table 3.1: Type I and II Errors in hypothesis testing.

In particular, we have the following definition.

Definition 3.6. *We say that a hypothesis test results in a* Type I Error *whenever the decision rule falsely rejects a true null hypothesis. Similarly, we say that a hypothesis test results in a* Type II Error *whenever the decision rule accepts a false null hypothesis.*

In other words, a Type I Error occurs whenever we reject a null hypothesis that is true, and a Type II Error occur whenever we fail to reject a null hypothesis that is false. Again, Table 3.1 is a luciferous remedy should these words seem at first opaque.

Example 3.2. Our magician friend flips his coin and it lands *heads*. Using the hypothesis test from Example 3.1, we accuse him of cheating. It turns out, much to our dismay, that he in fact used an ordinary coin.

In this example, the null hypothesis was that the coin was fair (or even favored us). We falsely rejected this, concluding that his coin was weighted, making a Type I Error.

Consider now a variation: our magician friend is using a weighted coin that is more likely to turn up heads than it is tails. Despite odds being in his favor, and much to his chagrin, the coin lands *tails*. In this case, we accept the null hypothesis, that the coin is fair. However, we do so erroneously, thereby committing a Type II Error.   ▷

Note 3.3. The type of error is determined entirely upon whether or not the null hypothesis is actually true, i.e., the column of Table 3.1. If the null hypothesis is true and our conclusion is wrong, it is a Type I error. If the alternative hypothesis is true and our conclusion is wrong, it is a Type II error.   ▷

3.1.3 The Power Function

We still need a method of evaluating *how good* a particular hypothesis test is. We have an intuitive sense that the hypothesis test from Examples 3.1 and 3.2 is a poor test, but how can we quantity this? The answer lies in understanding the probability of making different types of errors. Those probabilities, however, depend on the *actual* value of the parameter θ, which is typically unknown. A useful tool for discussing these probabilities is given in our next definition.

Definition 3.7. *The* power function *of a hypothesis test with test statistic T and rejection region R is defined as the function $\beta : \Theta \to [0,1]$ given by the relation*

$$\beta(\theta) = \mathbb{P}_\theta(T(\mathcal{X}) \in R). \tag{3.1}$$

Thus, the power of a test tells us the probability of rejecting the null hypothesis as a function of the true value of parameter θ. Note that the power function is determined by the test statistic and rejection region, and is implicitly dependent on the sample size of the data but not on the data itself.

We next show how the power function is related to Type I and II Errors. First, let us assume that the null hypothesis is true; i.e., let us assume that $\theta \in \Theta_0$. In this case, the power function $\beta(\theta)$, when evaluated at the true value for θ, represents the probability of making a Type I Error.

Next, let us assume that the null hypothesis is false; i.e., let us assume that $\theta \in \Theta_0^c$. Since the power function represents the probability of rejecting the null hypothesis, the quantity $1 - \beta(\theta)$ must therefore represent the probability of accepting the null hypothesis. Thus, when $\theta \in \Theta_0^c$, one minus the power function represents the probability of making a Type II Error.

We therefore have the equality

$$\beta(\theta) = \begin{cases} \mathbb{P}(\text{Type I Error}) & \text{if } \theta \in \Theta_0 \\ 1 - \mathbb{P}(\text{Type II Error}) & \text{if } \theta \in \Theta_0^c \end{cases}.$$

Of course, the preceding equation assumes that we can actually evaluate the power function at the *true* value of the parameter θ. This is hardly the case in practice, and, as such, we must quantify the worst case scenario. Moreover, when devising tests, we do not typically prescribe the precise amount of error, but rather seek to place an upper bound on the error. This gives rise to the following definition.

Definition 3.8. *Let $\beta(\theta)$ be the power function of a hypothesis test. Then, for $0 \leq \alpha \leq 1$, we say that the test has* significance level α *if*

$$\sup_{\theta \in \Theta_0} \beta(\theta) \leq \alpha. \tag{3.2}$$

The exact value of the supremum is referred to as the size *of the test. The number $(1 - \alpha)$ is sometimes referred to as the* confidence level *of the test.*

Note 3.4. As pointed out in Casella and Berger [2002], it is, in practice, often difficult to construct a test with an exact size. It is much simpler to require that a test is valid at a particular significance level; typical choices are $\alpha = 0.01$, 0.05, and 0.10. A hypothesis test at the 5% significance level is therefore guaranteed to have no more than a 5% probability of producing a Type I error.   ▷

Note 3.5. A test with a low significance level can still have a high probability of producing a Type II Error.   ▷

Example 3.3. So we've burnt through several magician friends and have decided that it's time for a modest improvement to our test. Our new test will require that we observe ten coin tosses, and if 7 or more are *heads*, we will reject the null hypothesis $H_0 : \mathbb{P}(\text{heads}) \leq 0.50$. We can characterize the test statistic as $T = \sum_{i=1}^{10} X_i$, and the rejection region is $R = \{7, 8, 9, 10\}$. Of course, if the true probability of heads is p, our test statistic will have a binomial distribution; i.e., $T \sim \text{Binom}(10, p)$. Moreover, given p, the probability of observing a value within the rejection region is simply

$$\mathbb{P}_p(T \in R) = S(6) = 1 - F(6),$$

where $F(x)$ is the CDF for Binom(10, p) and $S(x) = 1 - F(x)$ is the associated survival function[2].

---

[2] Note that $\mathbb{P}_p(T \in R) = S(6)$ and not $S(7)$. Since the CDF is defined as $F(x) = \mathbb{P}(X \leq x)$, its complement, the survival function, is the probability $S(x) = \mathbb{P}(X > x)$ Hence we must evaluate the survival function at 6 in order to obtain the probability that $T \in \{7, 8, 9, 10\}$.

Luckily, the plot of the survival code is easily achieved using Python, as shown in Code Block 3.1. The resulting output is shown in Figure 3.1.

```
1  p = np.linspace(0, 1, num=100)
2  beta = scipy.stats.binom.sf(6, 10, p)
3  alpha = scipy.stats.binom.sf(6, 10, 0.5)
4  plt.plot(p, beta)
5  plt.fill_between(p[:50], beta[:50], color='#1f77b4', alpha=0.5)
6  x = 0.5 * np.ones(50)
7  y = np.linspace(0, alpha)
8  plt.plot(x, y, '--', color='#1f77b4')
9  x = np.linspace(0, 0.5)
10 y = alpha * np.ones(50)
11 plt.plot(x, y, '--', color='#1f77b4')
```

Code Block 3.1: Survival code for coin-toss experiment.

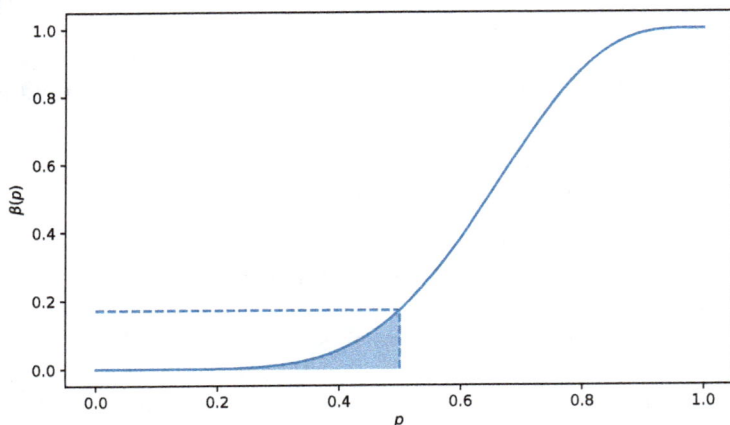

Fig. 3.1: Power function for Example 3.3.

Moreover, the size of our test is revealed by `print(alpha)`, and is (approximately) 17.2%. This point is shown with dotted lines in Figure 3.1, and the area under the power curve corresponding to our null hypothesis is shaded.

We conclude that if our null hypothesis is true, we still have up to a 17.2% probability of bearing false witness against our dear magician friend.

Moreover, it is likely that the probability of a Type I Error under the null hypothesis is exactly 17.2%, and not one of the lesser values. This is because our null hypothesis is actually a *composite hypothesis*: the null hypothesis is true if $\theta = 0.50$ (the coin is fair) *or if* $\theta < 0.50$ (the coin is weighted, but our friend is incompetent). Under the nulll hypothesis, we very well might assess that a fair coin is much more likely than a weighted coin being used the wrong way. In general, the size determines the worst-case scenario, under the null hypothesis, for the probability of committing a Type I Error. In this case, it is also the most likely value of that probability, if the null hypothesis is true.

Our magician friend proceeds to flip his coin ten times. Seven out of ten are heads. Our test says we should reject the null hypothesis, but we now know that, if the null hypothesis is true, we danger a 17.2% chance of rejecting the null hypothesis in error. *Should we reject the null hypothesis?* The answer, in this case, depends on our fondness for magicians and their scarcity.   ▷

As we might expect after examining Figure 3.1, the ideal power function will be as close as possible to zero for $\theta \in \Theta_0$ and as close as possible to one for $\theta \in \Theta_0^c$.

Essentially, the significance level is a bound we place on the probability of a Type I Error. We stress again that the power function, size, and significance level are a function of the test, not of the data. When we go to apply the test for an actual, realized set of data $\mathcal{X}_0 \in \mathbb{X}$, it is further useful to report the smallest level of significance that would result in rejection of the null hypothesis.

Definition 3.9. *Let T be a test statistic and R_c be a rejection region in standard form. For a given set of data $\mathcal{X}_0 \in \mathbb{X}$, the p-value is given by*

$$p_0 = \sup_{\theta \in \Theta_0} \mathbb{P}_\theta(T(\mathcal{X}) \in R_{c^*}), \tag{3.3}$$

where $c^* = T(\mathcal{X}_0)$.

Thus, the *p*-value is what the size of the test *would have been*, had the rejection region been chosen to be the smallest possible rejection region, consistent with the standard form, that contains the observed value of our test statistic. In other words, the *p*-value is the best-case probability under the null hypothesis of obtaining a sample at least as extreme, as measured by the test statistic T, as the data that were actually observed.

Note 3.6. Though one of the earliest and most familiar concepts the author is familiar with, we found the *p*-value difficult to define. It is often simply described as *the probability of observing data at least as extreme as your sample* and is illustrated as the tail-end probability under the distribution for the test statistic. While we admit that the definition we finally gravitated to has limits in the sense that it relies on the form of the rejection region, we feel the final result is closest to practice and therefore most intuitive.

Some authors define the p-value as any statistic $p : \mathbb{X} \to [0, 1]$ that satisfies the relation

$$\mathbb{P}_\theta(p(\mathcal{X}) \le \alpha) \le \alpha$$

for every $\theta \in \Theta_0$ and $\alpha \in [0, 1]$; see Casella and Berger [2002]. Thus, the p-value is immediately seen to be a test with significance level α when used in conjunction with the rejection region $R_\alpha = [0, \alpha]$. We felt that such a definition obfuscates where the p-value comes from in the first place: the original test statistic.

Other authors (Shao [2003], Wasserman [2004]) suppose that, for a given test statistic T, and for each $\alpha \in (0, 1)$, that a size α test can be constructed by choice of an appropriate rejection region R_α. They go on to define the p-value, given an observed set of data $\mathcal{X}_0 \in \mathbb{X}$, by the relation

$$p = \inf\{\alpha \in (0, 1) : T(\mathcal{X}_0) \in R_\alpha\}.$$

The p-value therefore represents the *smallest* size test that would have rejected the null hypothesis given the observed data. We were quite fond of this definition, but nevertheless felt that it suffered two shortcomings: there was no continuity in the family of rejection regions R_α and it is not always possible to construct such a region for *any* $\alpha \in (0, 1)$, in particular when the possible values of the test statistic are discrete. (Example 3.4 will constitute a counterexample to this.)

After constructing the language for Definition 3.9, our choice was later validated upon discovery that each of the above-mentioned authors ultimately came to the same formula for the special case of a standard rejection region. As this is the definition closest to practice, we ultimately decided to adopt it as our definition. We invite the reader to explore the alternative definitions and draw their own conclusions.   ▷

Example 3.4. Let us return to the hypothesis test discussed in Example 3.3. Recall that we are testing the null hypothesis $H_0 : p \le 0.5$ by performing a sequence of ten independent coin tosses. Our test statistic is $T(\mathcal{X}) = \sum_{i=1}^{10} X_i$, the number of observed *heads*, and our rejection region is of the form $R_c = [c, \infty)$; i.e., we will reject the null hypothesis if we observe more than c heads. Furthermore, we have shown that in order to have a test at the 20% significance level, we can set $c = 7$.

Thus, our test is set: we reject the null hypothesis if we observe at least 7 out of 10 *heads*. Having a 20% significance level means that, if the coin is fair, there is (up to) a 20% probability of falsely rejecting a true null hypothesis. The significance level depends on the *test*, not the *data*.

We perform our test and observe *nine* heads! Our p-value is 1.1%. This is because, under the null hypothesis, there is only a 1.1% probability of observing a result that is so extreme, which is computed by evaluating the survival function for a binomial distribution Binom$(10, 0.5)^3$ at $x = 8$.

```
1  for c in range(11):
2      alpha = scipy.stats.binom.sf(c-1, 10, 0.5)
3      print (c, round(alpha, 3))
```

Code Block 3.2: Generate all possible p-values of coin-toss experiment.

$\sum_{i=1}^{10} X_i$	p-value
0	1.000
1	0.999
2	0.989
3	0.945
4	0.828
5	0.623
6	0.377
7	0.172
8	0.055
9	0.011
10	0.001

Table 3.2: p-values for coin-toss experiment

In fact, since there are only a handful of possible outcomes of the test statistic T, it is not difficult to compute a p-value for all of them, as shown in Code Block 3.2. The output is given in Table 3.2.   ▷

3.1.4 Confidence Intervals

Hypothesis testing is closely related to the topic of confidence intervals, which we now define.

Definition 3.10. *Given two statistics* $L, U : \mathbb{X} \to \mathbb{R}$ *satisfying* $L(\mathcal{X}) \leq U(\mathcal{X})$ *for all* $\mathcal{X} \in \mathbb{X}$, *we say that the random interval* $[L(\mathcal{X}), U(\mathcal{X})]$ *is a* $(1-\alpha)$-*confidence interval for the parameter* $\theta \in \Theta$ *if it satisfies the relation*

$$\inf_{\theta \in \Theta} \mathbb{P}_\theta(\theta \in [L(\mathcal{X}), U(\mathcal{X})]) = 1 - \alpha. \tag{3.4}$$

Note 3.7. In Definition 3.10, the parameter θ is a *fixed* albeit unknown quantity; it is the interval $[L(\mathcal{X}), U(\mathcal{X})]$ that is random. Hence, the quantity

---

[3] Using a probability of $p = 0.5$ maximizes the probability of a given number of *heads* out of all $p \in \Theta_0 = [0, 0.5]$.

$$\mathbb{P}_\theta(\theta \in [L(\mathcal{X}), U(\mathcal{X})])$$

should be interpreted as the probability, given parameter θ, that the statistics L and U satisfy $L(\mathcal{X}) \le \theta$ and $U(\mathcal{X}) \ge \theta$.   ▷

In other words, a 95% confidence interval means that, regardless the true value of the parameter θ, as least 95% of samples $\mathcal{X}$ from our distribution will produce correct upper and lower bounds on the parameter θ. Since the parameter θ is *fixed*, and not *random*, this is *not* a probability statement on θ.

Confidence intervals are oftentimes closely connected with hypothesis tests, in that each implies each other. To see this, consider a simple hypothesis of the form $H_0 : \theta = \theta_0$. Suppose a size α test is devised with test statistic T and rejection region $R_c = \{t : |t| \ge c\}$. The *acceptance region* is therefore given by $A_c = R_c^c = (-c, c)$, so that we accept the null hypothesis whenever $-c < T(\mathcal{X}) < c$. Now, since the test is size α, and since there is but a single point in Θ_0, Equation (3.2) implies that

$$\mathbb{P}_\theta(T(\mathcal{X}) \in R_c) = \alpha,$$

and hence

$$\mathbb{P}_\theta(T(\mathcal{X}) \in A_c) = 1 - \alpha.$$

It is often the case (as we shall shortly see) that the test statistic depends on the value θ_0 and, moreover, that this relation can be inverted. This inversion creates a correspondence

$$-c < T(\mathcal{X}; \theta_0) < c \qquad \text{if and only if} \qquad L(\mathcal{X}) < \theta_0 < U(\mathcal{X}),$$

for some functions L and U. Thus, under the above assumptions, a size α hypothesis test is often equivalent to a $(1 - \alpha)$-confidence interval.

3.2 Tests for Estimates

In this section we discuss various tests related to the expected value of a random variable. Additionally, we hope to further illustrate each of the concepts in Section 3.1.

3.2.1 The Mean of a Normal Random Variable

We begin with a test for the mean of a normal random variable. We will treat the cases of known and unknown variance separately.

Proposition 3.1 (z-test). *Let* $X_1, \ldots, X_n \sim \mathrm{N}(\mu, \sigma^2)$, *with known variance* σ^2. *The hypothesis test consisting of null hypothesis*

$$H_0 : \mu = \mu_0,$$

test statistic

$$Z_n = \frac{\overline{X}_n - \mu_0}{\sigma/\sqrt{n}}, \tag{3.5}$$

and rejection region

$$R_\alpha = \{z : |z| \geq z_{\alpha/2}\},$$

where $z_{\alpha/2}$ is the solution to $\Psi(z_{\alpha/2}) = \alpha/2$, as given by Definition 2.11, constitutes a size α test.

Moreover, given an observed sample $\mathcal{X}$ and test statistic $Z_n = z$, the p-value of the test for the given data is given by $2\Psi(|z|)$.

Proof. Under the null hypothesis $\mu = \mu_0$, the sample mean has distribution $\overline{X}_n \sim \mathrm{N}(\mu_0, \sigma^2/n)$, according to Corollary 2.2. Therefore, our test statistic given by Equation (3.5) is a standard normal random variable, $Z_n \sim \mathrm{N}(0, 1)$, according to Corollary 2.1.

The size of our test is therefore given by

$$\mathbb{P}_{\mu_0}(Z_n \in R_\alpha) = \alpha.$$

This follows due to symmetry of the two tails: $\mathbb{P}_{\mu_0}(Z_n \geq z_{\alpha/2}) = \alpha/2$ and $\mathbb{P}_{\mu_0}(Z_n \leq -z_{\alpha/2}) = \alpha/2$.

For a given, observed value of the test statistic $Z_n = z$, the p-value is the probability, under the null hypothesis, of observing a result as extreme. Following Definition 3.9, the p-value is given by

$$p_0 = p_{\mu_0}\left(|Z_n| > |z|\right) = \mathbb{P}\left(Z > |z|\right) + \mathbb{P}\left(-Z < -|z|\right) = 2\mathbb{P}\left(Z > |z|\right),$$

for $Z \sim \mathrm{N}(0, 1)$. The result follows since $\Psi(z) = \mathbb{P}(Z > z)$. $\qquad \square$

Note 3.8. A straightforward variation of Proposition 3.1 is the one-sided test, with null hypothesis

$$H_0 : \mu \leq \mu_0,$$

same test statistic Equation (3.5), and rejection region

$$R_\alpha = [z_\alpha, \infty).$$

We would therefore reject the null hypothesis if $Z_n \geq z_\alpha$. Since this is a one-sided test, the full probability α of rejection must lie on one tail and not be divided among two tails. To show that this is a size α test, we must show that

$$\sup_{\mu \leq \mu_0} \mathbb{P}_\mu(Z_n \geq z_\alpha) = \alpha.$$

However, this probability is maximized by taking $\mu = \mu_0$, and so the result follows. (If the true value of μ is less than μ_0, it would shift the true probability distribution to the left, thereby decreasing the residual probability in the *right* tail $Z_n \geq z_\alpha$.)

Similarly, for the one-sided null hypothesis $H_0 : \mu \geq \mu_0$, we would use $R_\alpha = (-\infty, -z_\alpha]$ to obtain a size α test. $\qquad \triangleright$

We stated previously that under normal circumstances there is a correspondence between hypothesis tests and confidence intervals. The confidence interval corresponding to the z-test is given below.

Proposition 3.2. *Let* $X_1, \ldots, X_n \sim \mathrm{N}(\mu, \sigma^2)$, *with known variance* σ^2. *Then a* $(1 - \alpha)$ *confidence interval for the population mean* μ_0 *is given by the interval*

$$I_n = \left(\overline{X}_n - \frac{\sigma z_{\alpha/2}}{\sqrt{n}}, \overline{X}_n + \frac{\sigma z_{\alpha/2}}{\sqrt{n}} \right). \tag{3.6}$$

Proof. The acceptance region of the z-test is given by

$$A_\alpha = \left(-z_{\alpha/2}, z_{\alpha/2} \right),$$

so that the test statistic Z_n has a $(1 - \alpha)$ of being found within A_α. It is straightforward to show that

$$-z_{\alpha/2} < \frac{\overline{X}_n - \mu_0}{\sigma/\sqrt{n}} < z_{\alpha/2}$$

if and only if

$$\overline{X}_n - \frac{\sigma z_{\alpha/2}}{\sqrt{n}} < \mu_0 < \overline{X}_n + \frac{\sigma z_{\alpha/2}}{\sqrt{n}},$$

which completes the result. □

If the population variance is unknown, we can use the next best thing: the sample variance. The result is given below.

Proposition 3.3 (t-test). *Let* $X_1, \ldots, X_n \sim \mathrm{N}(\mu, \sigma^2)$, *with* unknown *variance* σ^2. *The hypothesis test consisting of null hypothesis*

$$H_0 : \mu = \mu_0,$$

test statistic

$$T_n = \frac{\overline{X}_n - \mu_0}{S_n/\sqrt{n}}, \tag{3.7}$$

and rejection region

$$R_\alpha = \{ t : |t| \geq \mathrm{t}_{n-1,\alpha/2} \},$$

where $\mathrm{t}_{n-1,\alpha/2}$ *is the solution to* $S(\mathrm{t}_{n-1,\alpha/2}) = \alpha/2$, *where S is the survival function of a* t_{n-1} *random variable, constitutes a size α test.*

Proof. Under the null hypothesis, the test statistic Equation (3.7) is a t_{n-1} random variable due to Proposition 2.21. The result follows. □

Note 3.9. The one-sided variations of this test correspond to the null hypothesis $H_0 : \mu \geq \mu_0$ with rejection region $R_\alpha = (-\infty, -\mathrm{t}_{n-1,\alpha}]$ and null hypothesis $H_0 : \mu \leq \mu_0$ with rejection region $R_\alpha = [\mathrm{t}_{n-1,\alpha}, \infty)$. ▷

Note 3.10. It is helpful to recall that the t_n distribution looks much like a standard normal distribution with enlarged tails, and that $t_n \to N(0, 1)$ as $n \to \infty$. Thus, for large sample sizes, say $n > 30$, the t-test given by Proposition 3.3 will not yield a different result than the z-test of Proposition 3.1. Thus, the t-test is best suited for dealing with inferences from samples from a normally distributed population with unknown variance and small sample size.   ▷

Example 3.5. The difference between the z-test and t-test is relevant largely for small samples sizes. In this example, we verify the validity of the t-test over the z-test when constructing the test statistic using the sample variance. In particular, suppose our true distribution is $N(3, 1)$, and consider the null hypothesis $\mu_0 = 3$. Because we are constructing this numerical simulation, we know that the null hypothesis is in fact true. What we want to capture is, therefore, the rate that we are wrong if we were to use either test; i.e., an incorrect application of the z-test vs. a correct application of the t-test. The code is shown in Code Block 3.3.

```
1   mu_0 = 3
2   var = 1
3   n_samples = 10
4   alpha = 0.05
5   z_crit = scipy.stats.norm.isf(alpha/2) #1.960
6   t_crit = scipy.stats.t.isf(alpha/2, n_samples-1) # 2.262
7   significance_z_crit = 2 * scipy.stats.t.sf(z_crit, 9) # 8.165%
8
9   n_trials = 10000
10  rejections_z = np.zeros(n_trials)
11  rejections_t = np.zeros(n_trials)
12  p_values = np.zeros(n_trials)
13  for i in range(n_trials):
14      samples = np.random.normal(mu, scale=var, size=n_samples)
15      sample_mean = samples.mean()
16      sample_var = samples.var() * n_samples / (n_samples - 1)
17      T = (sample_mean - mu_0) / (np.sqrt(sample_var) /
            np.sqrt(n_samples))
18      rejections_z[i] = int( abs(T) > z_crit )
19      rejections_t[i] = int( abs(T) > t_crit )
20      p_values[i] = 2 * scipy.stats.t.sf(abs(T), 9)
21
22  print ( 'Z_crit Rejection Rate', rejections_z.mean() ) # 0.0834
23  print ( 'T_crit Rejection Rate', rejections_t.mean() ) # 0.0505
24  plt.hist(p_values, bins=100, normed=True)
```

Code Block 3.3: t-test and z-test using sample variance

First, we define our (true) mean, variance, sample size (10), and significance level $\alpha = 0.05$ on lines 1–4. Then we compute the critical value $z_{0.025} \approx 1.960$ (incorrect) and $t_{9,0.025} \approx 2.262$ (correct), as shown in lines 5 and 6. Using the critical value for the t-distribution is correct since we will be computing our test statistic using the sample variance, as if the population variance were unknown. By incorrectly using the critical value $z_{0.025}$, we therefore expect an *actual* significance of approximately 8.165% (line 7). If our theory is correct, that our test statistic satisfies a t-distribution as opposed to a normal distribution, we should expect to see approximately 5% rejections when using the critical value $t_{9,0.025}$ and approximately 8.165% rejections when using the critical value $z_{0.025}$.

Next, we simulate 10,000 random samples, each sample having size $n = 10$, and compute the test statistic T_n (line 17) for each sample. We then record our decision on whether or not to reject the null hypothesis using either criterion (lines 18 and 19). As an added bonus, we record the observed p-value for each sample (line 20).

The results are given in lines 22 and 23. If we incorrectly apply the critical value $z_{0.025}$, we actually observe 8.34% false rejections, as opposed to the observed 5.05% false rejection rate using the correct value $t_{9,0.025}$. (These numbers will vary slightly, of course, each time we run the simulation.)

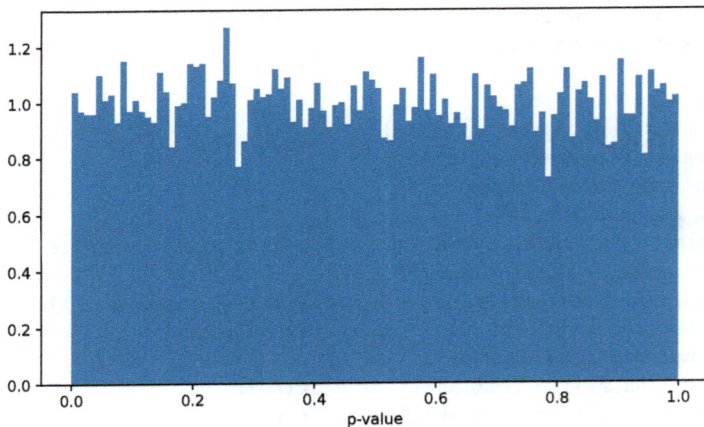

Fig. 3.2: Distribution of p-values for Example 3.5.

Finally, we can plot the distribution of our p-values, which is shown in Figure 3.2. The result is as expected: under the null hypothesis, the distribution of p-values is approximately Unif(0, 1). Our simulated data results in a distribution fairly close to this.   ▷

Note 3.11. It is true in general that whenever our test statistic has a continuous distribution, the distribution of p-values, under the null hypothesis, is Unif$(0, 1)$. Therefore, a test of the form p-value $< \alpha$ results in a size-α test. ▷

Note 3.12. Figure 3.2 further illustrates the distinction between p-value and significance level. The p-value is a measure of how unlikely a particular sample is, given the null hypothesis, and not a statement as to the Type I Error rate. For example, suppose our significance level was 20%. We would therefore reject each sample with a p-value less than 0.20. If the null hypothesis is true, we will falsely reject 20% of the samples. But, in practice, *we only get to observe one sample!* Suppose its p-value is 0.005. This does not mean that our test has a Type I Error rate of 0.5%! Our test has a Type I Error rate of 20%, because we *would have rejected* any p-value less than 0.20. It simply means that our *actual* observed sample is quite rare if the null hypothesis is true. In fact, the given sample of data *would have been rejected*, even by a test with a 1% significance level. ▷

3.2.2 The Wald Test

We have previously seen tests for comparing the sample mean from a normally distributed population against a hypothesized value for the population mean. Of course, when the population is not normally distributed, we may still use this test as an approximation, due to the central limit theorem. The result is known as the Wald test, which is stated below.

Definition 3.11 (Wald Test). *Let $\hat{\theta}_n$ be an asymptotically normal and unbiased estimator for a parameter $\theta \in \Theta$ and let $\mathrm{se}(\hat{\theta}_n)$ be the standard error. Then the size α Wald test consists of the null hypothesis*

$$H_0 : \theta = \theta_0,$$

the test statistic

$$Z_n = \frac{\hat{\theta}_n - \theta_0}{\mathrm{se}(\hat{\theta}_n)}, \tag{3.8}$$

and the rejection region

$$R_\alpha = \{Z_n : |Z_n| \geq z_{\alpha/2}\},$$

where $z_{\alpha/2}$ is the solution to $\Psi(z_{\alpha/2}) = \alpha/2$, as given by Definition 2.11. The function $W_n : \mathbb{X} \to \mathbb{R}$ given by $W_n = Z_n^2$ is known as the Wald statistic.

Theorem 3.1. *The size of a size-α Wald test converges to α, i.e.,*

$$\lim_{n \to \infty} \mathbb{P}_{\theta_0}(|Z_n| \geq z_{\alpha/2}) = \alpha.$$

Moreover, the result is still valid if we replace $\mathrm{se}(\hat{\theta}_n)$ in Z_n with any asymptotically unbiased estimate $\hat{\mathrm{se}}(\hat{\theta}_n)$.

Proof. Under the null hypothesis $\theta = \theta_0$, the statistic $Z_n = (\hat{\theta}_n - \theta_0)/\text{se}$ converges to a standard normal random variable $Z \sim N(0,1)$ as $n \to \infty$. Therefore, the probability of rejecting a true null hypothesis is given by

$$\mathbb{P}_{\theta_0}(|Z_n| > z_{\alpha/2}) = \mathbb{P}_{\theta_0}\left(\frac{|\hat{\theta}_n - \theta_0|}{\text{se}} > z_{\alpha/2}\right) \to \mathbb{P}(|Z| > z_{\alpha/2}) = \alpha.$$

Moreover, if $\hat{\text{se}}/\text{se} \to 1$ in probability, then, by Slutsky's theorem, it follows that

$$Z_n = \frac{\hat{\theta}_n - \theta_0}{\hat{\text{se}}} = \frac{\hat{\theta}_n - \theta_0}{\text{se}} \cdot \frac{\text{se}}{\hat{\text{se}}} \to N(0,1).$$

We therefore obtain the same result. $\square$

We mentioned in the discussion following Definition 3.10 that there is usually a correspondence between hypothesis tests and confidence intervals. We next turn to our first example of such a correspondence.

Corollary 3.1. *The Wald test rejects the null hypothesis if and only if the true parameter θ_0 is not contained within the interval*

$$I_n = (\hat{\theta}_n - \hat{\text{se}}\, z_{\alpha/2}, \hat{\theta}_n + \hat{\text{se}}\, z_{\alpha/2}). \tag{3.9}$$

Moreover, the interval I_n is an asymptotic $(1 - \alpha)$ confidence interval as $n \to \infty$.

Proof. A straightforward calculation shows that Z_n is in the acceptance region

$$-z_{\alpha/2} < \frac{\hat{\theta}_n - \theta_0}{\hat{\text{se}}} < z_{\alpha/2} \qquad \text{if and only if} \qquad \hat{\theta}_n - \hat{\text{se}}\, z_{\alpha/2} < \theta_0 < \hat{\theta}_n + \hat{\text{se}}\, z_{\alpha/2}.$$

The first inequality involves the test statistic whereas the second inequality involves the true value of the parameter. Since the Wald statistic is an asymptotic size-α test, it follows that there is a $(1 - \alpha)$ probability of being within the acceptance interval as $n \to \infty$, which validates the claim. $\square$

A straightforward generalization of the Wald statistic to higher dimensions is given in the following.

Definition 3.12 (Multivariate Wald Test). *Let $\hat{\theta}_n$ be an asymptotically normal and unbiased estimator for a k-dimensional parameter $\theta \in \Theta \subset \mathbb{R}^k$. Then the size α Wald test consists of the null hypothesis*

$$H_0 : \theta = \theta_0,$$

the test statistic

$$W_n = (\hat{\theta}_n - \theta_0)^T \mathbb{V}(\hat{\theta}_n)^{-1}(\hat{\theta}_n - \theta_0), \tag{3.10}$$

and the rejection region

$$R_\alpha = [\chi^2_{k,\alpha}, \infty),$$

where $\chi^2_{k,\alpha}$ is the solution to $S(\chi^2_{k,\alpha}) = \alpha$, where S is the survival function for a χ^2_k random variable. The function $W_n : \mathbb{X} \to \mathbb{R}$ is known as the Wald statistic.

Note 3.13. Definition 3.12 is consistent with Definition 3.11 since, for the special case of $k = 1$, we have

$$|Z| > z_{\alpha/2} \qquad \text{if and only if} \qquad W > \chi^2_{1,\alpha},$$

where $W = Z^2$. This follows since the square of a standard normal random variable is a chi-square random variable with one degrees of freedom. Moreover, the inequality $|Z| > z_{\alpha/2}$ captures $\alpha/2$ probability under $Z > z_{\alpha/2}$ and an additional $\alpha/2$ probability under $Z < -z_{\alpha/2}$, for a total probability of rejection equal to α. Since $W = Z^2$, we only have to worry about $W > \chi^2_{1,\alpha}$, which captures the full α amount of probability.   ▷

Theorem 3.2. *The size of a size-α multivariate Wald test converges to α, i.e.,*

$$\lim_{n \to \infty} \mathbb{P}_{\theta_0}(W_n \geq \chi^2_{k,\alpha}) = \alpha.$$

Moreover, the result is still valid if we replace $\mathbb{V}(\hat{\theta}_n)$ in Z_n with any asymptotically unbiased estimate $\hat{\mathbb{V}}(\hat{\theta}_n)$.

Proof. The result follows as long as $W_n \to \chi^2_k$ as $n \to \infty$. However, since the estimator is asymptotically normal and unbiased, it follows that, under the null hypothesis, $\hat{\theta}_n \to \mathbb{N}(\theta_0, \Sigma)$, for some limiting variance Σ. The Wald statistic therefore approaches the Mahalanobis distance, $W_n \to d_\Sigma(\hat{\theta}_n, \theta_0)$ as $n \to \infty$, which follows a χ^2_k distribution due to Corollary 2.3. This proves the result.   □

3.2.3 Test for Proportions

Our next test is concerned with inference regarding proportions. In this context, our sample is a sample of Bernoulli random variables $X_1, \ldots, X_n \sim \text{Bern}(p)$. Of course, such a sample can simply be described using a single binomial random variable.

Proposition 3.4. *Let $X \sim \text{Binom}(n, p)$ be a binomial random variable, and consider the hypothesis test consisting of the null hypothesis*

$$H_0 : p = p_0,$$

the test statistic

$$Z_n = \frac{X/n - p_0}{\sqrt{p_0(1 - p_0)/n}}, \tag{3.11}$$

and the rejection region

$$R_\alpha = \{z : |z| > z_{\alpha/2}\}.$$

The size of this hypothesis test approaches α as $n \to \infty$. Moreover, for large n, the size of this test is approximately α.

Essentially, Proposition 3.4 follows from the normal approximation to the binomial distribution; i.e., from a combination of the central-limit theorem and Proposition 3.1. It further constitutes an application of the Wald test, as $\overline{X}_n = X/n$ is simply the sample mean of n Bernoulli random variables, and, under the null hypothesis, $\mathbb{V}(\overline{X}_n) = p_0(1 - p_0)/n$.

One method for constructing a $(1 - \alpha)$ confidence interval for the true probability p_0 is to approximate the standard error $\mathrm{se}(X/n) = p_0(1 - p_0)/n$ with the estimated standard error $\hat{\mathrm{se}}(X/n) = \hat{p}(1 - \hat{p})/n$, where we define $\hat{p} = X/n$. This yields the confidence interval for p_0:

$$p_0 \in \left(\hat{p} - z_{\alpha/2}\sqrt{\hat{p}(1 - \hat{p})/n}, \hat{p} + z_{\alpha/2}\sqrt{\hat{p}(1 - \hat{p})/n} \right).$$

3.2.4 Power and Sample Size

We have thus far addressed the measurement of Type I Errors through devising tests with a fixed significance level and by measuring p-values. We have yet to discuss how to control for Type II Errors.

In particular, we have seen that by an appropriately selected rejection region, we can control the rate of false rejections, if the null hypothesis is true. So in order to control for Type I errors, we have specified a test statistic and a rejection region. What other lever do we have to further affect the likelihood of a Type II Error? The answer lies in the ability to specify a minimum required sample size for our test: the more data we collect, the more powerful our test will be.

Before proceeding, a slight shift in perspective is required. Instead of minimizing the Type II Error rate, we will seek to maximize the power of the test under the *alternative hypothesis*. Recall that the probability of a Type II Error is the probability of accepting a false null hypothesis. This is logically equivalent to the probability of correctly rejecting a null hypothesis that is false. This probability is often referred to as the *power* of a test. Thus, we seek to maximize the probability of detecting an effect when one is there.

There is, however, one complication with this approach, in that the power function varies across the parameter space θ. For bounding the probability of a Type I Error, we simply found the maximum value of the power function that was consistent with the null hypothesis. We cannot take the same approach when discussing the worst-case power under the alternative hypothesis, since the minimum power consistent with the alternative hypothesis is equivalent to the maximum power consistent with the null hypothesis, whenever the power function is continuous. This should be clear in Figure 3.1. To remedy this, we instead talk about the power of a test relative to some minimum detectable effect size. Referring back to Figure 3.1, we could argue that we have at least a 60% probability of correctly rejecting the null hypothesis, as long as the true probability is greater than 70%[4].

Power and Minimum Detectable Effect

In order to apply this to test design, we will invert the problem into the question: what is the minimum sample size needed to guarantee a certain power level for a minimum detectable effect. Formally, we proceed as follows.

Definition 3.13. *Given any metric space Θ and subspace $\Theta_0 \subset \Theta$, we define the* distance *between a point $\theta \in \Theta$ and the subspace Θ_0 as*

$$d(\theta, \Theta_0) = \inf_{\psi \in \Theta_0} ||\theta - \psi||.$$

In higher-dimensional spaces, distance is often specified relative to one of several norms.

Definition 3.14. *For $x \in \mathbb{R}^n$, common norms are the ℓ_1-norm (*Manhattan norm*)*

$$||x||_1 = \sum_{i=1}^{n} |x_i|,$$

*the ℓ_2 norm (*Euclidean norm*)*

$$||x||_2 = \sqrt{\sum_{i=1}^{n} x_i^2},$$

*and the ℓ_∞ norm (*infinity norm*)*

$$||x||_\infty = \max(|x_1|, \ldots, |x_n|).$$

Definition 3.15. *The* minimum detectable effect $\delta > 0$ *is a positive number that defines the* detectable effect space $\Theta_\delta \subset \Theta$ *as*

$$\Theta_\delta = \{\theta \in \Theta : d(\theta, \Theta_0) \geq \delta\}. \tag{3.12}$$

Note 3.14. The subspace Θ_δ is a subset of the alternative hypothesis, i.e., $\Theta_\delta \subset \Theta_0^c$, since $\delta > 0$ is positive and since $d(\theta, \Theta_0) = 0$ whenever $\theta \in \Theta_0$. ▷

---

[4] Note that this is well into the alternative-hypothesis territory.

Note 3.15. Unlike the size or significance level, the minimum detectable effect is not a function of the test alone, but rather a number that can be prescribed freely in conjunction with a test.   ▷

Definition 3.16. *The* power *of a hypothesis test with minimum detectable effect δ is given by*

$$\beta = \inf_{\theta \in \Theta_\delta} \beta(\theta). \tag{3.13}$$

Thus, given a minimum detectable effect, Equation (3.13) places a lower bound on the power of our test; i.e., it tells us the worst-case scenario probability of correctly rejecting a false null hypothesis at a given minimum detectable effect.

Note 3.16. In general, the power function for a particular test increases with distance to the null hypothesis space Θ_0. This makes sense, since the farther away our parameter is from the null hypothesis space, the more probable it will be that we reject the null hypothesis. Therefore, in practice, Equation (3.13) is typically evaluated for the $\theta \in \Theta_\delta$ that is closest to Θ_0. ▷

To see the relation between power, minimum detectable effect, and sample size, consider the case of a one sided null hypothesis $H_0 : \theta \leq \theta_0$ and minimum detectable effect δ. Let us further consider the case where the sampling distribution for an estimator $\hat{\theta}$ is normal. The most extreme sampling distribution under the null hypothesis is therefore $\mathrm{N}(\theta_0, \sigma^2/n)$, whereas the most extreme sampling distribution under the hypothesis of minimum effect $H_\delta : \theta \geq \theta_0 + \delta$ is $\mathrm{N}(\theta_0 + \delta, \sigma^2/n)$. The larger the sample size n, the more separated these two distributions become, and the more likely it is to detect an effect of size δ or larger.

Note 3.17. Some typical power levels for hypothesis tests are $\beta = 0.8$, 0.9, or 0.95. For example, a test with minimum detectable effect δ and power level $\beta = 0.8$ has at least an 80% probability of rejecting the null hypothesis if the null hypothesis is false and the effect size is at least δ.   ▷

Note 3.18. The minimum detectable effect is actually a bit of a misnomer: it does *not* mean that we cannot detect smaller effects. Rather, it is the minimum required effect that would need to be present to guarantee that our test has the desired power.   ▷

Power and the Coin Toss Example

To understand this in further detail, let us return to the example of the coin toss. We will assume that our sample size, once computed, is large enough to justify the normal approximation to the binomial distribution. We state our result in the following.

Theorem 3.3. *Let $X \sim \text{Binom}(n, p)$ be a binomial random variable with unknown probability p, and consider the test consisting of the single-sided null hypothesis $H_0 : p \leq p_0$, test statistic $\hat{p} = X/n$, and approximate rejection region*

$$R_\alpha = \left[p_0 + z_\alpha \sqrt{\frac{p_0(1 - p_0)}{n}}, 1 \right]. \tag{3.14}$$

Then the minimum sample size n required to achieve a power β test at minimum detectable effect δ is given by

$$n = \left(\frac{z_\alpha \sqrt{p_0(1 - p_0)} - z_\beta \sqrt{p_\delta(1 - p_\delta)}}{\delta} \right)^2, \tag{3.15}$$

where $p_\delta = p_0 + \delta$. The result holds as long as the sample size is large enough to justify the normal approximation at p_0 and p_δ; i.e., as long as $p_0 n, (1 - p_0)n, p_\delta n, (1 - p_\delta)n \geq 5$.

Proof. Under the null hypothesis, we have that $p \leq p_0$. In order to construct the rejection region for a size α test, we set $p = p_0$ and make the assumption that our sample size is large enough to justify the normal approximation $\hat{p} \sim N(p_0, p_0(1 - p_0)/n)$. The solution to $\Psi(z) = \alpha$ is z_α, so that a total of α probability lives in the tail $z \geq z_\alpha$. This yields the rejection region given by Equation (3.14). In particular, define the critical value

$$p_c = p_0 + z_\alpha \sqrt{\frac{p_0(1 - p_0)}{n}},$$

so that we reject the null hypothesis whenever $\hat{p} \geq p_c$.

Next, consider the hypothesis of minimum detectable effect, $p \geq p_0 + \delta$. The worst-case scenario here is the border case $p = p_0 + \delta$. Let us now assume that the null hypothesis is false, and that the effect size is δ. Hence, the normal approximation will yield $\hat{p} \sim N(p_\delta, p_\delta(1 - p_\delta)/n)$, where we define $p_\delta = p_0 + \delta$. Hence, the statistic

$$Z(\hat{p}) = \frac{\hat{p} - p_\delta}{\sqrt{p_\delta(1 - p_\delta)/n}}$$

is approximately a standard normal distribution.

Our next step is to ensure that under the hypothesis of minimum detectable effect, there is at least a probability β of rejecting the null hypothesis. The null hypothesis, however, is rejected whenever $Z \geq Z(p_c)$. The solution to $\mathbb{P}(Z \geq Z(p_c)) = \Psi(Z(p_c)) = \beta_0$ is simply $Z(p_c) = z_\beta$. Hence, we obtain

$$\frac{p_0 + z_\alpha \sqrt{\frac{p_0(1 - p_0)}{n}} - p_\delta}{\sqrt{p_\delta(1 - p_\delta)/n}} = z_\beta,$$

which is equivalent to

$$z_\alpha \sqrt{p_0(1 - p_0)} - \delta\sqrt{n} = z_\beta \sqrt{p_\delta(1 - p_\delta)}.$$

Solving for n yields the result.   □

Example 3.6. Let us return to our magician friend from Example 3.4. We will retain the null hypothesis $H_0 : p \le 0.5$ and test statistic $\hat{p}$, representing the frequency of observed *heads*. Let us devise a test with significance level $\alpha = 0.05$ and power $\beta = 0.80$ with minimum detectable effect $\delta = 0.20$. In other words, we desire a test that will produce a false positive 5% of the time when the coin is fair, and will correctly reject the null hypothesis 80% of the time, as long as the coin is biased by at least 20%, i.e., as long as the true probability of heads is at least 70%.

From Equation (3.15), our minimum sample size is $n = 37$, and from Equation (3.14), our critical point is $p_c = 0.635$. Thus, we will require 37 coin tosses, and we will reject the null hypothesis whenever we observer greater than 24 heads out of 37.

The extreme cases of the null hypothesis and hypothesis of minimum detectable effect are shown in Figure 3.3. The blue distribution is the sampling distribution for $\hat{p}$ under the edge case for the null hypothesis, $p = 0.50$. The shaded region is our $\alpha = 0.05$ significance rejection region; i.e., we reject the null hypothesis whenever $\hat{p} \ge 0.635$. The corresponding distribution under the edge case $p = 0.7$ for the hypothesis of minimum detectable effect is shown in orange; the shaded region constitutes $\beta = 0.80$ of the total probability. The code to produce Figure 3.3 is shown in Code Block 3.4.

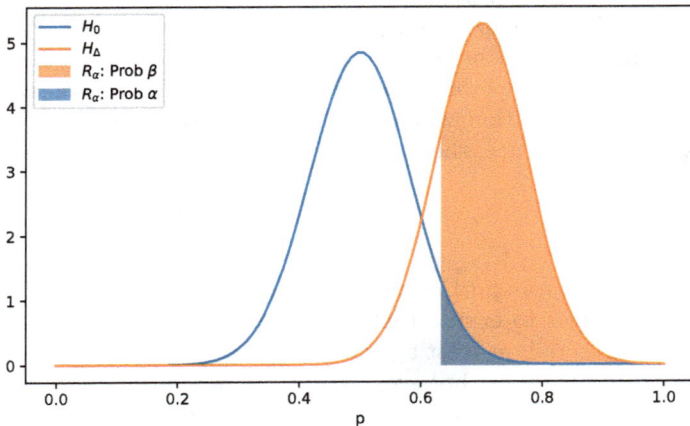

Fig. 3.3: Illustration of Theorem 3.3 for $p_0 = 0.5$, $\delta = 0.2$, $\alpha = 0.05$, and $\beta = 0.80$.

```
1  alpha = 0.05
2  beta = 0.8
3  p0 = 0.5
4  delta = 0.2
5  p1 = p0 + delta
6  za = scipy.stats.norm.isf(alpha)
7  zb = scipy.stats.norm.isf(beta)
8  n = np.ceil(((za*np.sqrt(p0*(1-p0)) - zb*np.sqrt(p1*(1-p1))) /
       delta)**2 )
9  p_crit = p0 + za*np.sqrt(p0*(1-p0)/n) # critical point
10 x = np.linspace(0, 1, num=100)
11 xr = np.linspace(p_crit, 1, num=50)
12 f_0 = scipy.stats.norm.pdf(x, p0, np.sqrt(p0*(1-p0)/n))
13 f_1 = scipy.stats.norm.pdf(x, p1, np.sqrt(p1*(1-p1)/n))
14 f_0r = scipy.stats.norm.pdf(xr, p0, np.sqrt(p0*(1-p0)/n))
15 f_1r = scipy.stats.norm.pdf(xr, p1, np.sqrt(p1*(1-p1)/n))
16 plt.plot(x, f_0, color='#1f77b4')
17 plt.plot(x, f_1, color='#ff7f0e')
18 plt.fill_between(xr, f_1r, color='#ff7f0e', alpha=0.7)
19 plt.fill_between(xr, f_0r, color='#1f77b4', alpha=0.7)
```

Code Block 3.4: Code to produce Figure 3.3.

Finally, we can verify our normal approximation by running a simulation. This is achieved by running Code Block 3.5 following Code Block 3.4. We ran the simulation twice, and observed rejection rates of 0.8000 and

```
1  n_trials = 10000
2  rejections = np.zeros(n_trials)
3  for i in range(n_trials):
4      p_hat = np.random.binomial(n, p1) / n
5      rejections[i] = int(p_hat > p_crit)
6
7  print('Rejection Rate', np.sum(rejections) / n_trials)
```

Code Block 3.5: Simulation of rejection rate under H_δ.

0.8039. This validates our theory that a sample size of 37 is required in order to correctly reject the null hypothesis 80% of the time, when the true probability of a heads is at least 70%.

This exercise was further repeated for various minimum detectable effects, holding the null hypothesis $p \leq 0.5$, significance level $\alpha = 0.05$, and power $\beta = 0.80$ fixed. The result is shown in Table 3.3.   ▷

δ	p_c	n
0.01	0.507	15,455
0.02	0.513	3,863
0.05	0.533	617
0.10	0.566	153
0.15	0.600	67
0.20	0.635	37

Table 3.3: Critical point and sample size for various minimum detectable effects; $p_0 = 0.50$, $\alpha = 0.05$, and $\beta = 0.80$.

Power and the Z-Test

Next, we turn to a power analysis of the classic z-test, as given in Proposition 3.1. This is relevant not only for the z-test, but also for similar tests, like the t-test, when there is sufficient sample size to approximate the distribution of the test statistic using the standard normal distribution.

We will consider the test statistic $T = (\overline{X}_n - \mu_0)/(\sigma/\sqrt{n})$, as defined in Equation (3.5), under three different null hypotheses: a simple two-sided null hypothesis $H_0^2 : \mu = \mu_0$, with rejection region

$$R_\alpha^2 = \{z : |z| \geq z_{\alpha/2}\}$$

and two one-sided null hypotheses, a right-tailed and a left-tailed test. For the right-tailed test, we will consider the null hypothesis $H_0^R : \mu \leq \mu_0$, with corresponding rejection region

$$R_\alpha^R = \{z : z \geq z_\alpha\}.$$

Similarly, for the left-tailed test, will will consider the null hypothesis $H_0^L : \mu \geq \mu_0$, with corresponding rejection region

$$R_\alpha^L = \{z : z \leq -z_\alpha\}.$$

For each of these, the superscript refers to the corresponding test: superscript 2 for the two-sided test, and superscripts L and R for the left-tailed and right-tailed tests, respectively. We will follow the convention that the directionality of the tail (i.e., *left-* or *right*-tailed) follows the direction of the corresponding rejection region.

Next, let us determine the power function $\beta(\mu; n)$, as a function of the hidden true value μ and the sample size n. Recall that

$$\beta(\mu; n) = \mathbb{P}_\mu(T \in R).$$

We must therefore obtain the distribution for T if the true value of the parameter is actually μ. If the true parameter value is μ, then the sample mean will be distributed as

$$X_n \sim \mathrm{N}\left(\mu, \frac{\sigma^2}{n}\right).$$

It follows that our test statistic is distributed as

$$T \sim \mathrm{N}\left(\frac{\mu - \mu_0}{\sigma/\sqrt{n}}, 1\right).$$

For a right- and left-sided test, we therefore have

$$\beta_\alpha^R(\mu; n) = \mathbb{P}(T \geq z_\alpha) = \Psi\left(z_\alpha - \frac{\mu - \mu_0}{\sigma/\sqrt{n}}\right), \tag{3.16}$$

$$\beta_\alpha^L(\mu; n) = \mathbb{P}(T \leq -z_\alpha) = \Phi\left(-z_\alpha - \frac{\mu - \mu_0}{\sigma/\sqrt{n}}\right), \tag{3.17}$$

respectively. For the two-sided test, we must consider the quantity

$$\beta_\alpha^2(\mu; n) = \mathbb{P}_\mu(T \leq -z_{\alpha/2} \text{ or } T \geq z_{\alpha/2}).$$

However, this can be found by combining Equations (3.16) and (3.17), while adjusting the significance level, to obtain

$$\beta_\alpha^2(\mu; n) = \Psi\left(z_{\alpha/2} - \frac{\mu - \mu_0}{\sigma/\sqrt{n}}\right) + \Phi\left(-z_{\alpha/2} - \frac{\mu - \mu_0}{\sigma/\sqrt{n}}\right). \tag{3.18}$$

Notice that for all three of these tests, we have the relationship

$$\beta_\alpha(\mu_0; n) = \alpha. \tag{3.19}$$

Furthermore, the power function of the two-sided test satisfies the following additive relation to the power functions (with reduced significance) of the two one-sided tests:

$$\beta_\alpha^2(\mu; n) = \beta_{\alpha/2}^L(\mu; n) + \beta_{\alpha/2}^R(\mu; n).$$

This relation is shown in Figure 3.4, in which Equations (3.16)–(3.18) are plotted using a significance level of $\alpha = 0.05$ for the two-sided test and $\alpha/2 = 0.025$ for each one-sided test. Note the value $z_{0.025} = 1.96$.

Next, let us determine the required sample size to achieve a two-sided test of a specific power β, given a minimum detectable effect δ. We are typically interested in constructing a test with a power of at least 80%. Notice in the right-hand side ($x > 0$) of Figure 3.4, that for values of $\beta \geq 0.20$, we have $\beta_\alpha^2(\mu; n) \approx \beta_{\alpha/2}^R(\mu; n)$. We can therefore determine our required sample size by solving the relation

$$\Psi\left(z_{\alpha/2} - \frac{\delta}{\sigma/\sqrt{n}}\right) \approx 0.80.$$

Recalling the relation $z_\alpha = \Psi^{-1}(\alpha)$, we can solve for the required sample size, obtaining

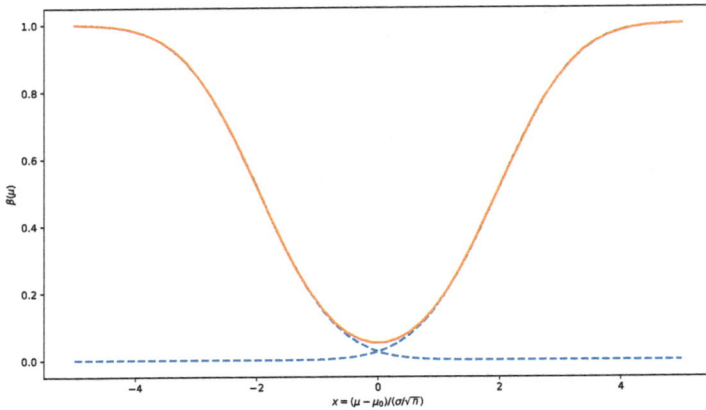

Fig. 3.4: Two-sided (β_α^2; orange solid) and both one-sided ($\beta_{\alpha/2}^L$ and $\beta_{\alpha/2}^R$; blue dashed) power functions, plotted as a function of the quantity $x = (\mu - \mu_0)/(\sigma/\sqrt{n})$; using significance level $\alpha = 0.05$.

$$n = \frac{\sigma^2}{\delta^2} \left[\Psi^{-1}(\alpha/2) - \Psi^{-1}(\beta) \right]^2 \approx \frac{7.85\sigma^2}{\delta^2}. \tag{3.20}$$

The value of 7.85 was obtained using a significance of $\alpha = 0.05$ and a power of $\beta = 0.80$. To achieve the sample size required for a one-sided test, we of course simply replace $\Psi^{-1}(\alpha/2)$ with $\Psi^{-1}(\alpha)$ and 7.85 with 6.18 in the above equation.

3.3 Tests for Dual Populations

We next turn to the case of comparisons between two groups. This is a common case to consider in A/B testing, where experimenters try to measure the effect of a treatment on a particular group. A common form of hypothesis test is therefore $\theta_1 < \theta_2$; i.e., did our treatment have a positive effect on a given parameter.

3.3.1 Test for Two Means; Equal Variances

We start by comparing the means of two normal populations. As we shall see, a fantastic simplification can be made by assuming the two populations share a common variance. We begin with this case.

Theorem 3.4 (t-test). *Let $X_1, \ldots, X_n \sim \mathrm{N}(\mu_X, \sigma_X^2)$ and $Y_1, \ldots, Y_m \sim \mathrm{N}(\mu_Y, \sigma_Y^2)$. Let us further assume that $\sigma_X^2 = \sigma_Y^2 = \sigma^2$. Finally, let*

$$\overline{X}_n = \frac{1}{n}\sum_{i=1}^{n} X_i \qquad and \qquad \overline{Y}_m = \frac{1}{m}\sum_{i=1}^{m} Y_i$$

be the sample means and

$$S_X^2 = \frac{1}{n-1}\sum_{i=1}^{n}(X_i - \overline{X}_n)^2 \qquad and \qquad S_Y^2 = \frac{1}{m-1}\sum_{i=1}^{m}(Y_i - \overline{Y}_m)^2$$

be the sample variances, as usual. Then the test consisting of the null hypothesis

$$H_0 : \mu_X = \mu_Y,$$

test statistic

$$T = \frac{\overline{X}_n - \overline{Y}_m}{\sqrt{S_P^2 \left(\frac{1}{n} + \frac{1}{m}\right)}}, \tag{3.21}$$

where S_P^2 is the pooled sample variance, defined by

$$S_P^2 = \frac{(n-1)S_X^2 + (m-1)S_Y^2}{n+m-2}, \tag{3.22}$$

and rejection region

$$R_\alpha = \{t : |t| > t_{n+m-2,\alpha/2}\}$$

constitutes a test of size α. Moreover, under the null hypothesis, the test statistic T is a Student's t-random variable with $n+m-2$ degrees of freedom; i.e., $T \sim t_{n+m-2}$.

Proof. Normality of the population implies normality of the sample means. In particular, Corollary 2.2 implies that

$$\overline{X}_n \sim \mathrm{N}(\mu_X, \sigma^2/n) \qquad and \qquad \overline{Y}_m \sim \mathrm{N}(\mu_Y, \sigma^2/m).$$

(Notice our use of the assumption of equal variances $\sigma_X^2 = \sigma_Y^2 = \sigma^2$.) Therefore, according to Theorem 2.3, the difference $\overline{X}_n - \overline{Y}_m$ satisfies

$$\overline{X}_n - \overline{Y}_m \sim \mathrm{N}\left(\mu_X - \mu_Y, \frac{\sigma^2}{n} + \frac{\sigma^2}{m}\right),$$

and, as such, the random variable

$$Z = \frac{\overline{X}_n - \overline{Y}_m - (\mu_X - \mu_Y)}{\sqrt{\sigma^2 \left(\frac{1}{n} + \frac{1}{m}\right)}}$$

is a standard normal random variable; i.e., $Z \sim \mathrm{N}(0,1)$.

Next, let us consider the sample variances. Theorem 2.6 provides that

$$\frac{(n-1)S_X^2}{\sigma^2} \sim \chi_{n-1}^2 \quad \text{and} \quad \frac{(m-1)S_Y^2}{\sigma^2} \sim \chi_{m-1}^2.$$

From Proposition 2.20, it thus follows that the random variable

$$U = \frac{(n-1)S_X^2}{\sigma^2} + \frac{(m-1)S_Y^2}{\sigma^2} \sim \chi_{n+m-2}^2$$

is a chi-squared random variable with $(n+m-2)$ degrees of freedom.
From Definition 2.14, it follows that

$$T = \frac{Z}{\sqrt{U/(n+m-2)}} \sim t_{n+m-2}$$

is a Student's t-distribution with $(n+m-2)$ degrees of freedom. Furthermore, under the null hypothesis, $\mu_X = \mu_Y$, which further simplifies our expression for Z. It is now a matter of simple algebra to show that the distribution defined above is equivalent to the test statistic given by Equations (3.21) and (3.22). The result follows. $\square$

Note 3.19. Due to the central limit theorem, Theorem 3.4 applies approximately for comparing the means from any populations with common variance, as $n, m \to \infty$. In practice, a good rule of thumb is to require $n > 30$ and $m > 30$. $\triangleright$

Confidence Intervals

An immediate consequence of Theorem 3.4 is the following. We leave the proof to the reader.

Corollary 3.2. *Given the setup of Theorem 3.4, the $(1-\alpha)$ confidence interval for the difference $\mu_X - \mu_Y$ is given by*

$$I_\alpha = (L(\mathcal{X}, \mathcal{Y}), U(\mathcal{X}, \mathcal{Y})),$$

where the lower and upper bounds are given by

$$L(\mathcal{X}, \mathcal{Y}) = \overline{X}_n - \overline{Y}_m - t_{n+m-2, \alpha/2} \sqrt{S_P^2 \left(\frac{1}{n} + \frac{1}{m}\right)}$$

$$U(\mathcal{X}, \mathcal{Y}) = \overline{X}_n - \overline{Y}_m + t_{n+m-2, \alpha/2} \sqrt{S_P^2 \left(\frac{1}{n} + \frac{1}{m}\right)},$$

respectively.

Example 3.7. In this example, we build a numerical simulation to validate the result of Theorem 3.4. In particular, we construct a simulation with true null hypothesis $\mu_X = \mu_Y = 10$ and equal variance $\sigma_X = \sigma_Y = 2$. Since

```
1   mu_x = mu_y = 10
2   sigma_x = sigma_y = 2
3   n = 3
4   m = 7
5   dof = n+m-2
6   alpha = 0.05
7   t_crit = scipy.stats.t.isf(alpha/2, dof) # 2.306
8
9   n_trials = 10000
10  rejections = np.zeros(n_trials)
11  p_values = np.zeros(n_trials)
12  for i in range(n_trials):
13      samples_x = np.random.normal(mu_x, scale=sigma_x, size=n)
14      samples_y = np.random.normal(mu_y, scale=sigma_y, size=m)
15      x_mean = samples_x.mean()
16      y_mean = samples_y.mean()
17      x_var = samples_x.var() * n / (n-1)
18      y_var = samples_y.var() * m / (m-1)
19      S_P2 = ( (n-1)*x_var + (m-1)*y_var ) / dof
20      T = (x_mean - y_mean) / np.sqrt( S_P2 * (1/n + 1/m) )
21      rejections[i] = int( abs(T) > t_crit )
22      p_values[i] = 2 * scipy.stats.t.sf(abs(T), dof)
23
24  print ('Rejection Rate', rejections.sum() / n_trials) # 0.0525
25  print ('Number of P-Values < 0.005', np.sum( p_values < 0.005 )) #
        48
```

Code Block 3.6: Simulation of Two Normal Populations with Equal Variance

the result is most interesting for small sample sizes, we select $n = 3$ and $m = 7$, with results in the degrees of freedom $(n + m - 2) = 8$. We further use a significance level of $\alpha = 0.05$, resulting in a critical value for our test $t_{8,0.025} \approx 2.306$. The code for the simulation is given in Code Block 3.6.

We run 10,000 trials. As expected, the number of rejections is approximately 5%. (For our particular simulation, the observed rejection rate was 5.25%.) It is also interesting to note that in 48 out of 10,000 simulations, we observed a p-value less than 0.5%. The expected number of simulations that would have such a low p-value is 50, since the p-value is uniformly distributed under the null hypothesis. In fact, the smallest p-value that was observed min(p_values) was 0.000247, or 0.0247%. It is a stark reminder that when conducting many experiments, such small p-values are destined to arise due to mere chance.   ▷

Power and Sample Size

We may easily modify our discussion leading up to Equation (3.20) to determine the required sample size for the t-test of Theorem 3.4. Assuming that $n = m$, the required sample size to obtain a power β test with minimum detectable effect δ at a significance level of α is given by

$$n = \frac{2S_P^2}{\delta^2} \left[\Psi^{-1}(\alpha/2) - \Psi^{-1}(\beta)\right]^2 \approx \frac{15.7S_P^2}{\delta^2}, \qquad (3.23)$$

assuming that n is large enough to justify use of the central limit theorem. In the event of unequal sample sizes, we can replace n with the harmonic mean of n and m in the preceding formula, thus obtaining

$$\frac{1}{1/n + 1/m} = \frac{S_P^2}{\delta^2} \left[\Psi^{-1}(\alpha/2) - \Psi^{-1}(\beta)\right]^2.$$

3.3.2 Test for Two Means; Unequal Variances

The case of comparing means from two normal populations with *different* variances is not as simple. In this case, the random variable Z, as defined in the proof of Theorem 3.4, would instead be given by

$$Z = \frac{\overline{X}_n - \overline{Y}_m - (\mu_X - \mu_Y)}{\sqrt{\sigma_X^2/n + \sigma_Y^2/m}}, \qquad (3.24)$$

which would constitute a standard normal random variable. To proceed, we will require a certain approximation for the weighted sum of chi-squared random variables due to Satterthwaite.

Satterthwaite's approximation

We describe the Satterthwaite's approximation as a method of approximating a weighted sum of random variables as a scaled random variable from the same family. If such an assumption is valid, we may match the first and second moments of the two expressions to obtain the following.

Definition 3.17 (Generalized Satterthwaite's Approximation). *Suppose $Y_1, \ldots, Y_k$ are independent random variables from the same one-parameter family, $Y_i \sim F_{\theta_i}$, for parameters $\theta_1, \ldots, \theta_k$. Then the generalized Satterthwaite's approximation for the sum $\sum_{i=1}^{k} a_i Y_i$, for arbitrary constants $a_1, \ldots, a_k$, is given by the Satterthwaite ansatz:*

$$\sum_{i=1}^{k} a_i Y_i = aY,$$

where $Y \sim F_\theta$, such that a and θ are approximated by solving the system of equations

$$\sum_{i=1}^{k} a_i \mathbb{E}[Y_i] = a\mathbb{E}[Y] \tag{3.25}$$

$$\sum_{i=1}^{k} a_i^2 \mathbb{V}(Y_i) = a^2 \mathbb{V}(Y). \tag{3.26}$$

Our first example is what is classically known as the *Satterthwaite's approximation*, i.e., his original use case. We have already seen that the sum of chi-squared random variables is also a chi-squared. The following provides us a way of approximated a *weighted* sum of chi-squared random variables.

Example 3.8 (Satterthwaite's Approximation). Let $Y_1, \ldots, Y_k$ be independent chi-squared random variables, with $Y_i \sim \chi_{r_i}^2$. Then the weighted sum $\sum_{i=1}^{k} a_i Y_i$ can be approximated by

$$\sum_{i=1}^{k} a_i Y_i = aY,$$

where $Y \sim \chi_\nu^2$, with parameters a and ν given by

$$a = \frac{\sum_{i=1}^{k} a_i^2 r_i}{\sum_{i=1}^{k} a_i r_i} \tag{3.27}$$

$$\nu = \frac{\left(\sum_{i=1}^{k} a_i r_i\right)^2}{\sum_{i=1}^{k} a_i^2 r_i}. \tag{3.28}$$

This follows by solving Equations (3.25) and (3.26) for a and ν, given the expected value and variance formulas for a chi-squared random variable: $\mathbb{E}[Y_i] = r_i$, $\mathbb{V}(Y_i) = 2r_i$, $\mathbb{E}[Y] = \nu$, and $\mathbb{V}(Y) = 2\nu$. ▷

Welch's T-test

Next, we proceed with our analysis of comparing means from two populations with unequal variances $\sigma_X^2 \neq \sigma_Y^2$. When the unequal variances are unknown, Equation (3.24) is best approximated by the statistic

$$W = \frac{\overline{X}_n - \overline{Y}_m - (\mu_X - \mu_Y)}{\sqrt{S_X^2/n + S_Y^2/m}}, \tag{3.29}$$

with population variances replaced by sample variances. Under the null hypothesis $H_0 : \mu_X = \mu_Y$, the distribution of the test statistic W can be approximated as follows.

Theorem 3.5 (Welch's t-test). *Let* $X_1, \ldots, X_n \sim N(\mu_X, \sigma_X^2)$ *and let* $Y_1, \ldots, Y_m \sim N(\mu_Y, \sigma_Y^2)$. *Then the distribution of the test statistic* W, *given by Equation (3.29), is approximately a Student's t-distribution with* $\hat{\nu}$ *degrees of freedom, where* $\hat{\nu}$ *is given by*

$$\hat{\nu} = \frac{\left(\dfrac{S_X^2}{n} + \dfrac{S_Y^2}{m} \right)^2}{\dfrac{1}{n-1}\left(\dfrac{S_X^2}{n} \right)^2 + \dfrac{1}{m-1}\left(\dfrac{S_Y^2}{m} \right)^2}. \tag{3.30}$$

Moreover, the null hypothesis $\mu_X = \mu_Y$ *can therefore be tested using the test statistic* W *and rejection region*

$$R_\alpha = \{ w : |w| > t_{\hat{\nu}, \alpha/2} \}.$$

The resulting test has approximate size α.

Proof. Comparing Equations (3.29) and (3.24), we seek to understand the distribution for

$$R = \frac{\dfrac{S_X^2}{n} + \dfrac{S_Y^2}{m}}{\dfrac{\sigma_X^2}{n} + \dfrac{\sigma_Y^2}{m}}.$$

Defining the random variables and constants

$$Y_1 = \frac{(n-1)S_X^2}{\sigma_X^2} \sim \chi_{n-1}^2$$

$$Y_2 = \frac{(m-1)S_Y^2}{\sigma_Y^2} \sim \chi_{m-1}^2$$

$$a_1 = \frac{\sigma_X^2}{n(n-1)} \left(\frac{\sigma_X^2}{n} + \frac{\sigma_Y^2}{m} \right)^{-1}$$

$$a_2 = \frac{\sigma_Y^2}{m(m-1)} \left(\frac{\sigma_X^2}{n} + \frac{\sigma_Y^2}{m} \right)^{-1},$$

we can express the statistic R as

$$R = a_1 Y_1 + a_2 Y_2.$$

Now let's make the Satterthwaite ansatz and assume that this sum may be approximated as $R = aY$, for $Y \sim \chi_\nu^2$, for some value of a and ν. In particular, we use Satterthwaite's approximation, which is given by Equations (3.27) and (3.28), with $r_1 = (n-1)$ and $r_2 = (m-1)$. In particular, note that

$$a_1(n-1) + a_2(m-1) = 1.$$

Thus, we see that

$$a_1 Y_1 + a_2 Y_2 \sim \frac{\chi_\nu^2}{\nu},$$

with

$$\frac{1}{\nu} = a_2^2(n-1) + a_2^2(m-1) = \left(\frac{\sigma_X^4}{n^2(n-1)} + \frac{\sigma_Y^4}{m^2(m-1)} \right) \left(\frac{\sigma_X^2}{n} + \frac{\sigma_Y^2}{m} \right)^{-2}.$$

Finally, we may approximate ν using the substitutions $\sigma_X^2 \to S_X^2$ and $\sigma_Y^2 \to S_Y^2$, thereby obtaining the desired result.   □

Corollary 3.3. *Given the setup of Theorem 3.5, an approximate $(1 - \alpha)$ confidence interval for the difference $\mu_X - \mu_Y$ is therefore given by*

$$\overline{X}_n - \overline{Y}_m \pm t_{\hat{\nu}, \alpha/2} \sqrt{\frac{S_X^2}{n} + \frac{S_Y^2}{m}}.$$

Note 3.20. The number of degrees of freedom given by Equation (3.30) goes to infinity like $\hat{\nu} \sim O(n)$ as $n, m \to \infty$ (commensurately). Thus, for large sample sizes, the test statistic W given by Equation (3.29) approximately follows a standard normal distribution. The Welch's t-test is therefore useful primarily in the context of small sample sizes. Moreover, for large sample sizes, the confidence interval given in Corollary 3.3 can be approximated by replacing $t_{\hat{\nu}, \alpha/2}$ with $z_{\alpha/2}$.   ▷

3.3.3 Test for Two Proportions

We next turn to the case of samples from two Bernoulli populations. We again rely on the normal approximation to the binomial distribution for large sample size. We will also use the sample variance $\hat{p}(1 - \hat{p})$ in lieu of the population variance $p(1 - p)$. In this context, the test statistic is a restatement of Equation (3.29), and we have the following.

Proposition 3.5. *Consider two independent binomial random variables $X \sim \text{Binom}(n_1, p_1)$ and $Y \sim \text{Binom}(n_2, p_2)$, and define the estimates $\hat{p}_1 = X/n_1$ and $\hat{p}_2 = Y/n_2$. Then the statistic*

$$Z = \frac{\hat{p}_1 - \hat{p}_2 - (p_1 - p_2)}{\sqrt{\hat{p}_1(1 - \hat{p}_1)/n_1 + \hat{p}_2(1 - \hat{p}_2)/n_2}} \tag{3.31}$$

is approximately a standard normal distribution for large n_1, n_2. Moreover, for large n_1, n_2, the null hypothesis $H_0 : p_1 = p_2$ can be tested using W, with $p_1 - p_2 = 0$, and with rejection region $R_\alpha = \{w : |w| \geq z_{\alpha/2}\}$. This test has approximate size α.

Finally, for large n_1, n_2, an approximate $(1 - \alpha)$ confidence interval for $p_1 - p_2$ is given by

$$\hat{p}_1 - \hat{p}_2 \pm z_{\alpha/2} \sqrt{\frac{\hat{p}_1(1 - \hat{p}_1)}{n_1} + \frac{\hat{p}_2(1 - \hat{p}_2)}{n_2}}.$$

Example 3.9. We can simulate an example of Proposition 3.5 in Code Block 3.7. We conduct simulations with a true null hypothesis $p_1 = p_2 = 0.30$, and using sample sizes $n_1 = 30$, $n_2 = 40$. The observed rejection rate is 6.39%, slightly higher than our approximate test size of 5%. To get a more accurate read on the size of this test, we repeated the simulation using 100,000 trials, obtaining a rejection rate of 5.989%. The simula-

```
1   p1 = p2 = 0.3
2   n1 = 30
3   n2 = 40
4   alpha = 0.05
5   z_crit = scipy.stats.norm.isf(alpha/2) # 1.960
6
7   n_trials = 10000
8   rejections = np.zeros(n_trials)
9   zs = np.zeros(n_trials)
10  for i in range(n_trials):
11      p1_hat = np.random.binomial(n1, p1) / n1
12      p2_hat = np.random.binomial(n2, p2) / n2
13      z = (p1_hat - p2_hat) / np.sqrt( p1_hat * (1-p1_hat) / n1 +
            p2_hat * (1-p2_hat) / n2 )
14
15      rejections[i] = int( abs(z) > z_crit )
16      zs[i] = abs(z)
17
18  print ('Rejection Rate', rejections.sum() / n_trials ) # 0.0639
```

Code Block 3.7: Simulation of two Bernoulli trials

tion can be repeated using various sample sizes. For example, when using $n_1 = 50$ and $n_2 = 100$, our observed false rejection rate was 5.511%. Using $n_1 = 200$ and $n_2 = 300$, our observed false rejection rate was 5.267%. As $n_1, n_2 \to \infty$, the test statistic Z in Equation (3.31) will approach a standard normal distribution, and our observed rejection rate will approach α.

Further note that in this simulation, we kept track of the statistic Z for each sample. We can actually use this to construct a better critical value than the one given by $z_{\alpha/2}$. By taking the quantile `np.quantile(zs, 0.95)`, we obtain a value $z^* = 2.0687$. Repeating the simulation with `z_crit=2.0687` yields a rejection rate of 4.988%, which is much closer to our desired significance level obtained from $z_{\alpha/2}$.   ▷

3.4 Tests for Categories

We next turn to the case of categorical data. By a *category*, we mean any one of several mutually exclusive and exhaustive outcomes for a given experiment. There are several flavors of tests that might be of interest: we shall consider the comparison of the outcome of a single experiment with a predicted set of values (Pearson's chi-squared); the comparison of the outcomes of two independent experiments (test for homogeneity); and tests concerning the independence of two or more classification attributes (test for independence). Additional details can be found in Agresti [2013, 2019].

3.4.1 Pearson's Chi-squared Test

We begin by considering an experiment that can produce k mutually exclusive and exhaustive outcomes, which we shall denote $A_1, \ldots, A_k$, and suppose that $\mathbb{P}(A_i) = p_i$, for $i = 1, \ldots, k$. We repeat our experiment n times, and count the number of times X_i that the experiment results in outcome A_i. Naturally, the random vector X is a multinomial random vector $X \sim \text{Multi}(n, p)$, with $p \in \Delta^{k-1}$. Based on an observed set of outcomes $X = x$, we would like to test the hypothesis $H_0 : p = p_0$ that the population probability vector is equal to some theoretical value p_0. Such a test is given in the following.

Definition 3.18 (Pearson's chi-squared test). *Let $X \sim \text{Multi}(n, p)$, where $p \in \Delta^{k-1}$. Pearson's chi-squared test consists of the null hypothesis $H_0 : p = p_0$, test statistic*

$$Q = \sum_{i=1}^{k} \frac{(X_i - np_{0i})^2}{np_{0i}}, \tag{3.32}$$

and rejection region

$$R_\alpha = [\chi_{k-1,\alpha}^2, \infty). \tag{3.33}$$

Theorem 3.6. *Under the null hypothesis $H_0 : p = p_0$ of Pearson's chi-squared test, the test statistic given by Equation (3.32) satisfies $Q \to \chi_{k-1}^2$ as $n \to \infty$. Therefore, for large n, Pearson's chi-squared test is approximately a size-α test.*

Note 3.21. A good rule of thumb is that Pearson's chi-squared test is safe to use as long as $np_{0i} \geq 5$, for $i = 1, \ldots, k$.   ▷

Before proceeding to the proof, let us first consider an easy warm-up: the case of $k = 2$. Let $X_1 \sim \text{Binom}(n, p_1)$, for $p_1 \in (0, 1)$. The normal approximation to the binomial, which follows due to the central limit theorem, implies that the statistic

$$Z = \frac{X_1 - np_1}{\sqrt{np_1(1 - p_1)}}$$

is approximately a standard normal random variable for large n, particularly when $\min(np_1, n(1 - p_1)) \geq 5$. It follows that the test statistic $Q = Z^2 \to \chi_1^2$ as $n \to \infty$. Now, if we define $X_2 = n - X_1$ and $p_2 = 1 - p_1$, a little algebra reveals

$$\begin{aligned}
Q &= \frac{(X_1 - np_1)^2}{np_1(1 - p_1)} \\
&= \frac{(X_1 - np_1)^2}{np_1} + \frac{(X_1 - np_1)^2}{n(1 - p_1)} \\
&= \frac{(X_1 - np_1)^2}{np_1} + \frac{(X_2 - np_2)^2}{np_2}.
\end{aligned}$$

This last line is equivalent to Equation (3.32) for the case $k = 2$. This proves that, for the case $k = 2$, $Q \to \chi_1^2$ as $n \to \infty$. Now, let us proceed to the the general proof.

Proof. Given $X \sim \text{Multi}(n, p)$ and $p \in \Delta^{k-1}$, we begin by defining the *reduced* vectors $\tilde{X} = (X_1, \ldots, X_{k-1})$ and $\tilde{p} = (p_1, \ldots, p_{k-1})$, consistent with Note 2.27. Moreover, the vector $\tilde{X}$ can be viewed as the sum of IID reduced Bernoulli random vectors $B_1, \ldots, B_n \sim \text{RedMultiBern}(p)$, such that $\tilde{X} = \sum_{i=1}^{n} B_i = n\bar{B}_n$.

The variance–covariance matrix $\tilde{\Sigma} = \mathbb{V}(B_i)$ is given by Equation (2.78) as

$$\tilde{\Sigma} = \text{diag}(\tilde{p}) - \tilde{p}\tilde{p}^T,$$

with inverse matrix given by Equation (2.79) as

$$\tilde{\Sigma}^{-1} = \text{diag}(\tilde{p})^{-1} + p_k^{-1} J_{k-1}.$$

Now, the central limit theorem (Theorem 2.11) tells us that

$$\sqrt{n}\tilde{\Sigma}^{-1/2}(\bar{B}_n - \tilde{p}) \to \text{N}(0_{k-1}, I_{k-1})$$

as $n \to \infty$. This in turn implies that the quantity

$$\begin{aligned}
Q &= n(\bar{B}_n - \tilde{p})^T \tilde{\Sigma}^{-1}(\bar{B}_n - \tilde{p}) \\
&= n(\bar{B}_n - \tilde{p})^T \left[\text{diag}(\tilde{p})^{-1} + p_k^{-1} J_{k-1}\right] (\bar{B}_n - \tilde{p}) \\
&= n(\bar{B}_n - \tilde{p})^T \text{diag}(\tilde{p})^{-1}(\bar{B}_n - \tilde{p}) + np_k^{-1}(\bar{B}_n - \tilde{p})^T J_{k-1}(\bar{B}_n - \tilde{p})
\end{aligned}$$

satisfies $Q \to \chi_{k-1}^2$ as $n \to \infty$. Let us examine each term separately.

Observing that $n(\bar{B}_n - \tilde{p}) = (\tilde{X} - n\tilde{p})$, we have, for the first term

$$n(\bar{B}_n - \tilde{p})^T \text{diag}(\tilde{p})^{-1}(\bar{B}_n - \tilde{p}) = \sum_{i=1}^{k-1} \frac{(X_i - np_i)^2}{np_i}.$$

For the second term, let us first observe that $x^T J x = \sum_{i=1}^{n} \sum_{j=1}^{n} J_{ij} x_i x_j = \left(\sum_{i=1}^{n} x_i\right)^2$, for $x \in \mathbb{R}^n$ and $J \in \mathbb{R}^{n \times n}$, with $J_{ij} = 1$. Thus, our second term is equivalent to

$$np_k^{-1}(\overline{B}_n - \tilde{p})^T J_{k-1}(\overline{B}_n - \tilde{p}) = \frac{1}{np_k}\left(\sum_{i=1}^{k-1}(X_i - np_i)\right)^2.$$

But,

$$\sum_{i=1}^{k-1}(X_i - np_i) = \sum_{i=1}^{k-1} X_i - n\sum_{i=1}^{n} p_i = (n - X_k) - n(1 - p_k) = -X_k + np_k.$$

Therefore,

$$np_k^{-1}(\overline{B}_n - \tilde{p})^T J_{k-1}(\overline{B}_n - \tilde{p}) = \frac{(X_k - np_k)^2}{np_k}.$$

Combining the above two results, we say that $Q = n(\overline{B}_n - \tilde{p})^T \tilde{\Sigma}^{-1}(\overline{B}_n - \tilde{p})$ is actually equivalent to Equation (3.32). The result follows. $\square$

Example 3.10. Consider an experiment that results in one of three different outcomes with probability vector $p = (0.2, 0.3, 0.5)$. We run the experiment ten times and use Pearson's chi-squared test to test the null hypothesis $p_0 = (0.2, 0.3, 0.5)$. We simulated such an experiment in Code Block 3.8. Under this simulation, the null hypothesis is always true, and we are tracking the fraction of tests in which the chi-squared test leads to a false rejection.

The output of the code is startling, with only a 0.3% rejection rate for a test that should supposedly produce a 5% rejection rate! We also print the critical value and the bootstrap critical value: the Q-value at which 5% of our simulated tests is greater than (same approach we did in Example 3.9). The theoretical and bootstrap critical values are *way off*! This is not surprising, as such a small fraction of cases landed in our rejection region.

Since the chi-squared test is supposed to be valid for "large" sample sizes, we run our simulation again with $n = 25$ and $n = 100$, and the results get worse each time. We begin to think that, maybe, there is a mistake somewhere in our code.

We will ask the reader to find the bug in Exercise 3.4. Upon correcting the bug, and running with n=10 and n_trials=100000, we received a more reassuring output:

                Observed Rejection Rate 0.05025.

Thus, the chi-squared test is valid even with such a relatively small sample size. $\triangleright$

```
1   n = 10
2   p = [0.2, 0.3, 0.5]
3   p_0 = [0.2, 0.3, 0.5]
4   alpha = 0.05
5   E_0 = n * np.array(p_0) # Expected frequencies; array REQUIRED!
6   chi_crit = scipy.stats.chi2.isf(alpha, n-1) # Critical Value
7   n_trials = 10000
8   rejections = np.zeros(n_trials)
9   Q_values = np.zeros(n_trials)
10  p_values = np.zeros(n_trials)
11
12  for i in range(n_trials):
13      x = np.random.multinomial(n, p)
14      Q = np.sum( (x - E_0)**2 / E_0 )
15      rejections[i] = int( Q > chi_crit )
16      Q_values[i] = Q
17      p_values[i] = scipy.stats.chi2.sf(Q, n-1)
18
19  print ('Observed Rejection Rate', np.sum(rejections) / n_trials)
20  print ('Critical Value', chi_crit.round(4) )
21  print ('95\% Quantile Test Statistic', np.quantile(Q_values,
            0.95).round(4))
22  # n=10, n_trials=100,000. OUTPUT ::
23  #     Observed Rejection Rate 0.00028
24  #     Critical Value 16.919
25  #     95\% Quantile Test Statistic 6.0333
```

Code Block 3.8: Chi-squared test simulation: CONTAINS A BUG!!

Note 3.22. We previously stated that Pearson's chi-squared test is valid for $\min(np_i) \geq 5$. Example 3.10, however, shows that it nevertheless produces good results for this case of $\min(np_i) = 2$. Some authors report that while $np_i \geq 5$ is a good rule-of-thumb, they have seen good results even when $np_i \geq 1$, for $i = 1, \ldots, k$; see Hogg, *et al.* [2015].   ▷

Note 3.23. The mistake in Code Block 3.8 is a very real example: this is the way I initially coded the simulation and, moreover, the discussion follows my initial thought process upon observing the results. (*Let's see what happens if we improve the sample size...*) Since it is a crucial skill in data science not only to code, but also to have a systematic approach for responding to bugs, I decided to leave the code as is and allow the reader to locate the mistake. Happy hunting!   ▷

3.4.2 Test for Homogeneity

We next turn to the case of *two* independent experiments, each capable of producing a set of k mutually exclusive and exhaustive outcomes $A_1, \ldots, A_k$. Consider two probability vectors $p_1, p_2 \in \Delta^{k-1}$ with components

$$p_{ij} = \mathbb{P}(A_j | \text{experiment } i),$$

for $i = 1, 2$ and $j = 1, \ldots, k$. Thus, we assume the outcome of experiment i may be described by $X_i \sim \text{Multi}(n_i, p_i)$, for $i = 1, 2$. We want to test whether or not these probabilities were affected by our experiment; i.e., whether or not $p_1 = p_2$.

Definition 3.19 (Two-experiment Test for Homogeneity). *Let $X_i \sim$ Multi(n_i, p_i), for $i = 1, 2$, and for $p_i \in \Delta^{k-1}$. The two-experiment test for homogeneity consists of the null hypothesis $p_1 = p_2$, the test statistic*

$$Q = \sum_{i=1}^{2} \sum_{j=1}^{k} \frac{(X_{ij} - n_i \hat{p}_j)^2}{n_i \hat{p}_j}, \tag{3.34}$$

where the quantity $\hat{p}_j$ is defined by

$$\hat{p}_j = \frac{X_{1j} + X_{2j}}{n_1 + n_2}, \tag{3.35}$$

for $j = 1, \ldots, k$, and the rejection region $R_\alpha = [\chi^2_{k-1,\alpha}, \infty)$.

Proposition 3.6. *The two-experiment test for homogeneity is approximately a size α test for large n. In particular, under the null hypothesis, the test statistic defined in Equation (3.34) satisfies $Q \to \chi^2_{k-1}$ as $n \to \infty$.*

Proof. From Theorem 3.6, we know that

$$\sum_{j=1}^{k} \frac{(X_{ij} - n_i p_{ij})^2}{n_i p_{ij}} \to \chi^2_{k-1}$$

as $n \to \infty$, for $i = 1, 2$. It follows that the quantity

$$\sum_{i=1}^{2} \sum_{j=1}^{k} \frac{(X_{ij} - n_i p_{ij})^2}{n_i p_{ij}} \to \chi^2_{2k-2}$$

as $n \to \infty$.

Now, under the null hypothesis, $p_{1j} = p_{2j}$, and we can therefore estimate this quantity using Equation (3.35). Substituting these estimates into the above results reduces the number of degrees of freedom by $k - 1$, as we are relying on k parameter estimates constrained by the single constraint $\sum_{j=1}^{k} \hat{p}_j = 1$. The result follows. $\square$

The above result can be generalized to more than two experiments, as follows.

Definition 3.20 (Many-experiment Test for Homogeneity). *Let $X_i \sim$ Multi(n_i, p_i), for $i = 1, \ldots, h$, and for $p_i \in \Delta^{k-1}$. The h-experiment test for homogeneity consists of the null hypothesis $p_1 = p_2 = \cdots = p_h$, the test statistic*

$$Q = \sum_{i=1}^{h} \sum_{j=1}^{k} \frac{(X_{ij} - n_i \hat{p}_j)^2}{n_i \hat{p}_j}, \tag{3.36}$$

where the quantity $\hat{p}_j$ is defined by

$$\hat{p}_j = \frac{\sum_{i=1}^{h} X_{ij}}{\sum_{i=1}^{h} n_i}, \tag{3.37}$$

for $j = 1, \ldots, k$, and the rejection region $R_\alpha = [\chi^2_{(k-1)(h-1),\alpha}, \infty)$.

Proposition 3.7. *The h-experiment test for homogeneity is approximately a size α test for large n. In particular, under the null hypothesis, the test statistic defined in Equation (3.36) satisfies $Q \to \chi^2_{(k-1)(h-1)}$ as $n \to \infty$.*

The proof is a straightforward generalization of Proposition 3.6.

3.4.3 Test for Independence

We return to the case of as single experiment, but add the following twist: the outcome of the experiment can be classified according to two different attributes. The first attribute assigns each outcome to one of k mutually exclusive and exhaustive events $A_1, \ldots, A_k$ whereas the second attribute assigns each outcome to one of h mutually exclusive and exhaustive events $B_1, \ldots, B_h$. We define the probability p_{ij} as

$$p_{ij} = \mathbb{P}(A_i, B_j),$$

for $i = 1, \ldots, k$ and $j = 1, \ldots, h$. Finally, let us repeat the experiment n times and define X_{ij} as the count of outcomes occurring in the bucket $A_i \cap B_j$. We may still say that $X \sim$ Multi(n, p), for $p \in \Delta^{kh-1}$, with the understanding that X and p are now *matrices* as opposed to *vectors*. We seek to test the independence of the two attributes, by testing the null hypothesis

$$H_0 : \mathbb{P}(A_i, B_j) = \mathbb{P}(A_i)\mathbb{P}(B_j),$$

for $i = 1, \ldots, k$ and $j = 1, \ldots, h$. Formally, we have the following.

Definition 3.21 (Test for Independence). *Let $X \sim$ Multi(n, p), where $p \in \Delta^{kh-1}$, be a multinomial random matrix arranged into k rows and h columns. Then the test for independence consists of the null hypothesis*

$$H_0 : p_{ij} = p_{i\cdot} p_{\cdot j},$$

where we have defined

$$p_{i\cdot} = \sum_{j=1}^{h} p_{ij} \quad and \quad p_{\cdot j} = \sum_{i=1}^{k} p_{ij}, \tag{3.38}$$

the test statistic

$$Q = \sum_{i=1}^{k} \sum_{j=1}^{h} \frac{(X_{ij} - n\hat{p}_{i\cdot}\hat{p}_{\cdot j})^2}{n\hat{p}_{i\cdot}\hat{p}_{\cdot j}}, \tag{3.39}$$

where we have defined

$$\hat{p}_{i\cdot} = \frac{1}{n} \sum_{j=1}^{h} X_{ij} \quad and \quad \hat{p}_{\cdot j} = \frac{1}{n} \sum_{i=1}^{k} X_{ij},$$

and the rejection region $R_\alpha = [\chi^2_{(k-1)(h-1),\alpha}, \infty)$.

Proposition 3.8. *The test for independence is approximately a size α test for large n. Moreover, the test statistic defined in Equation (3.39) satisfies $Q \to \chi^2_{(k-1)(h-1)}$ as $n \to \infty$.*

3.5 Tests for Categories II: Power Analysis

In this section, we will be exploring the power function for Pearson's chi-squared test. In particular, we will need to understand the distribution of the test statistic Q given by Equation (3.32) under the alternative hypothesis. Theorem 3.6 already tells us that Q has a χ^2_{k-1} distribution, for large n, when the null hypothesis is true. But what if $X \sim \mathrm{Multi}(n, p_A)$, for some alternative hypothesis $p_A \in \Theta_0^c$, and we use the test assuming a particular value of $p_0 \in \Theta_0$? This immediately creates a bias in the quantity $X_i - np_{0i}$, one which is best understood with an extension to the chi-squared distribution.

3.5.1 Noncentral Chi-squared Distribution

Our ultimate goal for the remainder of the section is to understand the power function for Pearson's chi-squared test. This understanding will require some familiarity of a new distribution: the non-central chi-squared distribution.

Definition 3.22. *Let $X_1, \ldots, X_k$ be a sequence of independent unit-variance normal random variables with $X_i \sim N(\mu_i, 1)$, for $i = 1, \ldots, k$. Then the random variable*

$$X = \sum_{i=1}^{p} X_i^2,$$

consisting of the sum of squares of the X_is, is said to constitute a noncentral chi-squared random variable, *denoted* $X \sim \text{NCX}_k^\lambda$, *with k degrees of freedom and* noncentrality parameter

$$\lambda = \sum_{i=1}^{k} \mu_i^2.$$

In order to derive the PDF for a noncentral chi-squared random variable, we will require a few preliminary results.

Proposition 3.9. *Let* $Y \sim N(\mu, \sigma^2)$ *be a normal random variable. The* PDF *of the random variable* $X = Y^2$ *is given by*

$$f_X(x) = \frac{1}{\sigma^2} \sum_{p=0}^{\infty} \frac{e^{-\lambda/2}(\lambda/2)^p}{p!} \frac{(x/\sigma^2)^{p-1/2}e^{-x/(2\sigma^2)}}{\Gamma(p+1/2)2^{p+1/2}}, \tag{3.40}$$

where $\lambda = \mu^2/\sigma^2$.

Proof. Consider the transformation $X = g(Y) = Y^2$, which is invertible on each of the domains $\mathcal{A}_1 = (-\infty, 0)$ and $\mathcal{A}_2 = (0, \infty)$. In fact, we may define $Y = g_1^{-1}(X) = -\sqrt{X}$ and $Y = g_2^{-1}(X) = \sqrt{X}$, with

$$\left| \frac{dg_i^{-1}(x)}{dx} \right| = \frac{1}{2\sqrt{x}},$$

for $i = 1, 2$. Additionally, the PDF for Y is given by

$$f_Y(y) = \frac{1}{\sigma} \phi\left(\frac{y-\mu}{\sigma} \right),$$

where $\phi(z) = e^{-z^2/2}/\sqrt{2\pi}$ is the PDF of a standard normal random variable. It therefore follows from Theorem 1.7 that the PDF of the random variable $X = Y^2$ is given by

$$\begin{aligned} f_X(x) &= \frac{1}{2\sqrt{x}} \left[f_Y(\sqrt{x}) + f_Y(-\sqrt{x}) \right] \\ &= \frac{1}{2\sigma\sqrt{x}} \left[\phi\left(\frac{\sqrt{x}-\mu}{\sigma} \right) + \phi\left(\frac{-\sqrt{x}-\mu}{\sigma} \right) \right] \\ &= \frac{1}{2\sigma\sqrt{x}\sqrt{2\pi}} \left[e^{-(\sqrt{x}-\mu)^2/(2\sigma^2)} + e^{-(\sqrt{x}+\mu)^2/(2\sigma^2)} \right] \\ &= \frac{e^{-x/(2\sigma^2)}e^{-\mu^2/(2\sigma^2)}}{\sigma\sqrt{x}\sqrt{2\pi}} \cosh\left(\frac{\sqrt{x}\mu}{\sigma^2} \right), \end{aligned}$$

where we used the definition for the hyperbolic cosine $\cosh(x) = (e^x + e^{-x})/2$. Next, let's replace the hyperbolic cosine with its Taylor series. We find

$$f_X(x) = \frac{e^{-x/(2\sigma^2)}e^{-\mu^2/(2\sigma^2)}}{\sigma\sqrt{x}\sqrt{2\pi}} \sum_{p=0}^{\infty} \frac{1}{(2p)!}\left(\frac{\sqrt{x}\mu}{\sigma^2}\right)^{2p}$$

$$= \sum_{p=0}^{\infty} \frac{e^{-x/(2\sigma^2)}e^{-\mu^2/(2\sigma^2)}(\mu^2/\sigma^2)^p}{\sigma^2\sqrt{2\pi}(2p)!}\left(\frac{x}{\sigma^2}\right)^{p-1/2}$$

Now, it might interest our readers to know that, in fact,

$$\sqrt{\pi}(2p)! = 2^{2p}\Gamma(p+1/2)p!.$$

Thus, we arrive at

$$f_X(x) = \sum_{p=0}^{\infty} \frac{e^{-x/(2\sigma^2)}e^{-\mu^2/(2\sigma^2)}(\mu^2/(2\sigma^2))^p}{\sigma^2 2^{p+1/2}\Gamma(p+1/2)p!}\left(\frac{x}{\sigma^2}\right)^{p-1/2}$$

Setting $\lambda = \mu^2/\sigma^2$ and rearranging yields our result. □

Now, there's an obvious corollary to all of this.

Corollary 3.4. *The* PDF *given by Equation* (3.40) *is represents Poisson-weighted mixture of Gamma random variables. In particular, if $Y \sim N(\mu, \sigma^2)$ is a normal random variable, then the random variable $X = Y^2$ is given by*

$$f_X(x) = \sum_{p=0}^{\infty} \text{Poiss}(p; \lambda/2)\text{Gamma}(x; p + 1/2, 2\sigma^2), \qquad (3.41)$$

where $\lambda = \mu^2/\sigma^2$.

Proof. From Equation (3.40), we have

$$f_X(x) = \frac{1}{\sigma^2}\sum_{p=0}^{\infty} \text{Poiss}(p; \lambda/2)\text{Gamma}(x/\sigma^2; p + 1/2, 2).$$

The result follows from Exercise 2.4. □

We may further restate Corollary 3.4 as follows.

Corollary 3.5. *Let $Y \sim N(\mu, \sigma^2)$ and define $\lambda = \mu^2/\sigma^2$. The random variable $X = Y^2$ is equivalent to the hierarchical model*

$$X|(P = p) \sim \text{Gamma}(p + 1/2, 2\sigma^2)$$
$$P \sim \text{Poiss}(\lambda/2).$$

Proof. Since $P \sim \text{Poiss}(\lambda/2)$, the PMF for P is given by

$$f_P(p) = \frac{e^{-\lambda/2}(\lambda/2)^p}{p!}.$$

Similarly, the PDF of the conditional random variable $X|P$ is given by

$$f_{X|P}(x|p) = \frac{1}{\sigma^2} \frac{(x/\sigma^2)^{p-1/2}e^{-x/(2\sigma^2)}}{\Gamma(p+1/2)2^{p+1/2}}.$$

The (unconditional) distribution of X is therefore obtained by marginalizing over P,

$$f_X(x) = \sum_{p=0}^{\infty} f_{X|P}(x|p)f_P(p),$$

which yields Equation (3.40).   □

An important variation of Corollary 3.5 is the following.

Corollary 3.6. *Let $Y \sim \text{N}(\mu, \sigma^2)$ and define $\lambda = \mu^2/\sigma^2$. The random variable $X = Y^2$ is equivalent to the hierarchical model*

$$X|(P = p) \sim \sigma^2 \chi^2_{1+2p}$$
$$P \sim \text{Poiss}(\lambda/2).$$

Proof. This follows since a $\text{Gamma}(p + 1/2, 2\sigma^2)$ distribution is equivalent to a scaled $\text{Gamma}(p + 1/2, 2)$ distribution, which is itself equivalent to a scaled χ^2_{1+2p} distribution. Specifically, we have

$$\text{Gamma}(x; p + 1/2, 2\sigma^2) = \frac{1}{\sigma^2} \text{Gamma}\left(\frac{x}{\sigma^2}; p + 1/2, 2\right)$$
$$= \frac{1}{\sigma^2} f\left(\frac{x}{\sigma^2}; 1 + 2p\right),$$

where $f(x; 1 + 2p)$ is the PDF of a χ^2_{1+2p} random variable. Finally, we recall from Theorem 1.10 that the random variable that has such a distribution is the scaled chi-squared random variable $\sigma^2 \chi^2_{1+2p}$.   □

And one final variation...

Corollary 3.7. *Let $Y \sim \text{N}(\mu, \sigma^2)$ and define $\lambda = \mu^2/\sigma^2$. Then random variable $Y^2 \sim \sigma^2 \text{NCX}_1^\lambda$.*

Proof. This follows immediately from Corollary 3.6, since the scale factor σ^2 multiplies the chi-squared random variable in the breakdown.   □

Next, let us apply these results to Definition 3.22 to obtain the distribution for the noncentral chi-squared random variable.

Theorem 3.7. *Let $X \sim \mathrm{NCX}_k^\lambda$ be a noncentral chi-squared random variable. Then the* PDF *for X is given by*

$$f_X(x) = \sum_{p=0}^{\infty} \frac{e^{-\lambda/2}(\lambda/2)^p}{p!} \frac{x^{k/2+p-1}e^{-x/2}}{\Gamma(k/2+p)2^{k/2+p}}, \tag{3.42}$$

Moreover, the random variable X can be alternatively described using the hierarchical model

$$X|(P = p) \sim \chi^2_{k+2p}$$
$$P \sim \mathrm{Poiss}(\lambda/2).$$

Proof. Definition 3.22 states that a noncentral chi-squared random variable X can always be expressed as the sum of squares of k normal random variables with unit variance, $X_1, \ldots, X_k$, with $X_i \sim \mathrm{N}(\mu_i, 1)$. Moreover, Corollary 3.6 tells us that the random variable X_i^2 is equivalent to the hierarchical model

$$X_i^2|(P_i = p_i) \sim \chi^2_{1+2p_i}$$
$$P_i \sim \mathrm{Poiss}(\mu_i^2/2),$$

since $\chi^2_{1+2p_i}$ is equivalent to $\mathrm{Gamma}((1+2p_i)/2, 2)$.

Since chi-squared random variables are additive (Proposition 2.20), it follows that our noncentral chi-squared random variable $X = \sum_{i=1}^{k} X_i^2$ is equivalent to

$$X|(P_1, \ldots, P_k) \sim \chi^2_{k+2(p_1+\cdots+p_k)},$$
$$P_1 \sim \mathrm{Poiss}(\mu_1^2/2),$$
$$\vdots$$
$$P_k \sim \mathrm{Poiss}(\mu_k^2/2),$$

Now, we observe that it is only the *sum* of the random variables $P_1, \ldots, P_k$ that matters, not their individual values. Given the additivity of the Poisson random variable (Proposition 2.14), this model therefore simplifies as

$$X|(P = p) \sim \chi^2_{k+2p}$$
$$P \sim \mathrm{Poiss}(\lambda/2),$$

where $\lambda = \sum_{i=1}^{k} \mu_i^2$. This model, however, is equivalent to Equation (3.42), which therefore completes our result. $\square$

Corollary 3.8. *The mean and variance of X are given by*

$$\mathbb{E}[X] = k + \lambda \quad and \quad \mathbb{V}(X) = 2(k + 2\lambda). \tag{3.43}$$

Proof. The mean of a noncentral chi-squared random variable is given by

$$\mathbb{E}[X] = \mathbb{E}[\mathbb{E}[X|P]] = \mathbb{E}[k + 2P] = k + \lambda.$$

The variance is obtained by applying Theorem 2.9. (See Exercise 3.5.)   □

Fortunately, we will not have occasion to revert to the analytical form of the noncentral chi-squared distribution, or its associated, unfathomably complex, CDF. Instead, basic usage in Python is given in Code Block 3.9.

```python
nc_lambda = 2 # noncentrality parameter
dof = 7       # degrees of freedom
n_samples = 100
samples = np.random.noncentral_chisquare(dof, nc_lambda,
    size=n_samples)

x = np.linspace(0, 20)
f = scipy.stats.ncx2.pdf(x, dof, nc_lambda)
F = scipy.stats.ncx2.cdf(x, dof, nc_lambda)
alpha = 0.05
crit_pt = scipy.stats.ncx2.isf(alpha, dof, nc_lambda) # Critical
    point
```

Code Block 3.9: Basic operations for a noncentral chi-squared random variable

3.5.2 Power Function for Chi-square Statistic

Our goal is to understand the distribution of the chi-squared statistic Q, as given by Equation (3.32), under the alternative hypothesis that $p = p_A$, for a particular $p_A \in \Delta^{k-1}$ with $p_A \neq p_0$. We begin with a slightly easier result.

Proposition 3.10. *Let* $X \sim \text{Multi}(n, p_A)$ *for some* $p_A \in \Delta^{k-1}$, *and let* $p_0 \in \Delta^{k-1}$ *be distinct from* p_A. *Then the statistic*

$$R = \sum_{i=1}^{k} \frac{(X_i - np_{0i})^2}{np_{Ai}} \tag{3.44}$$

tends towards $R \to \text{NCX}_{k-1}^{\lambda_n}$ *as* $n \to \infty$, *where the noncentrality parameter is given by*

$$\lambda_n = n \sum_{i=1}^{k} \frac{(p_{Ai} - p_{0i})^2}{p_{Ai}}. \tag{3.45}$$

Proof. We will follow the construction used in the proof of Theorem 3.6. In particular, consider a sequence of IID reduced Bernoulli random vectors $B_1, \ldots, B_n \sim \mathrm{RedMultiBern}(p_A)$, so that $\tilde{X} = (X_1, \ldots, X_{k-1}) = n\overline{B}_n$. The central limit theorem yields

$$Z_n = \sqrt{n}\tilde{\Sigma}_A^{-1/2}(\overline{B}_n - \tilde{p}_A) \to \mathrm{N}(0_{k-1}, I_{k-1})$$

as $n \to \infty$. Therefore, for large n, the exrpession $Z_n + \sqrt{n}\tilde{\Sigma}_A^{-1/2}(\tilde{p}_A - \tilde{p}_0)$ can be approximated by

$$\sqrt{n}\tilde{\Sigma}_A^{-1/2}(\overline{B}_n - \tilde{p}_0) \approx \mathrm{N}(\sqrt{n}\tilde{\Sigma}_A^{-1/2}\tilde{\delta}, I_{k-1}), \tag{3.46}$$

where we have defined the vector

$$\tilde{\delta} = \tilde{p}_A - \tilde{p}_0.$$

It follows that the quantity

$$R = n(\overline{B}_n - \tilde{p}_0)\tilde{\Sigma}_A^{-1}(\overline{B}_n - \tilde{p}_0) \approx \mathrm{NCX}_{k-1}^{\lambda_n}, \tag{3.47}$$

where $\lambda_n = n\tilde{\delta}^T\tilde{\Sigma}_A^{-1}\tilde{\delta}$. However, by following the same steps of the proof of Theorem 3.6, one can easily show that R is equivalent to the expression given in Equation (3.44). Moreover, the noncentrality parameter can be expressed as

$$\begin{aligned}
\lambda_n &= n\tilde{\delta}^T\tilde{\Sigma}_A^{-1}\tilde{\delta} \\
&= n\tilde{\delta}^T\left(\mathrm{diag}(\tilde{p}_A)^{-1} + p_{Ak}^{-1}J_{k-1}\right)\tilde{\delta} \\
&= n\tilde{\delta}^T\mathrm{diag}(\tilde{p}_A)^{-1}\tilde{\delta} + np_{Ak}^{-1}\tilde{\delta}^T J_{k-1}\tilde{\delta}
\end{aligned}$$

Now, the first term is simply

$$n\tilde{\delta}^T\mathrm{diag}(\tilde{p}_A)^{-1}\tilde{\delta} = n\sum_{i=1}^{k-1}\frac{(p_{Ai} - p_{0i})^2}{p_{Ai}}.$$

And the second term is

$$\begin{aligned}
np_{Ak}^{-1}\tilde{\delta}^T J_{k-1}\tilde{\delta} &= np_{Ak}^{-1}\left(\sum_{i=1}^{k-1}(p_{Ai} - p_{0i})\right)^2 \\
&= np_{Ak}^{-1}[(1 - p_{Ak}) - (1 - p_{0k})]^2 \\
&= \frac{n(p_{Ak} - p_{0k})^2}{p_{Ak}}.
\end{aligned}$$

It follows that $\lambda_n = n\tilde{\delta}^T\tilde{\Sigma}_A^{-1}\tilde{\delta}$ is equivalent to Equation (3.45), thereby completing the result. $\qquad\square$

The statistic R, defined in Proposition 3.10, is of course not equivalent to the chi-squared statistic Q, as defined by Equation (3.32), the difference being the denominator as the expected count under the null hypothesis (for the chi-squared statistic Q) or the expected count under the alternate hypothesis (for the statistic R). The remedy of such a blemish is the cause of much toil and strife. We begin by studying how Equations (3.46) and (3.47) must be modified to accommodate this change.

Lemma 3.1. *Under an alternate hypothesis $H_A : p = p_A$, with $p_A \neq p_0$, the chi-squared statistic Q, defined in Equation (3.32) is equivalent to*

$$Q = X^T X,$$

where

$$X = \sqrt{n} \tilde{\Sigma}_0^{-1/2}(\overline{B}_n - \tilde{p}_0) \approx \mathrm{N}(\sqrt{n}\tilde{\Sigma}_0^{-1/2}\tilde{\delta}, \tilde{\Sigma}_0^{-1/2}\tilde{\Sigma}_A\tilde{\Sigma}_0^{-1/2}), \qquad (3.48)$$

following the notation of the proof of Proposition 3.10.

Proof. It is straightforward to show, following the same steps done in the proof of Proposition 3.10, that

$$Q = n(\overline{B}_n - \tilde{p}_0)\tilde{\Sigma}_0^{-1}(\overline{B}_n - \tilde{p}_0)$$

is equivalent to Equation (3.32). Equation (3.48) is obtained by premultiplying Equation (3.46) by $\tilde{\Sigma}_0^{-1/2}\tilde{\Sigma}_A^{1/2}$, using the fact that $\tilde{\Sigma}_0^{-1/2}$ is symmetric.[5]
□

Lemma 3.2. *Under an alternate hypothesis $H_A : p = p_A$, with $p_A \neq p_0$, the chi-squared statistic Q, defined in Equation (3.32) is approximately distributed like*

$$Q \sim \sum_{i=1}^{k-1} \tau_i^2 \mathrm{NCX}_1^{n\lambda_i}, \qquad (3.49)$$

for large n, where $\lambda_i = \mu_i^2/\tau_i^2$, and where

$$\mu = V^T \tilde{\Sigma}_0^{-1/2}(\tilde{p}_A - \tilde{p}_0), \qquad (3.50)$$

and where D and V is the eigendecomposition of $\tilde{\Sigma}_0^{-1/2}\tilde{\Sigma}_A\tilde{\Sigma}_0^{-1/2} = VDV^T$, with $D = \mathrm{diag}(\tau_1^2, \ldots, \tau_{k-1}^2)$.

Proof. Let V be the orthogonal matrix of eigenvectors for $\tilde{\Sigma}_0^{-1/2}\tilde{\Sigma}_A\tilde{\Sigma}_0^{-1/2}$ and let $D = \mathrm{diag}(\tau_1^2, \ldots, \tau_{k-1}^2)$ be the corresponding (ordered) diagonal matrix of eigenvalues. Now consider the transformation $Y = V^T X$. First, since V is orthogonal, we have

---

[5] Note that if $X \sim \mathrm{N}(\mu, \Sigma)$, then $AX \sim \mathrm{N}(A\mu, A\Sigma A^T)$.

$$Y^T Y = X^T V V^T X = X^T X = Q.$$

Second, Equation (3.48) yields

$$Y \approx \mathrm{N}(\sqrt{n} V^T \tilde{\Sigma}_0^{-1/2} \tilde{\delta}, D).$$

If we define $\mu = V^T \tilde{\Sigma}_0^{-1/2} \tilde{\delta}$, then the ith component of Y is approximately

$$Y_i \approx \mathrm{N}(\sqrt{n} \mu_i, \tau_i^2),$$

since D is diagonal. The result follows from Corollary 3.7.   □

Our task is to approximate the expression for the chi-squared statistic, as given y Equation (3.49). To proceed, we shall require a number of preliminary results.

Lemma 3.3. *The matrix $\tilde{\Sigma}_A \tilde{\Sigma}_0^{-1}$ may be expressed as*

$$\tilde{\Sigma}_A \tilde{\Sigma}_0^{-1} = \mathrm{diag}\left(\frac{\tilde{p}_A}{\tilde{p}_0}\right) - \tilde{p}_A \left(\frac{\tilde{p}_A}{\tilde{p}_0}\right)^T + \frac{p_{Ak}}{p_{0k}} \tilde{p}_A \tilde{1}^T, \qquad (3.51)$$

where "vector division" is understood to be component-wise, and where $\tilde{1} = \langle 1, \ldots, 1 \rangle \in \mathbb{R}^{k-1}$.

Proof. Using a few matrix–vector properties,

$$\mathrm{diag}(x) J = x 1^T, \quad x^T J = \left(\sum x_i\right) 1^T, \quad x^T \mathrm{diag}(y) = (xy)^T,$$

we have the following

$$\tilde{\Sigma}_A \tilde{\Sigma}_0^{-1} = \left(\mathrm{diag}(\tilde{p}_A) - \tilde{p}_A \tilde{p}_A^T\right)\left(\mathrm{diag}(\tilde{p}_0)^{-1} + \frac{1}{p_{0k}} J_{k-1}\right)$$

$$= \mathrm{diag}\left(\frac{\tilde{p}_A}{\tilde{p}_0}\right) + \frac{\mathrm{diag}(\tilde{p}_A) J_{k-1}}{p_{0k}} - \tilde{p}_A \tilde{p}_A^T \mathrm{diag}(\tilde{p}_0)^{-1} - \frac{\tilde{p}_A \tilde{p}_A^T J_{k-1}}{p_{0k}}$$

$$= \mathrm{diag}\left(\frac{\tilde{p}_A}{\tilde{p}_0}\right) + \frac{\tilde{p}_A \tilde{1}^T}{p_{0k}} - \tilde{p}_A \left(\frac{\tilde{p}_A}{\tilde{p}_0}\right)^T - \frac{(1 - p_{Ak}) \tilde{p}_A \tilde{1}^T}{p_{0k}}.$$

The result follows.   □

Lemma 3.4. *The trace of the matrix $\tilde{\Sigma}_A \tilde{\Sigma}_0^{-1}$ is given by*

$$\mathrm{trace}\left(\tilde{\Sigma}_A \tilde{\Sigma}_0^{-1}\right) = \sum_{i=1}^{k} \frac{p_{Ai}(1 - p_{Ai})}{p_{0i}}.$$

Proof. Equation (3.51) yields the diagonal elements of $\tilde{\Sigma}_A \tilde{\Sigma}_0^{-1}$, which are

$$\left[\tilde{\Sigma}_A \tilde{\Sigma}_0^{-1}\right]_{ii} = \frac{p_{Ai}}{p_{0i}} - \frac{p_{Ai}^2}{p_{0i}} + \frac{p_{Ak}}{p_{0k}} p_{Ai},$$

for $i = 1, \ldots, (k-1)$. Therefore

$$
\begin{aligned}
\text{trace}\left(\tilde{\Sigma}_A \tilde{\Sigma}_0^{-1}\right) &= \sum_{i=1}^{k-1} \left[\tilde{\Sigma}_A \tilde{\Sigma}_0^{-1}\right]_{ii} \\
&= \sum_{i=1}^{k-1} \frac{p_{Ai}(1-p_{Ai})}{p_{0i}} + \sum_{i=1}^{k-1} \frac{p_{Ak}}{p_{0k}} p_{Ai} \\
&= \sum_{i=1}^{k-1} \frac{p_{Ai}(1-p_{Ai})}{p_{0i}} + \frac{p_{Ak}}{p_{0k}}(1-p_{Ak}).
\end{aligned}
$$

The result follows.   □

Lemma 3.5. *The trace of the matrix $(\tilde{\Sigma}_A \tilde{\Sigma}_0^{-1})^2$ is given by*

$$
\text{trace}\left((\tilde{\Sigma}_A \tilde{\Sigma}_0^{-1})^2\right) = \sum_{i=1}^{k} \frac{p_{Ai}^2}{p_{0i}^2}(1 - 2p_{Ai}) + \left(\sum_{i=1}^{k} \frac{p_{Ai}^2}{p_{0i}}\right)^2.
$$

Proof. Using the expression for $\tilde{\Sigma}_A \tilde{\Sigma}_0^{-1}$ given by Equation (3.51), we find that

$$
\begin{aligned}
(\tilde{\Sigma}_A \tilde{\Sigma}_0^{-1})^2 &= \text{diag}\left(\frac{\tilde{p}_A^2}{\tilde{p}_0^2}\right) + S\tilde{p}_A \left(\frac{\tilde{p}_A}{\tilde{p}_0}\right)^T + \frac{p_{Ak}^2}{p_{0k}^2}(1-p_{Ak})\tilde{p}_A \tilde{1}^T \\
&\quad -2\left(\frac{\tilde{p}_A^2}{\tilde{p}_0}\right)\left(\frac{\tilde{p}_A}{\tilde{p}_0}\right)^T + 2\frac{p_{Ak}}{p_{0k}}\left(\frac{\tilde{p}_A^2}{\tilde{p}_0}\right)\tilde{1}^T - 2\frac{p_{Ak}}{p_{0k}} S\tilde{p}_A \tilde{1}^T,
\end{aligned}
$$

where we define the sum S as

$$
S = \left(\frac{\tilde{p}_A}{\tilde{p}_0}\right)^T \tilde{p}_A = \sum_{i=1}^{k-1} \frac{p_{Ai}^2}{p_{0i}}.
$$

Summing the diagonal elements, we obtain

$$
\begin{aligned}
\text{trace}\left((\tilde{\Sigma}_A \tilde{\Sigma}_0^{-1})^2\right) &= \sum_{i=1}^{k-1} \left[(\tilde{\Sigma}_A \tilde{\Sigma}_0^{-1})^2\right]_{ii} \\
&= \sum_{i=1}^{k-1} \frac{p_{Ai}^2}{p_{0i}^2} + S^2 + \frac{p_{Ak}^2}{p_{0k}^2}(1-p_{Ak})^2 \\
&\quad -2\sum_{i=1}^{k-1} \frac{p_{Ai}^3}{p_{0i}^2} + 2\frac{p_{Ak}}{p_{0k}} S - 2\frac{p_{Ak}}{p_{0k}} S(1-p_{Ak}).
\end{aligned}
$$

Now, the third term may be expanded to find the kth term to complete the sum represented by the first and fourth term. Also, the fifth and sixth terms may be combined. This leaves us with

$$\text{trace}\left((\tilde{\Sigma}_A \tilde{\Sigma}_0^{-1})^2\right) = \sum_{i=1}^{k} \frac{p_{Ai}^2}{p_{0i}^2} + S^2 + \frac{p_{Ak}^4}{p_{0k}^2} - 2\sum_{i=1}^{k} \frac{p_{Ai}^3}{p_{0i}^2} + 2\frac{p_{Ak}^2}{p_{0k}} S$$

Noting that

$$\sum_{i=1}^{k} \frac{p_{Ai}^2}{p_{0i}^2} - 2\sum_{i=1}^{k} \frac{p_{Ai}^3}{p_{0i}^2} = \sum_{i=1}^{k} \frac{p_{Ai}^2}{p_{0i}^2}(1 - 2p_{Ai})$$

and

$$\left(\sum_{i=1}^{k} \frac{p_{Ai}^2}{p_{0i}}\right)^2 = \left(S + \frac{p_{Ak}^2}{p_{0k}}\right)^2 = S^2 + 2\frac{p_{Ak}^2}{p_{0k}} S + \frac{p_{Ak}^4}{p_{0k}^2},$$

we obtain our result. $\qquad\square$

Lemma 3.6. *Let μ and D be as in Lemma 3.2. Then the quantity $\mu^T D\mu$ is given by*

$$\mu^T D\mu = \sum_{i=1}^{k} \frac{p_{Ai}(p_{Ai} - p_{0i})^2}{p_{0i}^2} - \left(\sum_{i=1}^{k} \frac{p_{Ai}(p_{Ai} - p_{0i})}{p_{0i}}\right)^2.$$

Proof. From Equation (3.50), we have

$$\mu^T D\mu = \tilde{\delta}^T \tilde{\Sigma}_0^{-1/2} V D V^T \tilde{\Sigma}_0^{-1/2} \tilde{\delta}$$
$$= \tilde{\delta}^T \tilde{\Sigma}_0^{-1} \tilde{\Sigma}_A \tilde{\Sigma}_0^{-1} \tilde{\delta}$$

Now, in general, for $x \in \mathbb{R}^{k-1}$, we have

$$x^T \tilde{\Sigma}_A x = x^T \text{diag}(\tilde{p}_A) x - x^T \tilde{p}_A \tilde{p}_A^T x$$
$$= \sum_{i=1}^{k-1} p_{Ai} x_i^2 - \left(\sum_{i=1}^{k-1} p_{Ai} x_i\right)^2.$$

Therefore, since

$$\tilde{\Sigma}_0^{-1} \tilde{\delta} = \left(\text{diag}(\tilde{p}_0)^{-1} + p_{0k}^{-1} J\right)(\tilde{p}_A - \tilde{p}_0)$$
$$= \frac{\tilde{p}_A - \tilde{p}_0}{\tilde{p}_0} + \frac{p_{0k} - p_{Ak}}{p_{0k}} \tilde{1},$$

it follows that

$$\tilde{\delta}^T \tilde{\Sigma}_0^{-1} \tilde{\Sigma}_A \tilde{\Sigma}_0^{-1} \tilde{\delta} = \sum_{i=1}^{k-1} p_{Ai} \left[\tilde{\Sigma}_0^{-1}\tilde{\delta}\right]_i^2 - \left(\sum_{i=1}^{k-1} p_{Ai}\left[\tilde{\Sigma}_0^{-1}\tilde{\delta}\right]_i\right)^2$$
$$= \sum_{i=1}^{k-1} \frac{p_{Ai}(p_{Ai} - p_{0i})^2}{p_{0i}^2} + \frac{(p_{0k} - p_{Ak})^2}{p_{0k}^2}(1 - p_{Ak})$$
$$+ \frac{2(p_{0k} - p_{Ak})}{p_{0k}} \sum_{i=1}^{k-1} \frac{p_{Ai}(p_{Ai} - p_{0i})}{p_{0i}}$$
$$- \left[\sum_{i=1}^{k-1} \frac{p_{Ai}(p_{Ai} - p_{0i})}{p_{0i}} + \frac{(p_{0k} - p_{Ak})}{p_{0k}}(1 - p_{Ak})\right]^2$$

This last term is equivalent to

$$\left[\sum_{i=1}^{k} \frac{p_{Ai}(p_{Ai} - p_{0i})}{p_{0i}} + \frac{(p_{0k} - p_{Ak})}{p_{0k}}\right]^2 = \left(\sum_{i=1}^{k} \frac{p_{Ai}(p_{Ai} - p_{0i})}{p_{0i}}\right)^2 + \frac{(p_{0k} - p_{Ak})^2}{p_{0k}^2}$$

$$+2\frac{(p_{0k} - p_{Ak})}{p_{0k}} \sum_{i=1}^{k} \frac{p_{Ai}(p_{Ai} - p_{0i})}{p_{0i}}$$

Combining the above, we have

$$\mu^T D \mu = \sum_{i=1}^{k-1} \frac{p_{Ai}(p_{Ai} - p_{0i})^2}{p_{0i}^2} + \frac{(p_{0k} - p_{Ak})^2}{p_{0k}^2}(1 - p_{Ak})$$

$$+\frac{2(p_{0k} - p_{Ak})}{p_{0k}} \sum_{i=1}^{k-1} \frac{p_{Ai}(p_{Ai} - p_{0i})}{p_{0i}}$$

$$-\left(\sum_{i=1}^{k} \frac{p_{Ai}(p_{Ai} - p_{0i})}{p_{0i}}\right)^2 - \frac{(p_{0k} - p_{Ak})^2}{p_{0k}^2}$$

$$-2\frac{(p_{0k} - p_{Ak})}{p_{0k}} \sum_{i=1}^{k} \frac{p_{Ai}(p_{Ai} - p_{0i})}{p_{0i}}$$

Combining the third and sixth terms, we have

$$2\frac{p_{Ak}(p_{0k} - p_{Ak})^2}{p_{0k}^2}.$$

Similarly, the second and forth terms combine to yield

$$-\frac{p_{Ak}(p_{0k} - p_{Ak})^2}{p_{0k}^2}.$$

Altogether, the second, third, forth, and sixth terms combine to complete the sum represented by the first term. This completes the result.   □

Before moving further, let us collect the results from Lemmas 3.4–3.6 in the following.

Lemma 3.7. *Let $D = \text{diag}(\tau_1^2, \ldots, \tau_{k-1}^2)$ and μ be defined as in Lemma 3.2. Then*

$$S_1 = \sum_{i=1}^{k-1} \tau_i^2 = \sum_{i=1}^{k} \frac{p_{Ai}(1 - p_{Ai})}{p_{0i}}$$

$$S_2 = \sum_{i=1}^{k-1} \tau_i^4 = \sum_{i=1}^{k} \frac{p_{Ai}^2}{p_{0i}^2}(1 - 2p_{Ai}) + \left(\sum_{i=1}^{k} \frac{p_{Ai}^2}{p_{0i}}\right)^2$$

$$S_3 = \sum_{i=1}^{k-1} \mu_i^2 = \sum_{i=1}^{k} \frac{(p_{Ai} - p_{0i})^2}{p_{0i}}$$

$$S_4 = \sum_{i=1}^{k-1} \mu_i^2 \tau_i^2 = \sum_{i=1}^{k} \frac{p_{Ai}(p_{Ai} - p_{0i})^2}{p_{0i}^2} - \left(\sum_{i=1}^{k} \frac{p_{Ai}(p_{Ai} - p_{0i})}{p_{0i}}\right)^2.$$

Proof. Expression S_1 is given by

$$S_1 = \sum_{i=1}^{k-1} \tau_i^2 = \text{trace}(\tilde{\Sigma}_0^{-1/2}\tilde{\Sigma}_A\tilde{\Sigma}_0^{-1/2}) = \text{trace}(\tilde{\Sigma}_A\tilde{\Sigma}_0^{-1}),$$

and is therefore given by Lemma 3.4.

Similarly, expression $S_2 = \sum_{i=1}^{k-1} \tau_i^4$ is given by Lemma 3.5.

Expression S_3 is given by

$$\sum_{i=1}^{k-1} \mu_i^2 = \mu^T \mu$$

$$= (\tilde{p}_A - \tilde{p}_0)^T \tilde{\Sigma}_0^{-1/2} VV^T \tilde{\Sigma}_0^{-1/2}(\tilde{p}_A - \tilde{p}_0)$$
$$= (\tilde{p}_A - \tilde{p}_0)^T \tilde{\Sigma}_0^{-1}(\tilde{p}_A - \tilde{p}_0)$$

However, we already showed that such an expression simplifies to Equation (3.55) in the proof of Proposition 3.10.

Finally, Expression $S_4 = \mu^T D\mu$ is given by Lemma 3.6. This completes the result. □

Theorem 3.8. *Under an alternative hypothesis $H_A : p = p_A \in \Theta_0^c$, the chi-squared statistic, for null hypothesis $H_0 : p = p_0$, is approximately distributed for large n as a location–scale shifted noncentral chi-squared distribution*

$$Q \sim a\text{NCX}_\nu^{n\lambda} + c, \tag{3.52}$$

where parameters a, ν, λ, and c are given by

$$a = \frac{S_4}{S_3} \tag{3.53}$$

$$\lambda = \frac{S_3}{a} \tag{3.54}$$

$$\nu = \frac{S_2}{a^2}, \tag{3.55}$$

$$c = S_1 - a\nu \tag{3.56}$$

where we have defined

$$S_1 = \sum_{i=1}^{k} \frac{p_{Ai}(1 - p_{Ai})}{p_{0i}} \tag{3.57}$$

$$S_2 = \sum_{i=1}^{k} \frac{p_{Ai}^2}{p_{0i}^2}(1 - 2p_{Ai}) + \left(\sum_{i=1}^{k} \frac{p_{Ai}^2}{p_{0i}}\right)^2 \tag{3.58}$$

$$S_3 = \sum_{i=1}^{k} \frac{(p_{Ai} - p_{0i})^2}{p_{0i}} \tag{3.59}$$

$$S_4 = \sum_{i=1}^{k} \frac{p_{Ai}(p_{Ai} - p_{0i})^2}{p_{0i}^2} - \left(\sum_{i=1}^{k} \frac{p_{Ai}(p_{Ai} - p_{0i})}{p_{0i}}\right)^2. \tag{3.60}$$

Proof. From Lemma 3.2, we know that under an alternative hypothesis $p = p_A$, the chi-squared statistic is approximately distributed as a weighted sum of noncentral chi-squared random variables, such that

$$Q \sim \sum_{i=1}^{k-1} \tau_i^2 \mathrm{NCX}_1^{n\lambda_i},$$

where τ_i^2 is an eigenvalue of $\tilde{\Sigma}_0^{-1/2}\tilde{\Sigma}_A\tilde{\Sigma}_0^{-1/2}$, and $\mu = V^T\tilde{\Sigma}_0^{-1/2}(\tilde{p}_A - \tilde{p}_0)$, and $\lambda_i = \mu_i^2/\tau_i^2$. Let us make a Satterthwaite ansatz that this distribution may be approximated in the form of Equation (3.52); i.e.,

$$\sum_{i=1}^{k-1} \tau_i^2 \mathrm{NCX}_1^{n\lambda_i} \approx a\mathrm{NCX}_\nu^{n\lambda} + c,$$

for some a, ν, λ, and c. Comparing the first and second moments of both sides of this and applying Equations (3.25) and (3.26) we obtain

$$\sum_{i=1}^{k-1} \tau_i^2(1 + n\lambda_i) = a(\nu + n\lambda) + c$$

$$\sum_{i=1}^{k-1} \tau_i^4(1 + 2n\lambda_i) = a^2(\nu + 2n\lambda)$$

Separating the $O(1)$ and $O(n)$ terms, we can convert this into a system of four equations:

$$\sum_{i=1}^{k-1} \tau_i^2 = a\nu + c, \qquad\qquad \sum_{i=1}^{k-1} \tau_i^2\lambda_i = a\lambda,$$

$$\sum_{i=1}^{k-1} \tau_i^4 = a^2\nu, \quad \text{and} \quad \sum_{i=1}^{k-1} \tau_i^4\lambda_i = a^2\lambda.$$

Recognizing the left-hand sides as the quantities S_1–S_4, as per Lemma 3.7, we may solve this system for our parameters, thereby obtaining our result. □

We now arrive are our key result.

Corollary 3.9. *The power function of Pearson's chi-squared test, as defined in Definition 3.18, with sample size n, is given by*

$$\beta(p_A) = S(\chi^2_{k-1,\alpha}; n, a, \nu, \lambda, c),$$

where S is the survival function of the scaled noncentral chi-squared random variable defined in Theorem 3.8.

Example 3.11. Suppose we are running a five-category Pearson chi-squared test with null hypothesis

$$H_0 : p = p_0 = \langle 0.2, 0.2, 0.2, 0.2, 0.2 \rangle,$$

and we want to determine the required sample size n to ensure that we achieve a power of $\beta = 0.80$ for the particular alternate hypothesis

$$p_A = \langle 0.1, 0.3, 0.2, 0.3, 0.1 \rangle.$$

We proceed as shown in Code Block 3.10.   We begin by defining our null hypothesis and our particular alternate hypothesis (lines 1–3). Next, we compute the sums S_1–S_4 (lines 5–8), which we use to compute our parameters a, λ, ν, and c (lines 10–13). We obtain values of

$$S_1 = 3.8, \qquad S_2 = 4.24, \qquad S_3 = 0.2, \qquad S_4 = 0.16,$$

and parameter values

$$a = 0.8, \qquad \lambda = 0.25, \qquad \nu = 6.625, \qquad c = -1.5.$$

Next, we compute the critical value of our test, $Q_{crit} = 9.487729$, and then solve for the value of n that will yield an 80% survival value at Q_{crit} under the alternate hypothesis (lines 15–20).

Theoretically, we are done. But let's go a step further and validate that our answer is correct. To do this, we can run 10,000 simulations of our test, using data consistent with our particular alternate hypothesis, and count the frequency at which our test (correctly) rejects the null hypothesis.

Finally, we note that we can further validate the theory behind our expressions for S_1–S_4, from Equations (3.57)–(3.60), as shown in Code Block 3.11.[6] Here, we compute the reduced probability vectors $\tilde{p}_A$ and $\tilde{p}_0$,

---

[6] Note that in lines 3–4 of Code Block 3.11, we explicitly use `np.outer(ps, ps)` as opposed to `ps @ ps.T`. This is because the latter will not work correctly unless we reshape the array `ps` as a column vector.

```
1   p_0 = np.array([0.2, 0.2, 0.2, 0.2, 0.2])
2   p = np.array([0.1, 0.3, 0.2, 0.3, 0.1])
3   dof = len(p) - 1
4
5   S1 = np.sum( p*(1 - p) / p_0 ) #3.8
6   S2 = np.sum( p**2*(1-2*p)/p_0**2 ) + (np.sum( p**2/p_0 ))**2 #4.24
7   S3 = np.sum( (p-p_0)**2/p_0 ) #0.20
8   S4 = np.sum( p*(p-p_0)**2/p_0**2 ) - (np.sum( p*(p-p_0)/p_0 ))**2
        #0.16
9
10  a = S4 / S3     #0.800
11  la = S3 / a     #0.250
12  nu = S2 / a**2 #6.625
13  c = S1 - a * nu #-1.50
14
15  alpha = 0.05
16  Q_crit = scipy.stats.chi2.isf(alpha, dof) # Critical Value 9.487729
17  beta = 0.80
18  results = scipy.optimize.root(lambda x: scipy.stats.ncx2.sf(Q_crit,
        nu, x*la, scale=a, loc=c) - beta, 10)
19  n = int(results['x']) #57
20  beta = scipy.stats.ncx2.sf(Q_crit, nu, n*la, scale=a, loc=c)
21
22  n_trials = 10000
23  E_0 = n * np.array(p_0) # Expected frequencies; array REQUIRED!
24  E_A = n * np.array(p) # Expected frequencies; array REQUIRED!
25  rejections = np.zeros(n_trials)
26  for i in range(n_trials):
27      x = np.random.multinomial(n, p)
28      Q = np.sum( (x - E_0)**2 / E_0 )
29      rejections[i] = int( Q > Q_crit )
30
31  print ('Predicted Power', beta) #0.7964
32  print ('Observed Rejection Rate', np.sum(rejections)/n_trials)
        #0.7962
```

Code Block 3.10: Power of a chi-squared test

and compute the matrices $\tilde{\Sigma}_A$, $\tilde{\Sigma}_0$, and $\tilde{\Sigma}_0^{-1/2}$, from which we can compute the matrices eigenvalues D and eigenvectors V, from which we can further determine the vector μ. We can then compute the four expressions given in Lemma 3.7:

```
1   ps = p[:dof]
2   ps0 = p_0[:dof]
3   Sigma = np.diag(ps) - np.outer(ps, ps) # Outer product: ps @ ps.T
4   Sigma0 = np.diag(ps0) - np.outer(ps0, ps0)
5   Sigma0Rt = scipy.linalg.sqrtm(Sigma0)
6   Sigma0RtInv = np.linalg.inv(Sigma0Rt)
7   D, V = np.linalg.eig(Sigma0RtInv @ Sigma @ Sigma0RtInv)
8   mu = V.T @ Sigma0RtInv @ (ps - ps0)
9
10  S1 = np.sum( D ) # 3.8
11  S2 = np.sum( D**2 ) # 4.24
12  S3 = np.sum( mu**2 ) #0.20
13  S4 = np.sum( D * mu**2 ) # 0.16
```

Code Block 3.11: Validation of our expressions for S_1–S_4

$$S_1 = \sum_{i=1}^{k-1} \tau_i^2 = \text{np.sum(D)}, \qquad S_2 = \sum_{i=1}^{k-1} \tau_i^4 = \text{np.sum(D**2)}$$

$$S_3 = \sum_{i=1}^{k-1} \mu_i^2 = \text{np.sum(mu**2)} \qquad S_4 = \sum_{i=1}^{k-1} \tau_i^2 \mu_i^2 = \text{np.sum(D*mu**2)}.$$

As expected, the results produced by lines 10–13 of Code Block 3.11 produce identical results to the output of lines 5–8 of Code Block 3.10, thereby validating all of our work from Lemmas 3.3–3.6, which is further summarized in Lemma 3.7.   ▷

3.5.3 Sampling the Simplex

Let us take a moment to recap. Given a null hypothesis $H_0 : p = p_0 \in \Delta^{k-1}$ and an alternate hypothesis $p_A \in \Theta_0^c$, we have constructed an approximate power function as a function of the sample size $\beta = \beta(p, n; p_0)$ for Pearson's chi-squared test (Definition 3.18). However, this method computes the power of a test at a single point, whereas we are often interested to know the worst-case power for a given minimum detectable effect. One approach to modeling this is to sample the simplex Δ^{k-1}, and remove those samples within a δ-ball of the point p_0, i.e., consider only $p \in \Delta^{k-1}$ for which $||p - p_0|| \geq \delta$, relative to a given norm. For this reason, we briefly discuss a method for sampling the standard simplex.

Recall from Definition 2.23 that the standard simplex Δ^{k-1} is equivalent to the portion of the hyperplane $\sum_{i=1}^{k} x_i = 1$ that lies within the unit cube $[0, 1]^k \subset \mathbb{R}^k$. A naive approach to sample the simplex would therefore be to sample the hypercube and then normalize the result. This approach, however, does not produce a uniform sampling of the standard simplex.

The correct approach relies on the concept of a *random spacing* of the unit interval, defined as follows.

Definition 3.23. *Let $X_1, \ldots, X_{k-1} \sim \text{Unif}(0,1)$ be a IID uniform sample of the unit interval $[0,1]$, and let $X_{(1)} < \cdots < X_{(k-1)}$ be the corresponding order statistics. Further, let us define the endpoints $X_{(0)} = 0$ and $X_{(k)} = 1$. Then the random variables $S_1, \ldots, S_k$, defined by*

$$S_i = X_{(i)} - X_{(i-1)}$$

for $i = 1, \ldots, k$, are called a random spacing of size k *of the unit interval.*

We now present our result, which follows the approach contained in Devroye [1986].

Theorem 3.9. *The random vector $S = (S_1, \ldots, S_k)$ consisting of a random spacing of size k of the unit interval is distributed uniformly on the standard simplex Δ^{k-1}.*

We leave the proof to Devroye [1986]. However, note that due to the definition of a random spacing, it follows that each $S_i \geq 0$ and the sum

$$\sum_{i=1}^{k} S_i = 1,$$

which shows that the random vector $S \in \Delta^{k-1}$.

A simple function which generates a random point from the simplex is given by the first few lines of Code Block 3.12. This is then generalized to produce a sample of a given size in the subsequent lines of code. The random sample of the simplex Δ^2 is shown in Figure 3.5[7].

Sample Size for Pearson's Chi-squared Test

We can combine our sampling technique with our theory of the power function in order to approximate the power of the test as a function of the effect size

$$\delta(p) = ||p - p_0||.$$

We illustrate the approach with the following shining example.

Example 3.12. Determine the required sample size for a chi-squared test with 80% power and 5% significance level, given the null hypothesis

$$p_0 = \langle 0.2, 0.2, 0.2, 0.2, 0.2 \rangle.$$

The code is shown in Code Block 3.13.

---

[7] Make sure to import: `from mpl_toolkits.mplot3d import Axes3D`.

```python
def randomSimplex(k):
    """ Returns a random point on the
    simplex Delta^{k-1}
    """
    x = np.random.random(k-1)
    x = np.append(x, [0, 1])
    x.sort()
    return np.diff(x)

def randomMultiSimplex(k, size=10):
    """ Returns a random sample from the
    simplex Delta^{k-1}.
    Output is an array of shape (size, k)
    """
    X = np.ones((size, 2))
    X[:, 0] = 0 # First column is 0
    x = np.random.random((size, k-1))
    x = np.append(x, X, axis=1)
    x.sort(axis=1)
    return np.diff(x, axis=1)

## Generate and plot a random sample Delta^2.
fig = plt.figure(figsize=(8, 9/2))
ax = fig.add_subplot(111, projection='3d')
X = randomMultiSimplex(3, size=1000)
ax.scatter3D(X[:, 0], X[:, 1], X[:, 2], '.')
```

Code Block 3.12: Generate a random vector from the standard simplex

We begin by creating a random sample of our simplex, using the result of Theorem 3.9 (line 8). We can then sort this sample so that the individual samples are ordered by the effect size (lines 9–11). For each sample, we compute the parameters a, λ, ν, and c for the approximation given by Equation (3.52) of the distribution of the chi-squared statistic under the given sample alternate hypothesis (lines 19–26). We can use the built-in `scipy.optimize.root` algorithm for root-finding in order to determine the sample size that generates a power of 80%, for the given alternate hypothesis. We then record the result and a flag for instances in which the algorithm fails to converge (lines 28–30). We used an initial guess of 5, as a larger value would generate convergence issues for larger effect sizes. Finally, we can compute the required sample size for any given minimum detectable effect in lines 32–33. More interesting, however, is the plot of `delta` vs. `ns`, as shown in Figure 3.6. For instance, we see from this curve that a minimum sample size of 237 is required to guarantee a minimum detectable effect of $\delta = 0.1$.   ▷

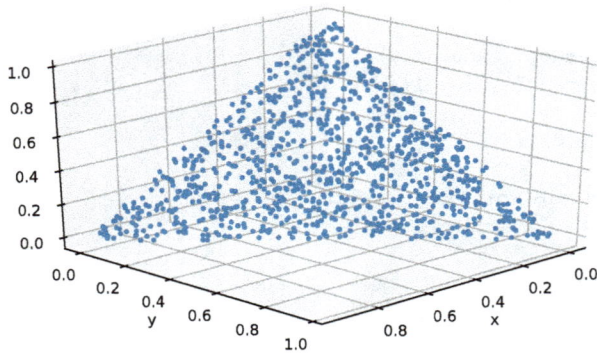

Fig. 3.5: Uniform random sample of the simplex Δ^2.

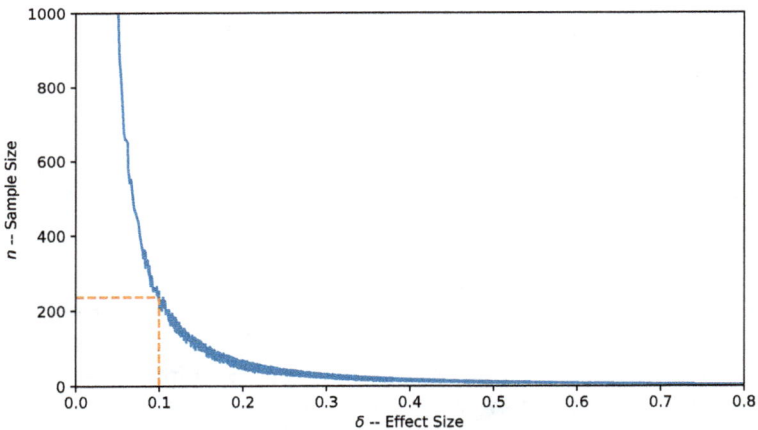

Fig. 3.6: Power function for chi-squared test.

3.6 Tests for Distributions

Often, in practice, we have a set of data. The distribution that generated the data—or even its form—is unknown. The problem of statistical inference therefore extends beyond the simple problem of estimating a set of parameters that match an observed set of data to a parametric family of distributions: we must also infer which family of distributions the data belong to. In this section, we will discuss various tests, collectively known

```
1   p_0 = np.array([0.2, 0.2, 0.2, 0.2, 0.2])
2   d = len(p_0)
3   dof = d - 1
4   alpha = 0.05
5   Q_crit = scipy.stats.chi2.isf(alpha, dof)
6
7   n_samples = 10000
8   X = randomMultiSimplex(d, size=n_samples)
9   i = np.argsort( np.linalg.norm(X-p_0, axis=1) )
10  X = X[i, :]
11  delta = np.linalg.norm(X-p_0, ord=order, axis=1) #Array of deltas
12
13  beta = 0.8
14  ns = np.zeros(n_samples) # array of sample sizes
15  flags = np.zeros(n_samples)
16  for i in range(n_samples):
17      p = X[i, :]
18
19      S1 = np.sum( p*(1 - p) / p_0 )
20      S2 = np.sum( p**2*(1-2*p)/p_0**2 ) + (np.sum( p**2/p_0 ))**2
21      S3 = np.sum( (p-p_0)**2/p_0 )
22      S4 = np.sum( p*(p-p_0)**2/p_0**2 ) - (np.sum( p*(p-p_0)/p_0 ))**2
23      a = S4 / S3
24      la = S3 / a
25      nu = S2 / a**2
26      c = S1 - a * nu
27
28      results = scipy.optimize.root(lambda x: scipy.stats.ncx2.sf(
            Q_crit, nu, x*la, scale=a, loc=c ) - beta, 5)
29      flags[i] = int( not results['success'] ) # Errors
30      ns[i] = int(results['x'])
31
32  MDE = 0.1
33  sample_size = int(max(ns[delta>MDE])) # 237
```

Code Block 3.13: Power as a Function of Effect Size; Sample Size for Minimum Detectable Effect

as goodness-of-fit tests, that seek to address this problem. These tests will each be of the following form.

Definition 3.24. *Given an* IID *sample* $X_1, \ldots, X_n \sim F$ *and a known* CDF F_0, *a* goodness-of-fit test *is any hypothesis test with a null hypothesis of the form*

$$H_0 : F = F_0.$$

3.6.1 Chi-squared Goodness-of-Fit Test

Pearson's chi-squared test (Definition 3.22) can readily be adopted as our first goodness-of-fit test, with the following natural modification.

Definition 3.25. *Let $X_1, \ldots, X_n \sim F$ be an* IID *sample from a continuous distribution F, and let F_0 be a known* CDF*. Fix $k \in \mathbb{Z}_+$, a positive integer (a typical value might be $k = 5$ or $k = 10$). Next, define the boundary points*

$$b_i = F_0^{-1}(i/k),$$

for $i = 1, \ldots, (k-1)$, which partition the real line into k buckets

$$A_i = (-b_{i-1}, b_i],$$

for $i = 1, \ldots, k$, where we have further defined $b_0 = -\infty$ and $b_k = \infty$. Finally, define the counts

$$Y_i = \sum_{j=1}^{n} \mathbb{I}[X_j \in A_i],$$

for $i = 1, \ldots, k$, of the number of occurrences of the random numbers X_i being found within each bucket. Then the chi-squared goodness-of-fit test *consists of the null hypothesis given by Definition 3.24, the test statistic*

$$Q = \sum_{i=1}^{k} \frac{(Y_i - np_0)^2}{np_0},$$

where $p_0 = 1/k$, and rejection region

$$R_\alpha = [\chi^2_{k-1,\alpha}, \infty).$$

Theorem 3.10. *Under the null hyptohesis, the chi-squared goodness-of-fit test is approximately a size-α test for large n. Moreover, under the null hyptohesis, the test statistic $Q \to \chi^2_{k-1}$, as $n \to \infty$.*

Proof. Under the null hypothesis, the sets $A_1, \ldots, A_k$ constitute equal-probability subsets of the real line, each with probability

$$\mathbb{P}(X \in A_i) = \frac{1}{k}.$$

Therefore, under the null hypothesis, $Y \sim \text{Multi}(k, p)$, where p is the equal-probability vector $p = (1/k)\langle 1, \ldots, 1 \rangle \in \mathbb{R}^k$. The result therefore follows from Theorem 3.6. □

Example 3.13. Given a sample of IID data $X_1, \ldots, X_n \sim F$, consider the null hypothesis $H_0 : F = F_0$, where $F_0 = \mathrm{Exp}(\beta_0)$, with $\beta_0 = 7$. We will follow Definition 3.25, using $k = 10$ buckets.

We will run a simulation to see what happens when our null hypothesis is correct. The setup is shown in Code Block 3.14. We define our null-hypothesis value β_0 in line 1, and the (hidden) true value in line 2. The samples are generated in line 4 using the true value, and the boundary points are determined using the null hypothesis distribution. The `bucketCounter` function takes as inputs the samples we generated in line 4 along with the boundary points we generated in line 6, and returns a vector of counts of the number of samples that fall within each of our ten buckets. An example output is shown on line 22. Our goal is then to use Pearson's chi-squared test against the null hypothesis that each bucket has a true probability of 10%.

```
1   beta_0 = 7 # null hypothesis
2   beta = 7   # true value
3   n_samples = 1000
4   samples = np.random.exponential(scale=beta, size=n_samples) #true
        value
5   percentiles = [0.1*i for i in range(1, 10)]
6   bs = scipy.stats.expon.ppf(percentiles, scale=beta_0) # bounaries
7
8   def bucketCounter(samples, boundaries=[0]):
9       boundaries = np.sort(boundaries) #Make sure they are ordered
10      n_buckets = len(boundaries) + 1
11      n_samples = len(samples)
12      counts = np.zeros(n_buckets)
13      n_counted = 0
14
15      for i, b in enumerate(boundaries):
16          counts[i] = (samples <= b).sum() - n_counted
17          n_counted += counts[i]
18
19      counts[i+1] = n_samples - n_counted
20      return counts
21      # example output:
22      # [ 96., 91., 102., 102., 121., 100., 85., 109., 86., 108.]
```

Code Block 3.14: Simulated draws from Exp(7) and the bucket counter

Next, let us repeat our simulation many times and compute the frequency at which we would (falsely) reject our (true) null hypothesis. This is shown in Code Block 3.15. Upon running the simulation 10,000 times,

```
1   beta_0 = 7 # null hypothesis
2   beta = 7   # true value
3   n_samples = 1000
4   n_trials = 10000
5   percentiles = [0.1*i for i in range(1, 10)]
6   bs = scipy.stats.expon.ppf(percentiles, scale=beta_0) # bounaries
7   alpha = 0.05
8   dof = 9
9   Q_crit = scipy.stats.chi2.isf(alpha, dof)
10  E_0 = n_samples / (dof+1) * np.ones(dof+1)
11  rejections = np.zeros(n_trials)
12
13  for i in range(n_trials):
14      samples = np.random.exponential(scale=beta, size=n_samples)
15      x = bucketCounter(samples, boundaries=bs)
16      Q = np.sum( (x - E_0)**2 / E_0 )
17      rejections[i] = int( Q > Q_crit )
18
19  print ('Observed Rejection Rate', np.sum(rejections) / n_trials)
        #0.0496
```

Code Block 3.15: Simulation of Chi-squared goodness of fit test

we see that the null hypothesis was rejected 4.96% of the time, in agreement with our value of $\alpha = 0.05$.   ▷

3.7 Analysis of Variance

We conclude this chapter with an introduction to the analysis of variance (ANOVA). Though our initial theoretical groundwork is laid from the perspective of hypothesis testing, we will later find that the theory has much broader implications that span experimentation, regression theory, and generalized linear models. In the oneway ANOVA framework, observations are classified into groups, or *treatments*, along with the following modeling assumptions.

Definition 3.26. *A oneway ANOVA model identifies each datum with one of m groups, called* treatments, *such that observations satisfy the model*

$$Y_{ij} = \theta_i + \epsilon_{ij}, \qquad i = 1, \ldots, m; \qquad j = 1, \ldots, r_i, \qquad (3.61)$$

according to the following ANOVA assumptions

1. *The* errors *have zero expected value and finite variance:* $\mathbb{E}[\epsilon_{ij}] = 0$ *and* $\mathbb{V}(\epsilon_{ij}) = \sigma_i^2 < \infty$, *for all* i, j;

2. the errors are uncorrelated $\mathrm{COV}(\epsilon_{ij}, \epsilon_{kl}) = 0$ whenever $i \neq k$ and $j \neq l$;
3. the errors ϵ_{ij} are independent and normally distributed;
4. the variance is independent of treatment, i.e., $\sigma_i^2 = \sigma^2$, for all i (also known as homoscedasticity).

Note 3.24. Sometimes the model given by Equation (3.61) is cast as

$$Y_{ij} = \mu + \tau_i + \epsilon_{ij},$$

where μ is a sort of *overall mean*, and τ_i is the *treatment effect*. This prescription, however, suffers identifiability issues, as the parameters cannot be uniquely solved for without further restraint. Nevertheless, this is the form will will predominantly use later during our discussion on experimentation.
▷

The assumption of independent and normally distributed error terms is important for defining confidence intervals. Further note that we do not require there to be an equal amount of data in each treatment. Though of limited interest in modern applications, the classic ANOVA test is given by the following.

Definition 3.27. *The* (classic) ANOVA null hypothesis *is the hypothesis*

$$H_0: \qquad \theta_1 = \theta_2 = \cdots = \theta_m. \tag{3.62}$$

In this section, we shall consider the classic ANOVA null hypothesis, which states that there is no difference between any of the treatments. The alternative hypothesis is therefore that there is *some* difference between the treatments, without regard as to where that difference may lie. In practice, however, one is typically more interested in estimating the performance of various treatments to determine which is superior. We shall discuss this in our next chapter when treating the theory of experimentation.

The main idea behind analysis of variance is to partition the observed variance, known as the sum of squares, into two components: one representing the *within*-group and one representing the *between*-groups sum of squares. This is captured in the following theorem.

Theorem 3.11 (Partitioning Theorem). *For any set of numbers y_{ij}, where $i = 1, \ldots, m$ and $j = 1, \ldots, r_i$,*

$$\sum_{i=1}^{m} \sum_{j=1}^{r_i} (y_{ij} - \bar{\bar{y}})^2 = \sum_{i=1}^{m} r_i \, (\bar{y}_{i\cdot} - \bar{\bar{y}})^2 + \sum_{i=1}^{m} \sum_{j=1}^{r_i} (y_{ij} - \bar{y}_{i\cdot})^2 ; \tag{3.63}$$

where we have followed the dot notation *for brevity:*

$$y_{i\cdot} = \sum_{j=1}^{r_i} y_{ij}, \qquad \bar{y}_{i\cdot} = \frac{1}{r_i} \sum_{j=1}^{r_i} y_{ij}, \qquad \bar{\bar{y}} = \frac{1}{n} \sum_{i=1}^{m} \sum_{j=1}^{r_i} y_{ij}, \tag{3.64}$$

where $n = \sum_{i=1}^{m} r_i$.

Proof. The proof is straightforward. Begin by expanding the left-hand side:

$$\sum_{i=1}^{m}\sum_{j=1}^{r_i}(y_{ij}-\bar{\bar{y}})^2 = \sum_{i=1}^{m}\sum_{j=1}^{r_i}[(y_{ij}-\bar{y}_{i\cdot})+(\bar{y}_{i\cdot}-\bar{\bar{y}})]^2 .$$

The result follows by expanding the right-hand side and recognizing that the cross terms vanish, which is verified in Exercise 3.6.   □

Theorem 3.11 relates the total sum of square to the between-group and within-group sums of squares, as defined in the following.

Definition 3.28. *For any set of numbers y_{ij}, where $i = 1,\ldots,m$ and $j = 1,\ldots,r_i$, the* between-group sum of squares *(SSB), the* within group sum of squares *(SSW), and the* total sum of squares *(SST) are defined by*

$$SSB = \sum_{i=1}^{m} r_i\,(\bar{y}_{i\cdot}-\bar{\bar{y}})^2 , \tag{3.65}$$

$$SSW = \sum_{i=1}^{m}\sum_{j=1}^{r_i}(y_{ij}-\bar{y}_{i\cdot})^2 , \tag{3.66}$$

$$SST = \sum_{i=1}^{m}\sum_{j=1}^{r_i}(y_{ij}-\bar{\bar{y}})^2 . \tag{3.67}$$

Moreover, the between-group (MSB) *and* within-group (MSW) *mean squares are defined by the relations*

$$MSB = \frac{SSB}{m-1} \quad and \quad MSW = \frac{SSW}{n-m}, \tag{3.68}$$

respectively.

Theorem 3.11 therefore states that the total sum of squares is equal to the sum of the between-group and within-group sums of squares.

The total sum of squares describes the overall variation of the data, without regard to classification into the various treatment groups. The between-group sum of squares, SSB, captures the variation by comparing the various groups; notice that the mean of each group is compared to the overall mean, and the sum is only over the m separate groups. Finally, the within-group sum of squares yields the average variation within each group; i.e., each data point within each group is compared to that group's mean,. Thus, the within-group sum of squares measures variation with respect to random error, whereas the between-group sum of squares measures variation due to treatment differences.

We next proceed to determine a test statistic for the ANOVA null hypothesis. We begin by considering two lemmas.

Lemma 3.8. *Under the ANOVA assumptions of Definition 3.26, the following scaled within-group sum of squares is a chi-squared random variable with $(n-m)$ degrees of freedom:*

$$\frac{1}{\sigma^2} \sum_{i=1}^{m} \sum_{j=1}^{r_i} \left(Y_{ij} - \bar{Y}_{i\cdot}\right)^2 \sim \chi^2_{n-m}.$$

Proof. Following the ANOVA assumptions, we have $Y_{ij} \sim \mathrm{N}(\theta_i, \sigma^2)$. Therefore, for each $i = 1, \ldots, m$, the quantities

$$\frac{1}{\sigma^2} \sum_{j=1}^{r_i} \left(Y_{ij} - \bar{Y}_{i\cdot}\right)^2 \sim \chi^2_{r_i-1}$$

are independent chi-square random variables. Due to their independence, their sum is also a chi-squared random variable with $\sum_{i=1}^{m}(r_i - 1) = n - m$ degrees of freedom, thus proving the result.   □

Lemma 3.9. *Under the ANOVA assumptions of Definition 3.26 and the ANOVA null hypothesis of Definition 3.27, the following scaled between-group and total sums of squares are chi-squared random variables with $(m-1)$ and $(n-1)$ degrees of freedom, respectively.*

$$\frac{1}{\sigma^2} \sum_{i=1}^{m} r_i \left(\bar{Y}_{i\cdot} - \bar{\bar{Y}}\right)^2 \sim \chi^2_{m-1} \qquad and \qquad \frac{1}{\sigma^2} \sum_{i=1}^{m} \sum_{j=1}^{r_i} \left(Y_{ij} - \bar{\bar{Y}}\right)^2 \sim \chi^2_{n-1}.$$

Proof. Under the ANOVA null hypothesis, we have $\theta_i = \theta_j$, for all i, j. Thus, each observation is an independent draw from

$$Y_{ij} \sim \mathrm{N}\left(\theta, \sigma^2\right).$$

We therefore recognize the total sum of squares as a scaled version of the sample variance. From Theorem 2.6, it therefore follows that $(1/\sigma^2)SST \sim \chi^2_{n-1}$. This proves the second relation.

For the first relation, we instead consider the sample means $\bar{Y}_{i\cdot}$. Under the null hypothesis, the sample mean is distributed according to

$$\bar{Y}_{i\cdot} \sim \mathrm{N}\left(\theta, \frac{\sigma^2}{r_i}\right).$$

We recognize the between-group sum of squares as a scaled version of the sample variance *of the sample means*. It therefore follows (again from Theorem 2.6) that $(1/\sigma^2)SSB \sim \chi^2_{m-1}$.   □

In light of Lemmas 3.8 and 3.9, the partitioning theorem, when scaled by σ^2, is actually a partitioning of a χ^2_{n-1} random variable into the sum of two independent chi-squared random variables χ^2_{m-1} and χ^2_{n-m}. Moreover, this decomposition leads us to the following theorem, which yields a test-statistic for the ANOVA null hypothesis.

Theorem 3.12. *Under the ANOVA assumptions of Definition 3.26 and the ANOVA null hypothesis of Definition 3.27, the statistic*

$$F = \frac{MSB}{MSW} = \frac{\dfrac{1}{m-1} \displaystyle\sum_{i=1}^{m} r_i \left(\bar{y}_{i\cdot} - \bar{\bar{y}}\right)^2}{\dfrac{1}{n-m} \displaystyle\sum_{i=1}^{m} \sum_{j=1}^{r_i} \left(y_{ij} - \bar{y}_{i\cdot}\right)^2}, \tag{3.69}$$

known as the F *statistic, has an F distribution with* $m-1$ *and* $n-m$ *degrees of freedom.*

Proof. This follows immediately from Lemmas 3.8 and 3.9. □

These definitions are summarized neatly in Table 3.4. In general, computation of Equation (3.69) is generally done by completion of such an ANOVA table.

	Degrees of freedom	Sum of squares	Mean square
Between groups	$m-1$	$SSB = \displaystyle\sum_{i=1}^{m} r_i(\bar{y}_{i\cdot} - \bar{\bar{y}})^2$	$MSB = \dfrac{SSB}{m-1}$
Within groups	$n-m$	$SSW = \displaystyle\sum_{i=1}^{m}\sum_{j=1}^{r_i}(y_{ij} - \bar{y}_{i\cdot})^2$	$MSW = \dfrac{SSW}{n-m}$
Total	$n-1$	$SST = \displaystyle\sum_{i=1}^{m}\sum_{j=1}^{r_i}(y_{ij} - \bar{\bar{y}})^2$	$F = MSB/MSW$

Table 3.4: ANOVA table.

Note that as a consequence of Theorem 3.12, we should reject the ANOVA null hypothesis whenever the F statistic exceeds the value

$$F > F_{m-1,n-m}^{\alpha},$$

where α is our significance level and $F_{m-1,n-m}^{\alpha}$ is the unique real number such that $\mathbb{P}(X > F_{m-1,n-m}^{\alpha}) = \alpha$ for $X \sim F_{m-1,n-m}$. Next, we will test our intuition versus solid statistical theory with a practical example.

Example 3.14. Consider data from five distinct treatment groups, as shown in Table 3.5. Each group consists of exactly eight data points, so that $r_i = 8$, for $i = 1, \ldots, 5$. The mean of each treatment group is given in the final column, and the overall mean of all the data is 97.175. What conclusions can we draw from these data? Is group 4 the clear and decisive victor, with groups 2 and 5 lagging sorely behind? Or is the variation in these means due to random chance?

Group	1	2	3	4	5	6	7	8	Mean
1	103	98	103	112	74	106	79	85	95
2	71	73	102	73	118	118	81	105	92.625
3	96	87	110	123	84	96	82	109	98.375
4	103	91	119	111	147	103	112	95	110.125
5	70	89	107	106	92	96	84	74	89.75

Table 3.5: Data for Example 3.14; overall mean 97.175

We begin by stating our classic ANOVA null hypothesis, that there is no difference between treatment groups:

$$H_0 : \theta_1 = \theta_2 = \theta_3 = \theta_4 = \theta_5.$$

The F statistic is computed in Table 3.6 to be $F = 1.914$. At a 5% sig-

	Degrees of freedom	Sum of squares	Mean square
Between groups	4	$SSB = 1997.65$	$MSB = 499.4125$
Within groups	35	$SSW = 9132.125$	$MSW = 260.917857$
Total	39	$SST = 11129.775$	$F = 1.914$

Table 3.6: ANOVA table for Example 3.14.

nificance level, $F_{4,35}^{0.05} = 2.64$. Therefore, we should *not* reject the null hypothesis. The data shown in Table 3.5 could have resulted from random chance. Indeed, the data *were* constructed from random chance.   Code Block 3.16 shows the Python code used to generate the data and perform the hypothesis test for this example. Lines 2–6 will load the data for the current example; to generate new examples, however, simply comment out these five lines of code.   ▷

Problems

3.1. Repeat Example 3.5 using the Z_n statistic on line 17 instead of the T_n statistic; i.e., with the sample variance replaced with the true variance. Given this modification, what is the expected rejection rate for each case (z_crit and t_crit)? Run the simulation and compare the observed rejection rates with your prediction.

3.2. Verify Equation (3.18).

3.3. Prove Corollary 3.2. *Hint*: How must Equation (3.21) be modified before proceeding?

```
1  X = np.random.normal(loc=100, scale=15, size=[5,8]).astype(int)
2  X = np.array([[103, 98, 103, 112, 74, 106, 79, 85],
3         [ 71, 73, 102, 73, 118, 118, 81, 105],
4         [ 96, 87, 110, 123, 84, 96, 82, 109],
5         [103, 91, 119, 111, 147, 103, 112, 95],
6         [ 70, 89, 107, 106, 92, 96, 84, 74]])
7
8  X_means = X.mean(axis=1) #mean for each group
9  X_bar = X.mean() # overall mean
10 r_i = 8
11
12 SSB = r_i * np.sum((X_means - X_bar)**2)
13 SSW = np.sum( (X - X_means.reshape((5,1)) * np.ones(8))**2 )
14 SST = np.sum( (X - X_bar)**2 )
15
16 MSB = SSB / 4
17 MSW = SSW / 35
18
19 F = MSB / MSW # 1.91406
20 alpha = 0.05
21 scipy.stats.f.isf(0.05, 4, 35) # 2.641465
```

Code Block 3.16: Simulation of an ANOVA test

3.4. Find the bug in Code Block 3.8. Validate your answer by reproducing the code, with the bug fixed, and showing that the rejection rate (even with $n = 10$) is close to 5%.

3.5. Use Theorem 2.9 to prove that the variance of a noncentral chi-squared random variable statisfies Equation (3.43).

3.6. Complete the proof to Theorem 3.11 by showing that

$$\sum_{i=1}^{m} \sum_{j=1}^{r_i} (y_{ij} - \bar{y}_{i\cdot}) (\bar{y}_{i\cdot} - \bar{\bar{y}}) = 0.$$

3.7. Explain how the ANOVA null hypothesis is used in Lemma 3.9.

3.8. Use the code from Code Block 3.16 to build a simulation that will generate 10,000 data sets and record the fraction of instances for which the ANOVA null hypothesis was falsely rejected. *Bonus*: Update the code block to use parameters m, r_i, and $n = m*r_i$ to be specified prior to the for-loop. Repeat the simulation using other values of sample size.

4

It's Alive!

Whereas the theory of hypothesis testing focuses on statistical inference, i.e., what one may infer about a population from a random sample, the theory of experimentation focuses on elements of *design*, i.e., how one may go about constructing an experiment that will produce statistically significant results. For additional references, we refer the reader to Berger, *et al.* [2018], Kohavi, Tang, and Xu [2020], Rosenbaum [2002], and Selvamuthu and Das [2018].

4.1 Principles of the Design of Experiments

In this section, we lay out the basic principles and definitions of experimental design and present some of the mathematical framework for our subsequent discussions. We approach the subject from the perspective of *product engineering*, where various treatment groups correspond to various versions of a given product (typically a digital product, such as a website or an app). Historically, however, much of the theory was developed in the context of *clinical trials*, designed to measure the efficacy of a given drug treatment on a given disease. These two applications are practically analogous as drugs are, after all, chemical products. A slight variation is sometimes also used, when the purpose of an experiment is to study the effect of a given covariate on a product's efficacy; e.g., the effect of age range on a shampoo's ability to treat split ends. Such an application departs from the product engineering and clinical trial approach—where treatments correspond to different versionings of a product—in that treatments correspond to different properties, i.e., covariates, of the populations being tested.

Experiments

We begin by laying out the ingredients that define an experiment. We will then proceed to discuss each in turn.

Definition 4.1. *An* experiment *is a six-tuple* $(\mathcal{T}, \mathcal{S}, \mathcal{P}, x, Z, R)$ *consisting of*

- *a finite set of m possible* treatments, $\mathcal{T}$,
- *a finite set of s known* covariates *or* strata $\mathcal{C}$ *of the population of interest,*
- *a population* $\mathcal{P}$, *consisting of a set of n experimental units,*
- *a classification function* $x : \mathcal{P} \to \mathcal{C}$ *that uniquely determines the stratum for each experimental unit,*
- *an* assigment mechanicsm $Z : \mathcal{P} \to \mathcal{T}$ *that assigns a treatment to each experimental unit, and*
- *a response* $R : \mathcal{T} \times \mathcal{P} \to \mathbb{R}$, *which records the experimental results of applying a particular treatment to a particular unit.*

The purpose of defining an experiment is to study the effect of the factors, broken up into individual treatments, on a response variable, while controlling for one or more covariates. The *treatments* are the individual variations of our product. Often, the set of treatments $\mathcal{T}$ is decomposed into the Cartesian product of one or more *factors*; i.e., $\mathcal{T} = \mathcal{F}_1 \times \cdots \times \mathcal{F}_k$, where $\mathcal{F}_i$ is the set of *levels* of factor i. When such a decomposition is deployed, we call the experiment a *k-factor experiment*. For example, suppose we are studying the effect of age and gender on a particular drug's efficacy. Then the factors would be age and gender, and a reasonable set of levels would be

$$\mathcal{F}_{\text{age}} = \{18\text{--}24, 25\text{--}34, 35\text{--}44, 45\text{--}54, 55\text{-}64, 65+, \text{unknown}\},$$
$$\mathcal{F}_{\text{gender}} = \{\text{male}, \text{female}, \text{unknown}\}.$$

Here, we assume an age restriction that participants must be at least eighteen years of age, and we further allow for the possibility that some participants may choose not to disclose their age or gender. The set of possible treatment groups is therefore given by $\mathcal{T} = \mathcal{F}_{\text{age}} \times \mathcal{F}_{\text{gender}}$, which is a set of size 21.

For a more product-centric example, suppose we are testing the effect of font and background color for a website or app. The result would be a two-factor experiment, comprised of a set of fonts and a set of background colors.

The purpose of defining the treatment set as a Cartesian product, for which all possible combinations are considered, is to capture *interaction effects*. For instance, New Times Roman might be the superior choice when considering variations in fonts alone, and a blue background might be the superior color choice when considering variations of color alone, but there might be something charmingly irresistible about the enchanting *combination* of Century Gothic with a chartreuse background. Only by testing all possible combinations of font plus background together can one discover such synergies.

A *factorial design* is any multi-factor design that involves various combinations of levels from different factors. A full (or complete) factorial design involves taking all combinations of levels from each factor, as described in the Cartesian product decomposition, above.

A special case of this is that of the so-called two-level factorial design[1]. Here, each level is binary, and so the total number of possible treatments for a k-factor experiment is given by 2^k, as each factor is comprised of two levels. The special case of a one-factor design with two levels is referred to as an *A/B test*. The levels of a binary factory are customarily referred to as the *control* and the *treatment*. In clinical trials, the control typically corresponds to a placebo whereas the treatment would correspond to the drug. In product design, the control could represent the current version of the product, whereas the treatment would represent a proposed change.

With our set of treatments in hand, we next turn to *experimental units*. A *unit* is an opportunity to withhold or apply any of the various treatments. The individual experimental units are drawn from our population $\mathcal{P}$.

Experimental units may further be described by a collection of one or more *covariates*, or *nuisance factors*. The only true difference between a covariate or nuisance factor (and their corresponding strata) and the experimental factors (and their corresponding treatments) is one of intent: it is the causal relation between the experimental factors and the response we are seeking to uncover. In the context of product engineering, however, there is a more natural distinction: the experimental factors typically represent variations in the product, whereas nuisance factors represent variations in certain features of the population.

To further complicate matters, there are both *known* and *unknown* nuisance factors. To control for unknown nuisance factors, i.e., those factors which *may* exist, but over which we have no direct insight or control, we employ a design principle known as *randomization*. Randomization is achieved through the choice of the assignment mechanism. The idea here is that by randomly assigning treatments to units, we will, on average, cancel out the effect of any unknown nuisance factor, thereby making the results of the experiment more robust. Randomization essentially breaks every causal link between unknown factors and the experiment's results.

On the flip side of the coin, the known nuisance factors are the factors we expressly define as *covariates*, which are characteristics of each unit which may be measured prior to assignment or experimentation. As covariates are properties of the units themselves, they cannot be influenced by the choice of treatment. For example, in a two-factor experiment designed to test various combinations of font and background color, age and gender might be considered covariates. Here, since we have direct access of age and gender prior to our assignment of treatment, we can design an assignment mechanism to balance out the effects of age and gender across the various

---

[1] sometimes referred to as 2^k-factorial design.

treatments, so that, for example, we don't accidentally assign one particular combination predominantly to one particular stratum. Such a balancing act is known as *blocking*. Here, the various covariate strata are thought of as *blocks*, and the idea is to balance the blocks over the given set of treatments.

Another principle of design is one of *replication*. By replication, we mean that for each treatment and stratum, the experiment should be repeated across several units. In this way, we can form a measure of statistical error. If, on the other hand, each stratum contains only one unit for each treatment, assigned randomly, then the experiment is referred to as *paired randomized experiment*.

Blocking, randomization, and replication all feed into the concept of an assignment mechanism. Typically, assignment mechanisms are not functional in nature but rather stochastic, in the sense that they incorporate random numbers to properly randomize treatments across experimental units. By placing various constraints on this random process, we can further achieve a blocked-design, where, though random, the various strata are balanced over the various treatment assignments.

There are two classic assignment mechanisms that earn special consideration in our studies. The first is called a *completely randomized design*. A completely randomized experiment has no covariates, so we can say that each unit belongs to the same stratum. To carry out a completely randomized experiment, the number of replicants for each treatment group is prescribed, and the available units are assigned randomly to each treatment group. It is important to note that the *order* that the experiments are to be carried out is also randomized. The second classic assignment mechanism is called a *randomized block design*. This is similar to a completely randomized design, except for the fact that there is more than one stratum. The experimental units are divided into their respective strata, and a completely randomized design is then implemented within each stratum. In other words, each stratum is constrained to posses an equal number of replicants for each treatment. We will explore these types of design over the pursuing pages.

Finally, the experiment is executed and a response is measured. Note that the response mapping is not completely known; we only know the response for a single treatment for each experimental unit, under a particular test condition. We cannot say what the response for a particular unit *would have been* had we applied a different treatment, or even had we applied the same treatment under different operating conditions (e.g., a different time of day or day of week or with a different operator). Thus, the response is simply a record of the results of the experiment and not a full functional description; it is censored by the fact that we live in but a single universe and have no access to the results that our counterparts in other universes may have achieved using a different seed for their random assignment mechanism.

For example, suppose that experimental unit i is assigned to treatment Z_i by our random assignment mechanism. We will record the response $R_{Z_i i}$. The other *potential outcomes* of the experiment,

$$R_{ji}, \qquad \text{for } j \neq Z_i,$$

are referred to as *counterfactuals*, as they do not factually exist in reality and, as such, cannot be observed or measured. We are capable of observing only the outcome $R_{Z_i i}$ obtained by applying our chosen treatment Z_i to experimental unit i, and not the other worldly outcomes observed by our brothers and sisters in parallel universes, whose random assignment mechanisms led them to choose other treatments $j \neq Z_i$ for unit i.

Analysis

Of course, we don't just run experiments for the fun of it. The purpose of running an experiment is to learn something. While Definition 4.1 lays out the definition of an experiment, this is only half of the picture. An experiment must be coupled with a statistical analysis which supports a conclusion. Since the conclusion is itself the goal of experimentation, this analysis must also be taken into account during the design phase of the experiment. For example, if one is attempting to measure the effect of polar-magnetic vortex flares on the choice of sock color, it would be most unfortunate if an experiment was devised that measured the blood pressure of the participants and not the color of their socks.

An *experimental analysis* is an experiment coupled with a hypothesis test, such that the test statistic is a function of the experimental data. The null hypothesis we consider is typically *the hypothesis of no effect*, which states that the choice of treatment had no effect on the outcome. Given the hypothesis of no effect, we then derive an appropriate test statistic. The null hypothesis of the experiment and the test statistic must therefore be considered as part of the experimental design, to ensure that the experiment produces data that can be used to confirm or reject the hypothesis of the experiment, and to ensure that enough data is collected to achieve a particular significance and power.

Direct and Index Notation

We will encounter two types of notation during our discussion on experimentation, one based directly on the enumeration of the individual units, and one based on the index of the assignment of units to various treatments and of the classification of units into covariates. We refer to these two systems as *direct notation* and *index notation* respectively.

In direct notation, each unit of the population is given a unique index i. This serves as an ID for individual units and is independent on the eventual assignment of units into treatment groups. Direct notation goes hand-in-hand with our Definition 4.1 and our discussion thus far. Unit i receives treatment Z_i, as determined by our random assignment mechanism, resulting in observation $R_{Z_i i}$.

In index notation, we reindex all of the units based on their assigned treatment group, their covariates, and a final enumeration to distinguish replicants within a given treatment group and stratum. This notation is necessarily fluid to allow for a variable number of experimental factors and nuisance factors. To achieve this, we use the following ordering for indices:

1. *treatment group*—the innermost index is reserved for the assigned treatment group. For a one-factor experiment, we use index i; for a two-factor experiment, we use indices (i, j), and so-on.
2. *strata*—the next index (or group of indices) is reserved for the covariate classification. Typically this will constitute a single index, though it is possible to use multiple indices to study interaction effects among the nuisance factors.
3. *replicants*—the final, outermost index is reserved for enumeration of the replicants within an individual treatment group and stratum.

For example, suppose a two-factor experiment is performed using replication and a single nuisance factor. The quantity Y_{ijkl} would refer to the result obtained from the lth replicant in treatment group (i, j) in stratum k. If the experiment is conducted *without replication*, so that only a single experimental unit for each strata is given each treatment, then we can dispense with the final index l and simply write Y_{ijk} for the result.

Similarly, consider the quantity Y_{ij}. This notation would be consistent with any of the possible experimental structures:

- A one-factor completely randomized experiment (no covariates) with replication. In this case, the quantity Y_{ij} corresponds to the result obtained from the jth replicant from treatment group i.
- A two-factor completely randomized experiment with no replication. (This constitutes a poor design choice, though it is consistent with our index notation.) Here, quantity Y_{ij} represents the result of applying treatment (i, j) to the sole replicant of that treatment group.
- A one-factor experiment with one covariate and no replication. Here, quantity Y_{ij} represents the result of applying treatment i to the sole replicant of stratum j.

Though the indices will take on different meanings in different contexts, we will follow the golden rule that: treatment groups are enumerated first, then the covariate strata, then the replicants within each treatment-group and stratum.

Each set of notation has its own advantages. The direct notation is conducive to discussions on counterfactuals, treatment effect, and causality. This is especially convenient for the analysis of observational studies.

Similarly, index notation is conducive to analysis of controlled experiments, especially experiments with many-level factors, and the analysis of variance. In particular, it is not possible to have a discussion about counterfactuals using index notation, as the notation itself enumerates the units *within their respective treatment groups*, i.e., it enumerates the units *after* the units have been assigned to their respective treatments.

When we are referring to particular units, it is possible that any particular index could exceed the value of 9, especially when using the direct notation. For example, the response of applying treatment 3 to unit 171 would be R_{3171}. In order to avoid such ambiguity, we will separate the index numbers with commas whenever one of them is not a single digit; for example, $R_{3,171}$ for $i = 3$ and $j = 171$ or $R_{3,17,1}$ for $i = 3$ and $j = 17$ and $k = 1$. The only rule is that whenever commas are deployed, we shall use commas throughout: we would write $R_{1,7,19}$ for $i = 1$, $j = 7$, and $k = 19$, and not $R_{17,19}$, which could be confused with $i = 17$ and $j = 19$ or $i = 17$ and $j = 1$ and $k = 9$.

In the context of database design, direct notation directly uses the "identity key" of each unit, whereas the index notation uses a multi-column key that represents each units assigned treatment group, properties, and replication number.

4.2 A/B Tests

In this section, we go into depth to explore the design and analysis of running A/B tests. For a practical guide that explores pitfalls and nuisances of A/B testing in practice, we refer the reader to the indispensable volume Kohavi, Tang, and Xu [2020]. Recall that an A/B test is a single-factor experiment with two levels, which are referred to as a control ($i = 0$) and a treatment ($i = 1$). Our attention is thus presently restricted to single factor experiments with two levels. In the following sections, we will extend this to single and multi-factor experiments with an arbitrary number of levels. We will conclude the chapter with a higher-dimensional generalization of A/B tests to more than a single factor, during our discussion of two-level factorial design.

4.2.1 Design

We begin by discussing the design and analysis of A/B tests with no co-variates; that is, we do not measure any characteristic of the units before the random assignment of the unit into a control or treatment group. We will suppose that unit i is assigned to treatment $Z_i \in \{0, 1\}$, resulting in observation $R_{Z_i,i}$. We begin with a discussion of the assignment mechanism.

Assignment

We will consider two possible assignment mechanisms.

Definition 4.2. *A* Bernoulli randomized experiment *is a randomized experiment in which each unit receives the treatment with the same probability; i.e.,*

$$\mathbb{P}\left[Z_i = 1\right] = \pi,$$

for all $i = 1, \ldots, n$, *where* $0 < \pi < 1$.

Typically, we use $\pi = 1/2$, so that each experimental unit is equally likely to be assigned to the treatment or control. Given a population of n-units, there are 2^n possible assignments that can be achieved, each occurring with equal probability. Since each unit has the same probability of being assigned to the treatment, one shortcoming of this approach is the possibility that *all* units receive the same treatment, though this possibility becomes vanishingly small as the number of units becomes large. To remedy this, however, we can choose a different assignment mechanism.

Definition 4.3. *A* completely randomized A/B test *is an A/B test for which the total number of replicants in the treatment group has a fixed value* r; *i.e., an experiment in which all assignments* Z_i *satisfying the constraint*

$$\sum_{i=1}^{n} Z_i = r$$

are equally likely.

Typically, one requires an even number of units n and fixes the number of treatments as $r = n/2$. In a completely randomized experiment, there are $\binom{n}{r}$ possible assignments, each being equally likely to occur; i.e., we randomly select a total of r units to be assigned to the treatment group.

A completely randomized experiment therefore restricts the number of possible assignments of the Bernoulli randomized experiment. For example, in an experiment with 10 available units, a Bernoulli randomized experiment will have $2^{10} = 1024$ possible assignments, whereas a completely randomized experiment will have $\binom{10}{5} = 252$ possible assignments.

There are two standard methods for carrying out a completely randomized experiment. The first method is to simply shuffle the n units and assign the first $(n - r)$ units to the control and the final r units to the treatment. An example of this is shown in Table 4.1. In this example, units 0, 2, 3, 5, and 6 are assigned to the control group, whereas units 1, 3, 4, 7, 8, and 9 are assigned to the treatment. Once the assignment has been made, we can now introduce index notation, which refers to each unit by its treatment-group and replicant number. For example, unit 5 is the fourth replicant of the control group, so unit 5 will correspond to Y_{03}. Finally, if the experiments

Unit	0	1	2	3	4	5	6	7	8	9
Shuffled	2	6	0	5	3	1	9	4	8	7
Assignment	0	0	0	0	0	1	1	1	1	1
Index Notation	Y_{00}	Y_{01}	Y_{02}	Y_{03}	Y_{04}	Y_{10}	Y_{11}	Y_{12}	Y_{13}	Y_{14}

Table 4.1: Assignment for a completely randomized experiment

are to be performed sequentially, we would typically shuffle the final row in order to perform the experiments in a random order; i.e., we would not want to perform all the control-group experiments first followed by all the treatment-group experiments second. As long as the *original* unit keys are random (i.e., the units were not originally ordered in any way), we could alternatively perform the experiments in the original order: Y_{02}, Y_{10}, Y_{00}, Y_{04}, Y_{12}, Y_{03}, Y_{01}, Y_{14}, Y_{13}, Y_{11}.

Alternatively, one can achieve a completely randomized design by using *one-at-a-time simple random sampling*. This assignment mechanism has the same effect as shuffling, but works on each unit *sequentially*. The idea is that each unit is assigned to a treatment group with a probability that is proportional to the number of "spots" left open in that treatment group. More formally, we have the following.

Definition 4.4. One-at-a-time simple random sampling *is a method for assigning n objects into m groups, such that group z has exactly n_i assignments, with $\sum_{z=0}^{m-1} n_i = n$, and such that the probability of allocating object i to group z is proportional to the number of remaining spaces in the group:*

$$\mathbb{P}[Z_i = z | Z_0, \ldots, Z_{i-1}] = \frac{n_z - \sum_{j=0}^{i-1} \mathbb{I}[Z_j = z]}{n - i}, \tag{4.1}$$

for $i = 0, \ldots, n - 1$.

It is straightforward to show that Equation (4.1) is a normalized probability mass function for each $i = 0, \ldots, n - 1$. For the special case of a completely randomized A/B test, we have the probability of assigning unit i to the treatment group is given by

$$\mathbb{P}[Z_i = 1 | Z_0, \ldots, Z_{i-1}] = \frac{r - \sum_{j=0}^{i-1} Z_j}{n - i}, \tag{4.2}$$

for $i = 0, \ldots, n - 1$.

Causal Effect

We are often interested in understanding the *causal effect* of a treatment, meaning, intuitively, the net effect of applying the treatment instead of the control. Before defining this, note that for the special case of an A/B test, we may write the *observed outcome* on unit i as

$$R_i = Z_i R_{1i} + (1 - Z_i) R_{0i}. \tag{4.3}$$

We call R_{0i} and R_{1i} the *potential outcomes* for unit i, as one of these two outcomes will come to pass. The observed outcome depends on the result of the random assignment: if unit i is assigned to the treatment $Z_i = 1$, then we will observe the outcome R_{1i}. If, on the other hand, unit i is assigned to the control $Z_i = 0$, then we will observe the outcome R_{0i}. The above formula accomplishes precisely this. It says that $R_i = R_{0i}$ if unit i is assigned to the control $Z_i = 0$ and that $R_i = R_{1i}$ if unit i is assigned to the treatment $Z_i = 1$. The potential outcome that is not observed is referred to as the *counterfactual* for unit i; i.e., we may write

$$C_i = (1 - Z_i) R_{1i} + Z_i R_{0i} \tag{4.4}$$

for the *counterfactual*. The counterfactual represents what the outcome of the experiment *would have been* for unit i, *had we made* the alternate assignment $1 - Z_i$ instead of Z_i. Counterfactuals exist only in our minds and in parallel universes. This handicap, however, does not prevent us from making inferences about them.

Definition 4.5. *The* causal effect *of applying the treatment over the control in an A/B test on unit i is the difference between the two potential outcomes*

$$\Delta_i = R_{1i} - R_{0i}. \tag{4.5}$$

Like counterfactuals, the causal effect *cannot* be observed *for any* unit. This follows since, for any unit i, we will *either* observe potential outcome R_{0i} *or* potential outcome R_{1i}, as we can *only* apply *either* the treatment or the control to unit i, but not both.

Though unobserved, we can estimate the average causal effect using

$$\hat{\Delta} = \frac{1}{r} \sum_{i=0}^{n-1} Z_i R_{1i} - \frac{1}{n-r} \sum_{i=0}^{n-1} (1 - Z_i) R_{0i}. \tag{4.6}$$

That is, we average the result of the treatment and control groups separately and take the difference. Notice that this estimate relies *only* on the observed outcomes, and not on any of the counterfactuals[2].

Randomization, of course, plays a crucial role in this conclusion. Only since the treatments have been properly randomized can we make this inference about *causality*. Without randomization, it is always possible that some other latent factor could have caused a certain set of units to end

---

[2] Even though there are counterfactuals within the two sums, their coefficients are always zero.

up in the control group and another different set of units to end up in the treatment. In such a case, the observed effect measured by Equation (4.6) could have been caused by this latent factor and not the actual experimental treatment itself. By randomizing, we are washing away all history and all unknown attributes of the experimental units, breaking their causal links to the observed outcome.

4.2.2 Analysis

Next, we discuss aspects of the analysis of A/B tests: choosing an appropriate test statistic and the roles of significance and power. As these necessarily play critical roles in the design, the analysis must be considered prior to the start of the experiment.

Test Statistic

Once we have randomized our units into treatment groups, we next switch to index notation. Our model for the response Y_{ij} of the jth replicant within the ith treatment group is

$$Y_{ij} = \mu + \tau_i + \epsilon_{ij}, \tag{4.7}$$

where $\mu = \mathbb{E}[Y_{ij}]$ is the overall average value, τ_i is the *differential effect* or *response* associated with the ith treatment, and ϵ_{ij} is the noise of the jth replicant in treatment group i.

Note 4.1. As mentioned in Note 3.24, Equation (4.7) is nonidentifiable, in the sense that the parameters cannot be uniquely determined. (In particular, the outcome, which is the true observable, is invariance under the shift $\tau_i' = \tau_i + \delta$, $\mu' = \mu - 2\delta$.) We must therefore impose an auxiliary condition so that the parameters are uniquely determined. There are two common choices: the first is to require that $\tau_0 = 0$. Under this choice, the treatment effect is identified with $\Delta = \tau_1$. This is a common choice in A/B testing, or whenever any one treatment group (the "control") is of particular interest, in that the treatment effect of the other groups is measured with respect to the former. More commonly, especially when we move beyond A/B tests, we will require

$$\sum_{i=0}^{m} \tau_i = 0,$$

so that μ is interpreted as the overall average. For the case of A/B tests, this equation reduces to $\tau_0 + \tau_1 = 0$. The average treatment effect in this case is given by $\Delta = \tau_1 - \tau_0$.   ▷

We further assume that this model satisfies the ANOVA assumptions from Definition 3.26. Since there are only two treatment groups, we soon realize that those assumptions imply the applicability of Theorem 3.4. As a result of these assumptions, we see that

$$Y_{ij} \sim N(\mu + \tau_i, \sigma^2), \qquad (4.8)$$

for $i = 0, 1$ and all $j = 1, \ldots, r$. Thus, the conditions of Theorem 3.4 are valid and we have the following theorem.

Theorem 4.1. *Given a completely randomized A/B test with r replicants, for which the outputs Equation (4.7) satisfy the ANOVA assumptions of Definition 3.26, then under the hypothesis of no effect $\tau_0 = \tau_1$, the test statistic*

$$T = \frac{\bar{y}_{1\cdot} - \bar{y}_{0\cdot}}{\sqrt{\lambda MSW}} \sim t_{2r-2}, \qquad (4.9)$$

where $\lambda = 2/r$ and MSW is the within-group mean squared error

$$MSW = \frac{1}{2r - 2} \sum_{i=0}^{1} \sum_{j=0}^{r-1} (y_{ij} - \bar{y}_{i\cdot})^2, \qquad (4.10)$$

is distributed as a Student's t-random variable with $2r - 2$ degrees of freedom. Moreover, the rejection region

$$R_\alpha = \{t : |t| > t_{2r-2, \alpha/2}\},$$

yields a test of size α.

Proof. The result follows as the conditions of Theorem 3.4 have been satisfied. □

If we were instead to make the inquiry as to whether the treatment was *better* than the control, our null hypothesis would be restated as $H_0 : \Delta = \tau_1 - \tau_0 \leq 0$. (The alternative hypothesis, in this case, would be $H_A : \Delta > 0$, that the treatment had a positive, non-zero effect.) In this case, we could still use the test statistic Equation (4.9), except with the alternate one-sided rejection region

$$R_\alpha = \{t : t > t_{2r-2, \alpha}\}.$$

Note 4.2. The within-group mean squared error of Equation (4.10) is analogous to the pooled sample variance of Equation (3.22). It makes intuitive sense to use the within-group error as opposed to the total error. Given the ANOVA assumption that $\sigma^2 = \sigma_0^2 = \sigma_1^2$, the within-group mean squared error is an unbiased estimate for σ^2. However, if the null hypothesis is false, and there are differences in the means, the total mean squared error (i.e., the sample variance of the entire data set, without regard to treatment group) can be biased by the between-group sum of squares.   ▷

Example 4.1. A professor, Dr. Austin, has two assistants who grade exams for her. Dr. Austin wants to determine whether her two assistants grade fairly. She therefore devises an experiment in which she gives half of the ten exams to one assistant and the other half to the other.

Let R_{0i} is the score that student i would obtain *if* his or her exam is graded by assistant 0, and R_{1i} is the score that student i would obtain *if* his or her exam is graded by assistant 1. For each student, we will have the ability to observe either R_{0i} or R_{1i}, but not both. For the purpose of illustration, however, we can claim omniscience.

Let's suppose that both assistants grade identically; i.e., $R_{0i} = R_{1i}$, for all $i = 0, \ldots, 9$. Let's further suppose that the scores are not random across the students, but correlated with the student's number. For example, suppose the exams are numbered in the order in which they were turned in, and suppose that better students tend to turn in exams sooner than the students who struggle. We can generate such a set of data by taking $R_{0i} \sim N(90 - 2i, 5^2)$. Random numbers pursuant to this scheme are given in Table 4.2.

i	R_{0i}	R_{1i}	Group NR	Observed NR	Group R	Observed R
0	96	96	0	96	1	96
1	89	89	0	89	1	89
2	79	79	0	79	0	79
3	81	81	0	81	0	81
4	82	82	0	82	0	82
5	82	82	1	82	1	82
6	72	72	1	72	0	72
7	69	69	1	69	1	69
8	80	80	1	80	0	80
9	73	73	1	73	1	73

Table 4.2: Nonrandom (NR) and Random (R) assignment for exam scores; both assistants grade identically.

At this point, let us suppose that Dr. Austin makes a faulty assumption, that the order in which she receives the exams is random. (This could even be an unconscious assumption.) She assigns the first five exams to assistant 0 and the second set of five exams to assistant 1. The results of this assignment are in the columns labeled *NR*. Upon obtaining the results, she first notices that the mean score for group 0 is 85.4, whereas the mean score for group 1 is 75.2. (Remember, she only sees R_{0i} for $i = 0, \ldots, 4$ and R_{1i} for $i = 5, \ldots, 9$; she does not see the counterfactuals.) She quickly computes $MSW = 40$ and $T = 2.55$. She then looks up the value $t_{8,0.025} = 2.306$, and rejects her null hypothesis (that both assistants grade identically), reporting a statistical significance of $\alpha = 0.05$. However, her conclusion is erroneous, as the assignments were not properly randomized. It was not the *choice of assistant* the *caused* the difference in exam scores between the two groups. It was the fact that the better students tended to finish earlier that caused their exam to be assigned to the earlier group that *caused* the difference in scores between the two groups.

Since we are omniscient beings in the universe of this example, let us like angels go back in time to see how things could have played out differently. The last exam lands on Dr. Austin's desk, abruptly awakening her from her daydream in which she realized the mistake in her plan. She then quickly randomizes the order of the exam and distributes the exams to the assistants following the mechanism in the columns labeled NR in Table 4.2. Upon reviewing the results, she finds that the average score of the exams graded by assistant 0 is 78.8, whereas the average score of the exams graded by assistant 1 is 81.8. She quickly calculates $MSW = 69.7$ and $T = 0.568$. She therefore fails to reject her null hypothesis and accepts that she does not have sufficient evidence to conclude that there is a difference in how the two assistants graded the exams. The assistants go happily on their ways, blissfully unaware of how close they came to mortal danger.   ▷

Example 4.2. Just down the hall, Dr. Powers has an identical group of ten students. His assistants, on the other hand, are not identical, in that assistant 1 always grades ten points higher than assistant 0. The potential outcomes for each student are given in Table 4.3.

i	R_{0i}	R_{1i}	Group NR	Observed NR	Group R	Observed R
0	96	100	0	96	1	100
1	89	99	0	89	1	99
2	79	89	0	79	0	79
3	81	91	0	81	0	81
4	82	92	0	82	0	82
5	82	92	1	92	1	92
6	72	82	1	82	0	72
7	69	79	1	79	1	79
8	80	90	1	90	0	80
9	73	83	1	83	1	83

Table 4.3: Nonrandom (NR) and Random (R) assignment for exam scores; assistant 1 always grades 10 points higher than assistant 0.

If Dr. Powers fails to randomize the exams before assigning them to his two assistants, the average score of group 0 will be 85.4 and the average score of group 1 will be 85.2. The *t*-score is $T = -0.05$. Dr. Powers fails to reject the null hypothesis and believes his two assistants to be grading fairly.

However, if Dr. Powers properly randomizes, the outcome is different. In this case, the average exam scores for assistants 0 and 1 are 78.8 and

90.6, respectively. This earns a t-score of $T = 2.587$, and Dr. Powers rejects the null hypothesis and concludes that his two assistants are not grading fairly.

In Exercise 4.1, we will simulate this and the preceding example to determine the probability of rejecting the null hypothesis in each of the four cases we discussed.   ▷

The Paradox of the One-Sided Test

The astute reader may have noticed the following oddity during our discussion on hypothesis testing: the rejection region for a standard two-sided test is

$$R_\alpha^2 = \{z < -z_{\alpha/2}\} \cup \{z > z_{\alpha/2}\},$$

and the rejection region for the corresponding one-sided (right-tailed) test is

$$R_\alpha^R = \{z > z_\alpha\}.$$

We therefore find ourselves with the peculiar outcome that if $z_\alpha < z < z_{\alpha/2}$, we would be able to reject the one-sided null hypothesis $H_0^R : \mu \leq \mu_0$, whereas we would not be able to reject the two-sided null hypothesis $H_0^2 : \mu = \mu_0$. (For $\alpha = 0.05$, this would happen if $1.65 < z < 1.96$.) From the same set of data, we would be able to conclude that the actual value of μ is greater than μ_0, since we are able to reject H_0^R, but we would not be able to conclude that μ is different than μ_0, as we are not able to reject H_0^2.

Both tests are producing the correct outcome. For the two-sided test, we will falsely reject a true null hypothesis H_0^2 exactly 5% of the time. This probability, however, is divided between the left and right tails, each of which hold 2.5% of the probability. For the one-sided test, on the other hand, we are not interested in cases for which $\mu < \mu_0$, as these cases have become incorporated into our null hypothesis. The full 5% probability is therefore entirely contained within the right tail, leading to the result that we will, again, falsely reject a true null hypothesis H_0^R exactly 5% of the time. So each test is valid for its given null hypothesis.

The conclusion, however, is still unsettling. How is it that we can conclude that μ is greater than μ_0, but yet we cannot conclude whether there is any effect at all? My view is this: the one-sided null hypothesis has done some of the lifting for us, as it has essentially removed the possibility that $\mu < \mu_0$ from our consideration. We are therefore no longer required to allocate some of our allotted $\alpha = 5\%$ significance in the pursuit of catching these cases. This results in the rejection region $\{z \geq z_\alpha = 1.65\}$, as opposed to the two-sided rejection region $\{|z| \geq z_{\alpha/2} = 1.96\}$. This is almost a Bayesian interpretation: the null hypothesis is, in some sense, our prior belief as to what the default state of the world should be. That prior belief therefore affects our posterior conclusions.

Next, let's address a practical matter: should we use one-sided tests? The answer, I think, is yes, possibly, with a caveat, but probably no. Let us suppose that we are running a test in which we are seeking to determine whether a variation in a product is beneficial, relative to some metric. (Or you could imagine we are conducting a clinical trial for a given drug against a placebo, if you like.) We are trying to improve the product, so we are only really interested in whether or not our change makes a positive impact on the product. So should we use a one-sided test? Not necessarily. It depends on our domain knowledge into the system we are studying. Consider the following two cases.

In the first case, suppose that, while we would like our product update to produce a positive result, we have no reason to rule out the possibility that it has a detrimental effect. In this case, we should run a two-sided test, and only proceed if we obtain a test statistic $z > z_{\alpha/2}$.

In the second case, suppose, on the other hand, that we have a good reason to rule out the possibility that our change can have a detrimental effect. That is, our prior belief about the system is that our change will either have no effect, or it will have a positive effect. Then we are safe to use a one-sided test with the less restrictive criterion $z > z_\alpha$.

In conclusion, our choice of whether or not to use a one- or two-sided test should not depend on our desired outcome, but on our most restrictive belief as to the possible range of outcomes. Even if we are only interested in catching positive outcomes, we would still run a two-sided test if a negative outcome is just as likely. We should not craft our null hypothesis to make our lives easier; we should craft our null hypothesis to mirror our understanding of the possible outcomes under consideration.

Example 4.3. Dr. Frankenstein is running an experiment to determine whether a corpse can be resurrected with an appropriate application of electricity. The outcome of this experiment is binary: either the corpse will remain deceased ($Y = 0$) or it will return to the living ($Y = 1$). Given the doctor's domain expertise, he is confident that cadavers randomly assigned to the control group are guaranteed to remain inanimate. It is therefore not possible for his treatment to produce a negative effect, as the patients in question are already deceased. He therefore constructs a one-sided test to determine whether or not his treatment has a positive effect. Furthermore, due to the vanishingly small variance, a single positive result will most likely be considered significant, especially to the test subject who finds himself stiffly awakened.   ▷

Power and Sample Size

The minimum sample size required to yield a power β test follows from Equation (3.23), which yields

$$r = \frac{2MSW}{\delta^2} \left[\Psi^{-1}(\alpha/2) - \Psi^{-1}(\beta) \right]^2 \approx \frac{15.7MSW}{\delta^2}, \qquad (4.11)$$

where the 15.7 corresponds to the choice of $\alpha = 0.05$ and $\beta = 0.80$. Here, the within-group mean-squared error has replaced the pooled sample variance, as expected.

4.2.3 Blocking

Finally, let us turn to the topic of *blocking*. Blocking is a design strategy used when different covariates are measured prior to the start of the experiment. The idea is that these covariates define certain strata (or blocks), in which units are uniform with respect to the observed set of covariates. A completely randomized design is then carried out within each stratum.

This approach is useful when the measured covariates, though not of direct interest in the study, are known or suspected to be correlated with the outcomes of the experiment. In particular, less populous strata have a higher tendency to posses an imbalance in treatment assignments, unless the assignment mechanism is devised to explicitly guard against it. This imbalance, coupled with the correlation between the covariates and the outcomes, could skew the results. We can therefore control for the covariates by ensuring that treatment and control assignments are balanced within each strata.

Example 4.4. To illustrate the importance of blocking, let us consider the following contrived example, in which the treatment has no effect, and the population is divided into two strata, as shown in Table 4.4. Units in stratum 0 have a fixed response of 150, whereas units in stratum 1 have a fixed response of 100. Since stratum 0 is small (only three units), it is

i	x_i	R_{0i}	R_{1i}	Group NB	Observed NB	Group B	Observed B
0	0	150	150	1	150	0	150
1	0	150	150	1	150	–	150
2	0	150	150	1	150	1	150
3	1	100	100	0	100	0	100
4	1	100	100	0	100	1	100
5	1	100	100	1	100	0	100
6	1	100	100	0	100	0	100
7	1	100	100	0	100	1	100
8	1	100	100	1	100	–	100
9	1	100	100	0	100	1	100

Table 4.4: Non-block (NB) and blocked (B) outcomes in a contrived two-strata problem with no treatment effect and no noise.

not hard to imagine the case when our random assignment places all three units from stratum 0 into the same treatment group, as shown in the NB (nonblocked) column in the table. In this case, our observed outcomes are

$\overline{Y}_{0\cdot} = 100$ and $\overline{Y}_{1\cdot} = 130$. In Exercise 4.2, we will show that the T-statistic is $T \approx 2.449$, so that the null hypothesis is rejected. We would therefore erroneously conclude that the treatment had an effect. However, this effect is in actuality entirely due to the fact that all units of stratum 0 have a higher response than the units of stratum 1. The treatment, by design, has no effect on the outcome.

With a block design, we would randomize within each stratum separately, enforcing the condition that we assign an equal number of units to the treatment and control groups within each stratum. (We randomly remove any excess units that would prevent us from assigning an equal number of units to each treatment group.) Such a randomized block design is shown in the final columns of the table. With this design, we see that observed outcomes between the two groups are identical: $\overline{Y}_{0\cdot} = \overline{Y}_{1\cdot} = 112.5$.

▷

Definition 4.6. *A block randomized A/B test is an A/B test consisting of a population with $s \geq 2$ strata (covariates) and an assignment mechanism that randomly assigns precisely half of the units of each stratum to the treatment group; i.e., if $2r_j$ units belong to stratum j, for $j = 0, \ldots, s - 1$, then the assignment mechanism will randomly assign r_j units to the control group and r_j units to the treatment group.*

In practice, we first group the units into their respective strata *horizontally*, determining the total number of units for each stratum. We then shuffle the units within each stratum in order to randomly assign half of those units to the treatment and half to the control group.

To express this using index notation, we let Y_{ijk} represent the observed result corresponding to treatment group $i = 0, 1$, stratum $j = 0, \ldots, s - 1$, and replicant $k = 0, \ldots, r_j - 1$. We may assume that the number $2r_j$ of units in stratum j is an even number, without loss of generality, as we can otherwise randomly omit one of the units, reducing that number to an even one. We may then model the response variable as

$$Y_{ijk} = \mu + \tau_i + \beta_j + \epsilon_{ijk}, \tag{4.12}$$

where $\epsilon_{ijk} \sim \mathrm{N}(0, \sigma^2)$ represents noise. Here, each stratum has its own "differential effect" β_j. By balancing units across the treatment and control within each stratum, we are therefore controlling for the effect produced by the different covariates.

Note 4.3. Whether or not one uses a randomized block design does not affect the model given by Equation (4.12). Rather, it directly affects the mechanism which assigns the units to the treatments, thereby affecting the *counting* of those units; i.e., without a block design, we would enumerate the replicants for treatment i and stratum j as $k = 0, \ldots, r_{ij} - 1$, as the total number of replicants r_{ij} for each stratum could vary treatment-to-treatment.

▷

Theorem 4.2. *In an A/B test with s strata, the average outcome* $\overline{Y}_{i..}$ *of treatment group* i *under the model prescribed by Equation (4.12) is a normally distributed random variable*

$$\overline{Y}_{i..} \sim N\left(\mu + \tau_i + \frac{1}{n_i}\sum_{j=0}^{s-1} r_{ij}\beta_j, \frac{\sigma^2}{n_i}\right),\tag{4.13}$$

where r_{ij} *is the number of replicants assigned to treatment group* i *and stratum* j *and* $n_i = \sum_{j=0}^{s-1} r_{ij}$ *is the total number of units assigned to treatment group* i *across all strata.*

Proof. According to the model given by Equation (4.12), each observation is normally distributed according to

$$Y_{ijk} \sim N\left(\mu + \tau_i + \beta_j, \sigma^2\right).$$

The sample mean of treatment group i and stratum j is therefore also normally distributed as

$$\overline{Y}_{ij.} \sim N\left(\mu + \tau_i + \beta_j, \frac{\sigma^2}{r_{ij}}\right).$$

The average value of treatment group i is obtained by averaging

$$\overline{Y}_{i..} = \frac{1}{n_i}\sum_{j=0}^{s-1} r_{ij}\overline{Y}_{ij..}$$

The result follows from Theorem 2.3.   □

Corollary 4.1. *In a randomized block A/B test, the expected observed effect under the model prescribed by Equation (4.12) is independent of the factors* β_j *determined by the covariates.*

Proof. In general, we see from Equation (4.13) that the expected treatment effect is

$$\mathbb{E}[\Delta] = \mathbb{E}\left[\overline{Y}_{1..} - \overline{Y}_{0..}\right] = (\tau_1 - \tau_0) + \sum_{j=0}^{s-1}\left(\frac{r_{1j}}{n_1} - \frac{r_{0j}}{n_0}\right)\beta_j.\tag{4.14}$$

However, under a randomized block A/B test, the number of replicants in each stratum are evenly divided between the control and treatment, yielding

$$r_{0j} = r_{1j}.$$

From this, it naturally follows that $n_0 = n_1$. For a randomized block design, we therefore conclude that the expected observed effect reduces to the actual causal effect of treatment,

$$\mathbb{E}[\Delta] = (\tau_1 - \tau_0).$$

We have therefore essentially *blocked* the effect of the covariates on the final observed effect of the experiment, by ensuring that units are allocated evenly, within each stratum, to both the treatment and control.   □

Note 4.4. Without blocking, we see that the final observed effect given by Equation (4.14) is *caused* by two factors: the treatment assignment and the covariates. When there are copious quantities of data within each stratum, the effect due to the covariates is likely to balance out, all by itself. This is, after all, the principle behind randomization. When covariates are available to the experimenter, however, and are known to be produce a significant effect on the outcome of the experiment, it is always prudent to control for those nuisance effects with an block design. Blocking is especially important when there is a pronounced affect β_j for one or more smaller strata, in which the imbalance of data can be large if not explicitly controlled for.   ▷

When carrying out a randomized block A/B test, we cannot apply the t-test of Theorem 4.1 directly. The problem is that Equation (4.10) estimates the error within each treatment group. However, since each treatment group has experimental units from multiple strata, this is not an unbiased estimate of the variance σ^2. In other words, despite having "blocked" the effect of the nuisance factor, one must nevertheless take into account both the treatment groups and strata when evaluating the t-statistic. The result is given in our next theorem.

Theorem 4.3. *Given a randomized block A/B test with s strata, following Equation (4.12), in which $r_j = r_{0j} = r_{1j}$ replicants in stratum j have been assigned to both treatment groups, define the test statistic*

$$T = \frac{\overline{y}_{1..} - \overline{y}_{0..}}{\sqrt{\lambda MSW}} \sim t_{n-2s}, \qquad (4.15)$$

where $\lambda = 1/n_0 + 1/n_1$, $n = n_0 + n_1$ is the total number of experimental units, and where MSW is the within-group mean squared error

$$MSW = \frac{1}{n - 2s} \sum_{i=0}^{1} \sum_{j=0}^{s-1} \sum_{k=0}^{r_j-1} \left(y_{ijk} - \overline{y}_{ij.}\right)^2. \qquad (4.16)$$

Then, under the hypothesis of no effect $\tau_0 = \tau_1$, the test statistic T is distributed as a Student's t-random variable with $n - 2s$ degrees of freedom. Moreover, the rejection region

$$R_\alpha = \{t : |t| > t_{n-2s,\alpha/2}\},$$

yields a test of size α.

Proof. Under the hypothesis of no effect, $\tau_0 = \tau_1$, and for a randomized block design, $r_{0j} = r_{1j}$. Under these assumptions, Equation (4.13) yields

$$\overline{Y}_{1..} - \overline{Y}_{0..} \sim N(0, \lambda\sigma^2).$$

Thus,

$$Z = \frac{\overline{Y}_{1..} - \overline{Y}_{0..}}{\sigma\sqrt{\lambda}} \sim N(0,1)$$

is a standard normal random variable.

Next, let

$$S_{ij}^2 = \frac{1}{r_{ij} - 1} \sum_{k=0}^{r_{ij}-1} \left(Y_{ijk} - \overline{Y}_{ij.}\right)^2 \tag{4.17}$$

represent the sample variance of treatment group i stratum j. Recall from Theorem 2.6 that

$$\frac{(r_{ij} - 1)S_{ij}^2}{\sigma^2} \sim \chi^2_{r_{ij}-1}.$$

It follows that the sum

$$\sum_{i=0}^{1}\sum_{j=0}^{s-1} \frac{(r_{ij} - 1)S_{ij}^2}{\sigma^2} = \frac{1}{\sigma^2}\sum_{i=0}^{1}\sum_{j=0}^{s-1}\sum_{k=0}^{r_{ij}-1} \left(Y_{ijk} - \overline{Y}_{ij.}\right)^2 \sim \chi^2_{n-2s}.$$

The result follows.   □

Example 4.5. Let's construct an example of a randomized block A/B test with 100 experimental units and three strata with frequency distribution $\langle 0.7, 0.2, 0.1\rangle$. Let us suppose that the hypothesis of no effect is true, so that $\delta = 0$. Further, suppose that $\mu = 100$ and $\beta = \langle 0, 100, 300\rangle$, so that the expected value of an observation from stratum 0, 1, and 2 is 100, 200, and 400, respectively. Finally, suppose that $\sigma = 20$.

As we are following a randomized block design, the test and control group will each receive 35 units from stratum 0, 10 units from stratum 1, and 5 units from stratum 2.

Following our experiment, we make the following observations. First, the overall mean of the control and treatment groups are given by

$$\overline{Y}_{0..} = 144.3 \quad \text{and} \quad \overline{Y}_{1..} = 150.3,$$

respectively. Furthermore, the six individual sums of squares (i.e., the sum portion of Equation (4.17)) are given in Table 4.5. Since there are three strata and $n = 100$ experimental units, the total within-group mean-squared error is given by the sum of these values divided by 94, which works out to be

$$MSW = 313.4.$$

Since there are fifty units assigned to treatment and control, the factor $\lambda = 2/50 = 0.04$. Thus, the T-statistic is

	Stratum 0	Stratum 1	Stratum 2
Treatment 0	8859	2714	303
Treatment 1	9870	5550	2166

Table 4.5: Sum of squares for each treatment-stratum combination.

$$T = \frac{150.3 - 144.3}{\sqrt{0.04 \cdot 313.4}} \approx 1.69.$$

The critical value for this test statistic is given by

$$T_{crit} = t_{94,0.025} \approx 1.9855.$$

Thus, we fail to reject the null hypothesis.

The data for this example were simulated using Code Block 4.1. By running this simulation many times, we find that the null hypothesis is rejected in approximately 5% of the simulations. (See Exercise 4.3.)   ▷

4.2.4 Online Randomized Block Design

We next consider a variation of randomized block design for the context of *online experiments*. An online experiment is an experiment in which information regarding the units is not known in advanced. For example, a clinical trial may be conducted over the period of several years, and units (i.e., patients) may arrive sporadically. Another example is in ad tech: an advertising campaign is run in order to test the effect of a particular creative (ad). In both of these situations, it is not known in advanced who the participants of the experiment will be. The participants show up from time to time as the experiment is underway. The covariates of each participant (unit) are only known once they arrive. Then a decision is made whether to assign each participant to the treatment or control. Though this assignment can depend on the particular covariates of each unit and the totals recorded *to date*, it cannot depend on the ultimate totals, as this number has yet to materialize. We will follow an approach presented in Efron [1971], which we call *online randomized block design*.

Definition 4.7. *An* online randomized block algorithm for A/B tests *for an online A/B test with s strata proceeds as follows:*

1. *Initialize a zero vector $b_j = 0$, for $j = 0, \ldots, s - 1$ and choose a fixed constant $p \in (1/2, 1)$. This vector will track the* imbalance *between the treatment and control as the experiment progresses.*
2. *Upon the arrival of each subsequent unit:*
   a) *Determine the current unit's stratum j;*
   b) *Assign unit to the treatment group with probability*
      - *$1/2$, if $b_j = 0$,*

```
 1  n = 100
 2  mu = 100
 3  beta = np.array([0, 100, 300])
 4  delta = 0
 5  sigma = 20
 6  r_j = np.array([0.35*n, 0.10*n, 0.05*n]).astype(int)
 7  T_crit = scipy.stats.t.isf(0.025,n-6)
 8
 9  # Create simulated data set.
10  Y = {}
11  for i in range(2):
12      Y[i] = {}
13      mu_delta = mu + delta * (i == 1)
14      for j in range(3):
15          Y[i][j] = np.random.normal(mu_delta + beta[j], scale=sigma,
                size=r_j[j])
16
17  # Create a 2 x 3 array of lists
18  Y = np.array([ list(Y[0].values()), list(Y[1].values()) ])
19
20  # All Control and Treatment Observations
21  Y0 = np.concatenate(Y[0, :])
22  Y1 = np.concatenate(Y[1, :])
23
24  MSW = 0
25  for i in range(2):
26      for j in range(3):
27          MSW += np.sum((Y[i,j] - Y[i,j].mean())**2)
28
29  MSW /= (n - 6)
30  la = 1 / len(Y0) + 1 / len(Y1)
31
32  T = ( Y1.mean() - Y0.mean() ) / np.sqrt(la*MSW)
```

Code Block 4.1: Simulated data for the randomized block A/B of Example 4.5

- p, if $b_j < 0$, and
- $(1 - p)$, if $b_j > 0$.

c) Increment b_j by $+1$ if the assignment was to the treatment group and increment b_j by -1 if the assignment was to the control group.

We may interpret b_j as the running total number of treatment assignments minus the running total number of control assignments. Whenever, for a given stratum, there have been more control assignments than treatments $(b_j < 0)$, we assign the unit to the treatment group with probability $p > 1/2$. Conversely, if there have been more treatment assignments

$(b_j > 0)$, we instead favor assignment to the control group. If there have been an equal number of treatment and control assignments $(b_j = 0)$, we do not favor one or the other, and instead assign the unit to the treatment group with a 50% probability.

Though the online randomized block design does not guarantee that the same number of units will be assigned to each treatment within each stratum, it creates a purposeful pressure towards balancing out units among treatment groups within each stratum. For example, suppose that $p = 0.80$. Then whenever there are more units assigned to the control, for a given stratum, subsequent units will receive an 80% probability of being assigned to the treatment, until that imbalance is remedied.

It should be cautioned that Theorem 4.3 does not apply directly, unless the condition is met that the experimental units of each stratum are equally divided among the treatment groups.

4.3 Single-Factor Experiments

In this section, we extend extend the classic A/B test to the more general case of multi-level (single-factor) experiments. For example, consider a three-level clinical trial in which patients receive either a placebo, a low dose, or a high dose of a certain drug. The generalization from a two-level experiment is mostly straightforward, with the analysis of variance replacing the t-test.

4.3.1 Design

Here, we consider both completely randomized experiments and randomized block experiments for multi-level experiments of a single factor.

Completely randomized experiments

For a completely randomized experiment, we fix the number of replicants r required for each treatment group, and then randomly assign units to the treatment groups while respecting this constraint. To achieve this, we may either shuffle the data or use one-at-a-time sampling as given by Definition 4.4.

Definition 4.8. *A* single-factor completely randomized experiment *is a single factor experiment in which units are randomly assigned to the m treatment groups, subject to the condition that each treatment group receive the same number r of replicants.*

Index notation for a multi-level experiment is practically identical to the index notation for A/B tests, except that the index i ranges over a larger set of possible treatment groups. Following the assignment of our population into treatment groups, we may let Y_{ij}, for $i = 0, \ldots, m - 1$ and $j = 0, \ldots, r - 1$, represent the observed outcome of the jth replicant of the ith treatment group. This follows the model given by Equation (4.7), with the only distinction being that the index i ranges from $i = 0, \ldots, m - 1$.

Randomized Block Design

As before, a randomized block design is achieved by implementing a completely randomized design within each of several strata. Formally, we have the following.

Definition 4.9. *A single-factor randomized block experiment is a single-factor experiment consisting of a population with $s \geq 2$ strata (covariates) and an assignment mechanism that randomly assigns an equal number of the units of each stratum to each of the m treatment groups; i.e., if mr_j units belong to stratum j, for $j = 0, \ldots, s - 1$, then the assignment mechanism will randomly assign r_j units to each of the m treatment groups.*

Following our assignment of units into their respective treatment groups, such that the units of each stratum are equally divided among treatments, we may make the switch to index notation. As before, we let Y_{ijk}, for $i = 0, \ldots, m - 1$, $j = 0, \ldots, s - 1$, and $k = 0, \ldots, r_j - 1$ represent the observed outcome of the kth replicant assigned to the ith treatment group and jth stratum.

Theorem 4.2 is still valid in this general context, as nothing in the proof relied on the fact that there were only two treatment groups; the same proof applies if the index i is allowed to run over additional groups.

As before, when using a randomized block design, we have $r_{0j} = \cdots = r_{(m-1)j} = r_j$; i.e., within stratum j each treatment group receives the same number r_j of replicants, and $n_0 = \cdots = n_{m-1} = r$, i.e., each treatment group overall receives the same number of replicants. Thus, as before, we see the benefit of the randomized block design, in that the expected observed effect between any two treatment groups no longer depends on the covariates β_j.

We will table our discussion on the analysis of a single-factor experiment with randomized block design until Section 4.4.3.

Online Randomized Block Design

We next generalize the algorithm presented in Section 4.2.3 for the case of an experiment with more than two levels. To this end, we combine elements of online randomized block design for A/B tests with one-at-a-time simple random sampling (Definition 4.4) to devise an approach suitable to a multi-level experiment.

Definition 4.10. *The* online randomized block algorithm *for an online experiment with m levels and s strata proceeds as follows:*

1. *Initialize a* count matrix $C_{ij} = 0$, *for* $i = 0, \ldots, m-1$ *and* $j = 0, \ldots, s-1$;
2. *Upon the arrival of each subsequent unit:*
   a) *Determine the current unit's stratum j;*
   b) *Assign unit to treatment group i with probability*

$$\mathbb{P}(Z = i | C_{ij}, x = j) = \frac{||C_{\cdot j}||_\infty + 1 - C_{ij}}{m \left(||C_{\cdot j}||_\infty + 1 - \overline{C}_{\cdot j} \right)}, \quad (4.18)$$

   where $C_{\cdot j}$ is the jth column vector of C, and $|| \cdot ||_\infty$ is the infinity norm[3] of $C_{\cdot j}$;
   c) *Increment the ijth component C_{ij} by 1 based on the final treatment assignment i and stratum j for the given unit.*

One can easily verify that Equation (4.18) is a normalized probability mass function over the treatment groups by summing over the index $i = 0, \ldots, m - 1$ (See Exercise 4.4). This assignment mechanism has the following interpretation: upon the arrival of a unit, we first identify the stratum j to which that unit belongs. We then look for the treatment group that has the greatest count of assignments $||C_{\cdot j}||_\infty$ out of all the units that have arrived *thus far* belonging to stratum j. We then assign the current unit to treatment group i with a probability proportional to the difference between this greatest value (plus one) and the number of units from that stratum already assigned to that treatment group. By adding 1, we ensure that there is always a non-vanishing probability of assigning each unit to any of the treatment groups.

Example 4.6. An online experiment with three levels and four strata is well underway. At a given point in time, the count matrix reads as follows:

$$C = \begin{bmatrix} 3 & 0 & 1 & 2 \\ 5 & 0 & 2 & 1 \\ 4 & 1 & 0 & 1 \end{bmatrix}.$$

What is the probability of assigning the next unit to treatment group i if the next arriving unit belongs to stratum 0?

To proceed, we first compute the infinity norm of column 0, which is given by

$$||C_{\cdot 0}||_\infty = 5.$$

We then construct the difference

---

[3] The *infinity norm* of a vector $x \in \mathbb{R}^m$ is defined as $||x||_\infty = \max\{|x_1|, \ldots, |x_m|\}$.

$$||C._0||_\infty + 1 - C_{i0} = \begin{bmatrix} 3 \\ 1 \\ 2 \end{bmatrix}.$$

By normalizing this vector, we should assign this unit to each treatment group with probability

$$\mathbb{P}(Z = i | C, x = 0) = \begin{bmatrix} 1/2 \\ 1/6 \\ 1/3 \end{bmatrix};$$

that is, a 50% probability for treatment group 0, a 16.7% probability for treatment group 1, and a 33.3% probability for treatment group 2. Suppose that we then choose a random number from this distribution, and find that our unit should be assigned to treatment group 2. We then update our count matrix to reflect this:

$$C_{20} = C_{20} + 1.$$

Finally, for purpose of illustration, we can compute the assignment probability for the next unit, as a function of the stratum. (Based on our original count matrix.) We have already computed the probability vector for stratum 0. A simple computation yields similar results for strata 1–3, which we record in the following matrix

$$\pi = \begin{bmatrix} 1/2 & 2/5 & 1/3 & 1/5 \\ 1/6 & 2/5 & 1/6 & 2/5 \\ 1/3 & 1/5 & 1/2 & 2/5 \end{bmatrix}.$$

Thus, if the next arriving unit belongs to stratum j, we use the assignment vector corresponding to column j of the probability matrix π.   ▷

Example 4.7. Write a simulation in Python that simulates an online randomized block design for three levels and four strata. Assume that the population is distributed over the strata according to the ratios $\langle 0.6, 0.2, 0.1, 0.1 \rangle$.

We accomplish this in Code Block 4.2.

We begin by initializing our 3×4 matrix of zeros and the number of units to simulate (lines 1–2). The assignment is accomplished term-by-term within our for-loop. Upon the arrival of each subsequent unit, we first observe the stratum to which the unit belongs. This process of observation is simulated on line 5. Next, we compute $||C._j||_\infty$, represented in code as Cj_inf, as the maximum value (thus far) in column j. We then form our probability mass function on lines 7–8, and carry out a random draw, according to the given probability vector, on line 9. Since unit u is assigned to treatment group i, we thus increment our count for C_{ij} by 1. This is exactly the set of code we used to simulate the count matrix in Example 4.6, except with $n = 20$.

Finally, note that this code can be modified to perform a simple random assignment, by replacing p=p in line 9 with p=[1/3,1/3,1/3].   ▷

```
1   C = np.zeros((3,4))
2   n = 100
3
4   for u in range(n):
5       j = np.random.choice(4, p=[0.6, 0.2, 0.1, 0.1])
6       Cj_inf = max( C[:, j] )
7       p = Cj_inf + 1 - C[:, j]
8       p /= p.sum()
9       i = np.random.choice(3, p=p)
10      C[i,j] += 1
```

Code Block 4.2: Simulation of an online randomized block assignment mechanism.

4.3.2 Analysis

For the remainder of this section, we will restrict our attention to single-factor completely randomized experiments; i.e., single-factor experiments with no covariates. This task will be itself broken into two tasks: an analysis of variance and an analysis of pairwise comparisons. The first task uses ANOVA to determine whether or not we can reject the hypothesis of no effect. If we reject the hypothesis of no effect, we then seek to determine which specific groups show a significant effect.

We will postpone our discussion of single-factor randomized block design until Section 4.4.3, as this analysis can be constructed as a special case of a two-factor completely randomized experiment.

Analysis of Variance

When analyzing the results of a single-factor completely randomized experiment, the first question is whether or not there is any effect at all. A single-factor completely randomized experiment that satisfies the ANOVA assumptions of Definition 3.26 can be analyzed using the analysis of variance method prescribed in Theorem 3.12, following the key formulas that are summarized in Table 3.4.

Example 4.8. Run a simulation using $\alpha = 0.05$, $\mu = 100$, $\sigma = 10$, $m = 4$, and $r = 12$ under the true null hypothesis. Determine the frequency of times in which the null hypothesis is incorrectly rejected.

This is achieved in Code Block 4.3. The observed rejection rate in 10,000 simulations was 5.03%, as expected. Note our use of **reshape** to convert arrays into column vectors (lines 7 and 16). This makes the computation in lines 13 and 16 possible.   ▷

```
1   alpha = 0.05
2   mu = 100
3   sigma = 10
4   r = 12
5   m = 4
6   n = r * m
7   tau = np.zeros(m).reshape((m,1))
8   F_crit = scipy.stats.f.isf(alpha, m-1, n - m) #2.866
9
10  rejections = 0
11  n_sims = 10000
12  for i in range(n_sims):
13      Y = np.random.normal(loc=mu+tau, scale=sigma, size=(m, r))
14
15      SSB = np.sum( (Y.mean(axis=1) - Y.mean())**2 ) * r
16      SSW = np.sum( (Y - Y.mean(axis=1).reshape((m,1)))**2 )
17      SST = np.sum( (Y - Y.mean())**2 )
18      MSB = SSB / (m-1)
19      MSW = SSW / (n-m)
20      F = MSB / MSW
21      rejections += 1 if F > F_crit else 0
22
23  print(rejections / n_sims) #0.0503
```

Code Block 4.3: ANOVA simulation for Example 4.8.

Power Analysis

In Chapter 3, we saw the importance of non-central distributions in the power analysis of statistical tests. The analysis of variance is no exception. Before we begin our discussion of power analysis, we first define a generalization to the F-distribution (Definition 2.15).

Definition 4.11. *Let $X \sim \text{NCX}_{\nu_1}^{\lambda}$ be a noncentral chi-squared random variable with ν_1 degrees of freedom and noncentrality parameter λ, and let $Y \sim \chi_{\nu_2}^2$ be a (central) chi-squared random variable with ν_2 degrees of freedom. Further, suppose X and Y are independent; i.e., $X \perp\!\!\!\perp Y$. Then the random variable F defined by the ratio*

$$F = \frac{X/\nu_1}{Y/\nu_2} \tag{4.19}$$

is a noncentral F random variable with degrees of freedom ν_1 and ν_2 and noncentrality parameter λ. Symbolically, we say that

$$F \sim \text{NCF}_{\nu_1,\nu_2}^{\lambda}.$$

Note that $\mathrm{NCF}^0_{\nu_1,\nu_2} = \mathrm{F}_{\nu_1,\nu_2}$.

Now, in order to perform a power analysis for single-factor experiments, we must determine the distribution of the F-statistic (Equation (3.69)) in the case that the null hypothesis is not true. To achieve this, we revisit Lemmas 3.8 and 3.9. However, we immediately realize that Lemma 3.8 is still valid, as its result is independent of the truth of the null hypothesis. Thus,

$$\frac{SSW}{\sigma^2} = \frac{1}{\sigma^2} \sum_{i=1}^{m} \sum_{j=1}^{r} \left(Y_{ij} - \overline{Y}_{i\cdot}\right)^2 \sim \chi^2_{n-m}. \tag{4.20}$$

However, Lemma 3.9 does rely on the truth of the null hypothesis. We therefore need to determine how the between-group sum of squares is distributed when the null hypothesis is false. (We restrict our attention to the case when $r_1 = \cdots = r_m$, which is true for completely randomized experiments.) The answer is provided in the following lemma.

Lemma 4.1. *Suppose the random variables Y_{ij} are distributed according to*

$$Y_{ij} \sim \mathrm{N}(\mu + \tau_i, \sigma^2), \tag{4.21}$$

for $i = 1, \ldots, m$ and $j = 1, \ldots, r$. (Let $n = mr$ be the total number of variables.) Without loss of generality, suppose that

$$\sum_{i=1}^{m} \tau_i = 0.$$

Then the scaled between-group sum of squares is distributed as a noncentral chi-squared random variable. In particular,

$$\frac{SSB}{\sigma^2} = \frac{r}{\sigma^2} \sum_{i=1}^{m} \left(\overline{Y}_{i\cdot} - \overline{\overline{Y}}\right)^2 \sim \mathrm{NCX}^{\lambda}_{m-1}, \tag{4.22}$$

where the noncentrality parameter is given by $\lambda = r\phi$, where we define the noncentrality effect by the relation

$$\phi = \frac{1}{\sigma^2} \sum_{i=1}^{m} \tau_i^2. \tag{4.23}$$

Note 4.5. The introduction of the noncentrality effect ϕ, as opposed to simply defining λ as r times the right-hand side of Equation (4.23), may at first pass seem superfluous. However, this notation has the benefit that ϕ is independent of the sample size r, and therefore more closely resembles the concept of a *minimum detectable effect*. Since it is a scaled version of the noncentrality parameter, we name it the noncentrality effect.   ▷

Proof. From Equation (4.21), we have

$$\overline{Y}_{i\cdot} \sim \mathrm{N}\left(\mu + \tau_i, \sigma^2/r\right) \qquad \text{and} \qquad \overline{\overline{Y}} \sim \mathrm{N}\left(\mu, \sigma^2/n\right).$$

Therefore,

$$\frac{\overline{Y}_{i\cdot} - \mu}{\sigma/\sqrt{r}} \sim \mathrm{N}\left(\frac{\tau_i\sqrt{r}}{\sigma}, 1\right).$$

It follows that the quantity W, defined by

$$W = \frac{r}{\sigma^2} \sum_{i=1}^{m} \left(\overline{Y}_{i\cdot} - \mu\right)^2 \sim \mathrm{NCX}_m^{r\phi}$$

is a noncentral chi-squared random variable with m degrees of freedom and noncentrality parameter $\lambda = r\phi$, with ϕ as defined in Equation (4.23). However,

$$W = \frac{r}{\sigma^2} \sum_{i=1}^{m} \left[\left(\overline{Y}_{i\cdot} - \overline{\overline{Y}}\right) + \left(\overline{\overline{Y}} - \mu\right)\right]^2$$

$$= \frac{r}{\sigma^2} \sum_{i=1}^{m} \left(\overline{Y}_{i\cdot} - \overline{\overline{Y}}\right)^2 + \frac{n}{\sigma^2} \left(\overline{\overline{Y}} - \mu\right)^2$$

The second line follows as the cross terms sum to zero. Now, the second term on the right is distributed as a χ_1^2 random variable. But $W \sim \mathrm{NCX}_m^{r\phi}$. We conclude that the first term on the right is distributed as $\mathrm{NCX}_{m-1}^{r\phi}$. This completes the proof. □

Theorem 4.4. *Under the model Equation (4.21), the F-statistic, defined as $F = MSB/MSW$ in Equation (3.69) is distributed as a noncentral F distribution:*

$$F \sim \mathrm{NCF}_{m-1,n-m}^{r\phi} \tag{4.24}$$

with noncentrality effect ϕ given by Equation (4.23)

Proof. This follows immediately from Lemmas 3.8 and 4.1 and Definition 4.9. □

We know that in an ANOVA test with significance level α, we reject the null hypothesis whenever

$$F > \mathrm{F}_{m-1,n-m}^{\alpha},$$

where $\mathrm{F}_{m-1,n-m}^{\alpha}$ is the unique real number that satisfies the relation

$$\mathbb{P}(X > \mathrm{F}_{m-1,n-m}^{\alpha}) = \alpha \qquad \text{whenever} \qquad X \sim \mathrm{F}_{m-1,n-m}.$$

In other words, $\mathrm{F}_{m-1,n-m}^{\alpha}$ is the inverse survival function of F with residual probability α. The power of our experiment, for a specific noncentrality effect ϕ, is therefore the probability of triggering this condition if the null hypothesis is false. In other words, the power is simply the probability

$$1 - \beta = \mathbb{P}(F > F^{\alpha}_{m-1,n-m} | F \sim \text{NCF}^{r\phi}_{m-1,n-m}). \tag{4.25}$$

In other terms, the power is the value of the survival function of the non-central F distribution evaluated at the critical value for the test:

$$1 - \beta = \text{NCF}^{r\phi}_{m-1,n-m} \cdot sf(F^{\alpha}_{m-1,n-m}).$$

(We borrow Pythonic syntax here; think of the ".sf" as the survival function method for the distribution $\text{NCF}^{r\phi}_{m-1,n-m}$.)

Example 4.9. Determine the power of the experiment described in Example 4.8, assuming a noncentrality effect of $\phi = 1$.

```python
def power(r, phi=1):
    m = 4
    n = r * m
    F_crit = scipy.stats.f.isf(alpha, m-1, n-m)
    return scipy.stats.ncf.sf(F_crit, m-1, n-m, r*phi)

r=12
print(f"The power is {power(r, phi=1)}") #802956
```

Code Block 4.4: Power calculation for the noncentral F-test

Equation (4.25) can easily be encoded as done in Code Block 4.4. We find that the power of the test is 80.2956%, for a noncentrality effect of $\phi = 1$.   ▷

Example 4.10. In Example 4.8, determine a value of τ that would yield a noncentrality effect of $\phi = 1$. Run a simulation, under this alternate hypothesis, to determine the fraction of times in which the null hypothesis is correctly rejected. Does the result agree with the theoretical result obtained in Example 4.9?

```python
tau = np.array([0,-np.sqrt(50),+np.sqrt(50),0]).reshape((m,1))
```

Code Block 4.5: Modification of Code Block 4.3 for power simulation

We can achieve this by simply modifying line 7 of Code Block 4.3 with the expression provided in Code Block 4.5. After rerunning the simulation with this modification, we obtained an output of 0.7999, which closely agrees with our result from Example 4.9.   ▷

4.3.3 Pairwise Comparisons

With the F-test, we are able to determine when to reject our null hypothesis, that there are no differences among the various treatment groups. In practice, however, we do not run experiments to confirm that everything is the same; rather, we run experiments to find out what is different and, in particular, what is better. The theory we developed thus far can only tell us when there is *some* difference among *some* of the treatment groups. But it fails to inform us which treatment groups are actually significantly different from each other. That is our goal for this section: to determine a test that will inform us when pairwise differences are significant.

The tools of this section should only be deployed when the null hypothesis has already been rejected. That is, it does not make sense to try to determine whether there is a significant difference between two particular treatment groups when we cannot even conclude that there is *any* significant difference in the first place.

Experimentwise Error

Consider a single-factor experiment with m treatment groups. If the ANOVA null hypothesis is rejected, we conclude that there is a significant difference between at least one pair of the m treatment groups. A *pairwise comparison analysis* seeks to determine which specific pairs of treatment groups are significantly different from each other. When performing a *pairwise* comparison, we seek to determine whether the observed effect

$$\Delta_{ii'} = \overline{Y}_{i'\cdot} - \overline{Y}_{i\cdot}$$

is significantly different, for each of the $c = m(m-1)$ pairs, for $i' \not\leq i$; i.e., for $i = 0, \ldots, m-1$ and $i' = i+1, \ldots, m-1$.

We know that a Type I error occurs when a true null hypothesis is falsely rejected. With multiple comparisons, however, we must carefully define whether we mean the error of a single comparison test or the overall error, a concept we now define.

Definition 4.12. *Consider an experiment that is comprised of a set of c null hypotheses $H_0^1, \ldots, H_0^c$ that are tested individually with separate test statistics. Let the event A_i be defined as accepting the ith hypothesis, and define the composite null hypothesis as $H_0 = \bigcap_{i=1}^{c} H_0^i$. Then the experimentwise error rate a is the probability of committing at least one Type I error among all of the tests, given that each null hypothesis is true; i.e.,*

$$a = \mathbb{P}\left(\sum_{i=1}^{c} \mathbb{I}[\neg A_i] \geq 1 \,\bigg|\, H_0 \right),$$

where $\neg A_i$ is the event that the ith null hypothesis is rejected.

If each of the c test statistics are *independent*, we have the following.

Proposition 4.1. *Suppose an experiment is comprised of a set of c individual null hypotheses $H_0^1, \ldots, H_0^c$ with independent test statistics, each with a probability α of a Type I error. Then the experimentwise error rate is*

$$a = 1 - (1 - \alpha)^c.$$

Proof. The probability that at we reject at least one null hypothesis is the complement of accepting each null hypothesis, so that

$$a = 1 - \mathbb{P}\left(\bigcap_{i=1}^c A_i \,\middle|\, H_0\right) = 1 - \prod_{i=1}^c \mathbb{P}(A_i | H_0) = 1 - (1 - \alpha)^c.$$

The second equality follows from the independence of the tests.   □

Example 4.11. Suppose an experiment consists of four independent hypothesis tests, each with significance level $\alpha = 0.05$. Then the experimentwise error rate is given by
$$a = 1 - 0.95^4 \approx 0.185.$$

▷

When we are performing pairwise comparisons for a single-factor experiment, however, the individual test hypotheses are *not* independent! The relationship between the comparisonwise and experimentwise error rates is generally not as simply stated as the expression in Proposition 4.1.

Example 4.12. Consider the null hypotheses for the pairwise comparisons

$$\begin{aligned}
H_0^1 : & \quad \tau_1 = \tau_2 \\
H_0^2 : & \quad \tau_1 = \tau_3 \\
H_0^3 : & \quad \tau_2 = \tau_3.
\end{aligned}$$

Suppose that we fail to reject H_0^1, but we do reject H_0^2. Given this information, what do you suspect about H_0^3?

From an intuitive perspective, it is clearly now more likely that H_0^3 should be rejected than it was before we knew anything about the outcome of the first two tests. This intuition is, however, not conclusive, as pairwise comparisons are not transitive. Our intuition nevertheless illustrates that the three tests are not independent.   ▷

In general, it is not straightforward to determine the experimentwise error rate a, given the individual comparison significance levels α, nor vice versa. However, if our goal is to ensure that the experimentwise error rate is below a given threshold, we can make use of the following correction.

Proposition 4.2 (Bonferroni Correction). *By setting the significance level of each comparison test at $\alpha = a^*/c$, the experimentwise error rate is bounded above by the value a^*; i.e., $a \leq a^*$.*

Proof. In the proof of Proposition 4.1, we saw that

$$a = 1 - \mathbb{P}\left(\bigcap_{i=1}^{c} A_i \,\middle|\, H_0 \right).$$

By applying the Bonferroni inequality, we may write

$$a \leq 1 - \left[\sum_{i=1}^{c} \mathbb{P}(A_i | H_0) - (c-1) \right] = c\alpha.$$

If we select $\alpha = a^*/c$, the result follows.   □

For any value of m greater than a few, this method, however, becomes inefficient, as seen in our next example.

Example 4.13. Consider the case of $m = 10$ treatment groups. In this case, there will be a total of $c = m(m-1)/2 = 45$ pairwise comparisons. In order to achieve an experimentwise error rate of $a = 0.05$, so that there is only a 5% chance of one of the pairwise comparisons of being falsely rejected, we would have to use a significance level of $\alpha = 0.05/45 \approx 0.00111$ for each comparison test.   ▷

Fisher's Least Significant Difference Test

Fisher's least significant difference (LSD) test controls for the comparison-wise significance levels α, without regard for the overall experimentwise error rate a. The philosophy here is to specify the significance level α for each pairwise comparison, and let the overall experimentwise error rate be as it may. The philosophy behind Fisher's LSD test is to simply apply the standard t-test (at fixed level α) to each pairwise comparison, with the following caveat: the pooled sample variance is replaced by the overall within-group mean squared error MSW. That is, we use a single estimate for σ^2 for each comparison. This correction allows for a larger number of degrees of freedom for each individual comparison, which, in turn, increases the power of each individual test. We now state Fisher's test as follows.

Theorem 4.5 (Fisher's least significant difference test). *Given a single-factor completely randomized experiment with m treatment groups and r replicants per treatment, for which the ANOVA null hypothesis has been rejected, define the* least significant difference (LSD) *as*

$$LSD = t_{n-m,\alpha/2}\sqrt{2MSW/r}, \tag{4.26}$$

for fixed significance level α and within-group mean squared error

$$MSW = \frac{1}{n-m} \sum_{i=0}^{m-1} \sum_{j=0}^{r-1} \left(Y_{ij} - \overline{Y}_{i\cdot} \right)^2.$$

Then, by rejecting any of the $m(m-1)/2$ null hypotheses

$$H_0^{i,i'} : \tau_i = \tau_{i'},$$

for $i' \not\leq i$, whenever the observed difference

$$\Delta_{ii'} = \left| \overline{Y}_{i\cdot} - \overline{Y}_{i'\cdot} \right| > LSD$$

constitutes a hypothesis test with significance level α.

In other words, once the ANOVA null hypothesis (i.e., the hypothesis of no effect) has been rejected, Fisher's LSD test entails classifying any pair $\{i, i'\}$ of treatment groups as significantly difference, at level α, if the observed difference Δ is greater than the least significant difference given by Equation (4.26).

Note 4.6. Fisher's LSD test differs from the t-test of Theorem 4.1 only in that the within-group mean squared error MSW is computed using all of the data, resulting in a total of $(n - m)$ degrees of freedom. The result follows by restating the test in terms of the observed difference as opposed to the T-statistic.   ▷

Example 4.14. Consider a single-factor experiment with five-levels, and six replicants per level. The observed data are given by

$$Y = \begin{bmatrix} 52 & 65 & 89 & 60 & 93 & 56 \\ 95 & 80 & 63 & 90 & 105 & 101 \\ 72 & 107 & 103 & 117 & 99 & 71 \\ 92 & 89 & 128 & 110 & 127 & 87 \\ 113 & 91 & 138 & 104 & 160 & 101 \end{bmatrix}.$$

For these data, we reject the ANOVA null hypothesis at the $\alpha = 0.05$ significance level. (See Exercise 4.5.) For Fisher's LSD test, we first compute the treatment group means:

$$\overline{Y}_{i\cdot} = \begin{bmatrix} 69.2 & 89.0 & 94.8 & 105.5 & 117.8 \end{bmatrix}.$$

The LSD can be easily computed, and works out to be

$$LSD = t_{25,0.025} \sqrt{2MSW/r} \approx 1.708\sqrt{129.68} = 19.45.$$

Fisher's LSD test therefore generates the following statistically significant comparisons:

Group 0 and Groups: 1, 2, 3, 4
Group 1 and Groups: 4
Group 2 and Groups: 4

Notice the failure of transitivity: groups 1, 2, and 3 are not significantly different from one another. Yet groups 1 and 2 are each significantly different from group 4, although group 3 is not.   ▷

Tukey's Honestly Significant Difference Test

Tukey's honestly significant difference (HSD) test is similar to Fisher's LSD test, except that it primarily controls for the experimentwise error rate a. Its derivation makes use of the following generalization of a T-random variable.

Definition 4.13. *Suppose that m samples of size r are collected from a normal distribution $N(\mu, \sigma^2)$, and define $\overline{Y}_{max}$ and $\overline{Y}_{min}$ by the relations*

$$\overline{Y}_{max} = \max_{i=1,\ldots,m} \overline{Y}_{i\cdot}. \quad and \quad \overline{Y}_{min} = \max_{i=1,\ldots,m} \overline{Y}_{i\cdot},$$

respectively. Then the random variable

$$Q = \frac{\overline{Y}_{max} - \overline{Y}_{min}}{\sqrt{s^2/r}}, \tag{4.27}$$

where s^2 is the pooled sample variance, is called a studentized range random variable *and its distribution the* studentized range distribution.

Note 4.7. For the case $m = 2$, the studentized range random variable is a scaled version of the T-statistic given in Equation (4.9), in that $Q = \sqrt{2}|T|$. This follows since when there are only two groups, one mean must be the maximum and the other must be the minimum.   ▷

Note 4.8. The studentized range distribution depends on two parameters: the number of groups m and the number of degrees of freedom ν of the pooled sample variance. If Q is such a random variable, we write $Q \sim Q_\nu^m$.
▷

The idea behind the studentized range distribution is as follows: the condition that there is at least one significant difference among the various pairwise comparisons is equivalent to the condition that there is a significant difference between the two most extreme group means. The logic is that if the difference between the largest and the smallest sample means is not significant, then none of the differences are significant.

Theorem 4.6 (Tukey's honestly significant difference test). *Given a single-factor completely randomized experiment with m treatment groups and r replicants per treatment, for which the ANOVA null hypothesis has been rejected, define the* honestly significant difference (HSD) *as*

$$HSD = q_\alpha(m, n - m)\sqrt{MSW/r}, \tag{4.28}$$

for fixed significance level α and within-group mean squared error

$$MSW = \frac{1}{n - m} \sum_{i=0}^{m-1} \sum_{j=0}^{r-1} \left(Y_{ij} - \overline{Y}_{i\cdot}\right)^2,$$

where the quantity $q_\alpha(m, \nu)$ is defined as the unique value such that

$$\mathbb{P}(Q > q_\alpha(m, \nu)) = a,$$

for $Q \sim Q_\nu^m$. Then, by rejecting any of the $m(m-1)/2$ null hypotheses

$$H_0^{i,i'} : \tau_i = \tau_{i'},$$

for $i' \not\leq i$, whenever the observed difference

$$\Delta_{ii'} = \left|\overline{Y}_{i\cdot} - \overline{Y}_{i'\cdot}\right| > HSD$$

constitutes a hypothesis test with experimentwise error rate a.

Example 4.15. Let us consider once again the experiment described in Example 4.14. Instead of specifying a pairwise significance level α, let us instead require that the experimentwise error rate be set at $a = 0.05$. If all of the groups are equivalent, there is only a 5% probability of obtaining at least one false rejection. The honestly significant difference is computed as

$$HSD = q_{0.05}(5, 25)\sqrt{MSW/r} \approx 4.15\sqrt{64.84} = 33.4.$$

Unfortunately, the studentized range distribution is not part of the `scipy.stats` library. It is nevertheless available in Python. We compute the above line using the `qsturng` method, as shown in Code Block 4.6.

Using this value for the Tukey test, we obtain the following reduced list of significant differences:

Group 0 and Groups: 3 and 4.

That is, the following pairs are statistically significant: $\{0, 3\}$ and $\{0, 4\}$. It makes sense that we get a smaller selection than we did previously, as we are requiring the experimentwise error rate—not the individual pairwise error rate–to be capped at 5%.

It is interesting to compare this result with the Bonferroni correction established in Proposition 4.2. In order to guarantee an experimentwise error rate of $a = 0.05$, we would use a value of $\alpha = a/[m(m-1)/2] = 0.005$ for Fisher's LSD test. The value of the LSD with this more strict level of significance is $LSD = 31.74$. It is interesting that the Fisher's LSD test with Bonferroni correction is actually slightly less strict than the Tukey HSD test.   ▷

```
from statsmodels.stats.libqsturng import psturng, qsturng
a=0.05
m = 5
r = 6
n = r * m
qsturng(1-a, m, n-m) #inverse cdf 4.15357
psturng(4.15357, m, n-m) # survival 0.05
```

Code Block 4.6: Inverse CDF and survival function for the studentized range distribution

4.4 Two-Factor Experiments

4.4.1 Design

A two-factor experiment is not unlike a single-factor experiment with blocking, except instead of the blocks constituting a nuisance factor, they are an orthogonal set of treatment groups arranged for testing. The other distinction is that, unlike the case of a single-factor experiment with blocking, units can be arranged at will. This makes the design a simpler process, as units are randomly allocated across both rows and columns of the design matrix.

Suppose we have m-levels for the first factor and s-levels for the second factor. A natural question is: why not simply arrange the ms total combinations as a single factor, as opposed to burning an additional index—and additional computational complexity—on this arrangement. The answer is that a two-factor experiment allows for the study not only of the response to each combination, but also for the study of the response due to *interaction effects*.

Example 4.16. A firm manages three separate factories, each which operate at different efficiencies, and wants to test five different machines (say, machines from five different brands) for a certain production process. A completely randomized two-factor experiment is devised, which uses the factory as one factor and the machine brand as the second factor. There are thus $m = 3$ levels for the first factor, and $s = 5$ levels for the second factor, yielding 15 different treatment groups. In addition, four separate machines of each type at each location are ordered, so that the experiment can be conducted using a replication factor of $r = 4$. There are thus 60 individual outcomes that will be measured.

In this example, the factory location and type of machine represent two intrinsically different factors for the experiment, which is why all enumerations are not simply lumped together in a single row.   ▷

A completely randomized two-factor experiment is conducted in the natural way.

Definition 4.14. *A two-factor experiment with m and s levels and r replicants is said to be* completely randomized *if the total $n = msr$ experimental units are randomly allocated such that each of the ms experimental cells receive a total of r units.*

4.4.2 Analysis: Two-way ANOVA

We model the response of a completely randomized two-factor experiment using

$$Y_{ijk} = \mu + \tau_i + \beta_j + \iota_{ij} + \epsilon_{ijk}, \tag{4.29}$$

for $i = 1, \ldots, m$; $j = 1, \ldots, s$; and $k = 1, \ldots, r_{ij}$. The term ι_{ij} represents the *interaction effect* between the two factors. We usually require the added constraints to make the equation well defined:

$$\sum_{i=1}^{m} \tau_i = 0, \qquad \sum_{j=1}^{s} \beta_j = 0, \qquad \text{and} \qquad \sum_{i=1}^{m} \iota_{ij} = \sum_{j=1}^{s} \iota_{ij} = 0.$$

In the following, we will consider the general case where there are r_{ij} replicants in the ijth cell. For a completely randomized experiment, however, $r_{ij} = r$ will not vary cell to cell. We further require the ANOVA assumption $\epsilon_{ijk} \sim \mathrm{N}(0, \sigma^2)$.

   To begin, let us define the various sums of squares, analogous of Equations (3.65)–(3.67).

Definition 4.15. *For any sets of numbers Y_{ijk}, with $i = 1, \ldots, m$; $j = 1, \ldots, s$; and $k = 1, \ldots, r_{ij}$, we define the following sums of squares*

$$SST = \sum_{i=1}^{m} \sum_{j=1}^{s} \sum_{k=1}^{r_{ij}} \left(Y_{ijk} - \overline{Y}_{...}\right)^2 \tag{4.30}$$

$$SSB_\tau = \sum_{i=1}^{m} s\hat{r}_{i \cdot} \left(\overline{Y}_{i \cdot \cdot} - \overline{Y}_{...}\right)^2 \tag{4.31}$$

$$SSB_\beta = \sum_{j=1}^{s} m\hat{r}_{\cdot j} \left(\overline{Y}_{\cdot j \cdot} - \overline{Y}_{...}\right)^2 \tag{4.32}$$

$$SSI = \sum_{i=1}^{m} \sum_{j=1}^{s} r_{ij} \left(\overline{Y}_{ij \cdot} - \overline{Y}_{i \cdot \cdot} - \overline{Y}_{\cdot j \cdot} + \overline{Y}_{...}\right)^2 \tag{4.33}$$

$$SSW = \sum_{i=1}^{m} \sum_{j=1}^{s} \sum_{k=1}^{r_{ij}} \left(Y_{ijk} - \overline{Y}_{ij \cdot}\right)^2, \tag{4.34}$$

as the total sum of squares, a between-group sum of squares for each factor, an interaction sum of squares, and, finally, a within-group sum of squares (also referred to as the error estimate, or the pooled sample variance), respectively. In the above, the "hat" operator represents the harmonic mean, *so that*

$$\hat{r}_{i\cdot} = \left(\frac{1}{s}\sum_{j=1}^{s} r_{ij}^{-1}\right)^{-1} \quad and \quad \hat{r}_{\cdot j} = \left(\frac{1}{m}\sum_{i=1}^{m} r_{ij}^{-1}\right)^{-1}.$$

For the case when we have the same number of replicants $r_{ij} = r$ in each cell, $\hat{r}_{i\cdot} = \hat{r}_{\cdot j} = r$.

Naturally, we have the following.

Theorem 4.7 (Partitioning Theorem). *For Y_{ijk} as in Definition 4.15, we have*

$$SST = SSB_\tau + SSB_\beta + SSI + SSW. \tag{4.35}$$

Proof. We begin by expressing the identity

$$\begin{aligned}
Y_{ijk} - \overline{Y}_{\cdots} &= \left(\overline{Y}_{i\cdot\cdot} - \overline{Y}_{\cdots}\right) \\
&\quad + \left(\overline{Y}_{\cdot j\cdot} - \overline{Y}_{\cdots}\right) \\
&\quad + \left(\overline{Y}_{ij\cdot} - \overline{Y}_{i\cdot\cdot} - \overline{Y}_{\cdot j\cdot} + \overline{Y}_{\cdots}\right) \\
&\quad + \left(Y_{ijk} - \overline{Y}_{ij\cdot}\right).
\end{aligned}$$

Next, we square both sides and sum over all indices. As was the case in Theorem 3.11, all the cross terms cancel (see Exercise 4.6). The result follows.   □

Given the sums of squares Equations (4.30)–(4.34), we next define the mean squares, analogous to Equation (3.68).

Definition 4.16. *Let Y_{ijk} be as in Definition 4.15. We define the following mean squares*

$$MSB_\tau = \frac{1}{m-1}SSB_\tau \tag{4.36}$$

$$MSB_\beta = \frac{1}{s-1}SSB_\beta \tag{4.37}$$

$$MSI = \frac{1}{(m-1)(s-1)}SSI \tag{4.38}$$

$$MSW = \frac{1}{n-ms}SSW. \tag{4.39}$$

where $n = \sum_{i=1}^{m}\sum_{j=1}^{s} r_{ij}$ is the total number of units. Note: for the case $r_{ij} = r$, the denominator $(n-ms)$ in MSW can be expressed as $ms(r-1)$.

Now, in a two-way analysis of variance (two-way ANOVA), there are three separate null hypotheses open for consideration.

Definition 4.17. *Under the model Equation (4.29), we define the three separate null hypotheses*

$$H_0^\tau : \qquad \tau_i = 0 \text{ and } \iota_{ij} = 0;$$
$$H_0^\beta : \qquad \beta_j = 0 \text{ and } \iota_{ij} = 0;$$
$$H_0^\iota : \qquad \iota_{ij} = 0.$$

For each of the ANOVA null hypotheses, we will conduct an F-test. We must therefore uncover the distributions of the various sums of squares, as we do in the following lemmas.

Lemma 4.2. *Under the ANOVA assumptions,*

$$\frac{1}{\sigma^2} SSW \sim \chi^2_{n-ms}. \tag{4.40}$$

Proof. Under the ANOVA assumptions, we have

$$Y_{ijk} \sim N(\mu + \tau_i + \beta_j + \iota_{ij}, \sigma^2).$$

Therefore, for each $i = 1, \ldots, m$ and $j = 1, \ldots, s$, the scaled sample variance for the ijth cell satisfies

$$\frac{1}{\sigma^2} \sum_{k=1}^{r_{ij}} \left(Y_{ijk} - \overline{Y}_{ij\cdot} \right)^2 \sim \chi^2_{r_{ij}-1}.$$

Summing over indices i and j yields our result. $\qquad\qquad\square$

Lemma 4.3. *Under the ANOVA assumptions and the null hypothesis H_0^τ from Definition 4.17, and under the additional assumption that r_{ij} is independent of i, then*

$$\frac{1}{\sigma^2} SSB_\tau \sim \chi^2_{m-1}. \tag{4.41}$$

Proof. Under the null hypothesis H_0^τ, we have $Y_{ijk} \sim N(\mu + \beta_j, \sigma^2)$. Therefore, the sample mean

$$\overline{Y}_{ij\cdot} \sim N \left(\mu + \beta_j, \frac{\sigma^2}{r_{ij}} \right).$$

Now, by averaging this equation over j, we obtain

$$\overline{Y}_{i\cdot\cdot} \sim N \left(\mu, \frac{\sigma^2}{s\hat{r}_{i\cdot}} \right),$$

where $\hat{r}_{i\cdot}$ is the harmonic mean defined in Definition 4.15. Now, if r_{ij} is independent of i, then so is $\hat{r}_{i\cdot}$, and the random variables $\overline{Y}_{i\cdot\cdot}$ are independent and identically distributed normal random variables. In this case, the quantity SSB_τ is therefore a scaled version of the sample variance of the random variable $\overline{Y}_{i\cdot\cdot}$. We therefore obtain from Theorem 2.6 that $(1/\sigma^2)SSB_\tau \sim \chi^2_{m-1}$. $\qquad\qquad\square$

Lemma 4.4. *Under the ANOVA assumptions and the null hypothesis H_0^β from Definition 4.17, and under the additional assumption that r_{ij} is independent of j, then*

$$\frac{1}{\sigma^2}SSB_\beta \sim \chi^2_{s-1}. \tag{4.42}$$

Proof. This follows from Lemma 4.3, by symmetry. $\square$

Lemma 4.5. *Under the ANOVA assumptions and the null hypothesis H_0^ι from Definition 4.17,*

$$\frac{1}{\sigma^2}SSI \sim \chi^2_{(m-1)(s-1)}. \tag{4.43}$$

We leave this final proof as an exercise for the reader. In general, however, the number of degrees of freedom for interaction terms (in two-way ANOVA as well as in higher-order ANOVA) is the product of the number of degrees of freedom in each factor. A quick way to see this is to recognize that the total sum of squares SST has $(n-1)$ degrees of freedom, and then solve for df_ι:

$$(n-1) = (m-1) + (s-1) + ms(r-1) + df_\iota.$$

We obtain $df_\iota = (m-1)(s-1)$.

Theorem 4.8. *In a completely randomized two-factor experiment, with an equal allocation of replicants $r_{ij} = r$ to each cell, the test statistics*

$$F_\tau = \frac{MSB_\tau}{MSW}, \qquad F_\beta = \frac{MSB_\beta}{MSW}, \qquad and \qquad F_\iota = \frac{MSI}{MSW} \tag{4.44}$$

are distributed as

$$F_\tau \sim \mathrm{F}_{m-1,n-ms}$$
$$F_\beta \sim \mathrm{F}_{s-1,n-ms}$$
$$F_\iota \sim \mathrm{F}_{(m-1)(s-1),n-ms},$$

under the null hypothesis H_0^τ, H_0^β, and H_0^ι, respectively.

Proof. This result follows from the preceding three lemmas. Note that since the number of replicants per cell is independent of both i and j, both Lemmas 4.3 and 4.4 apply. $\square$

Example 4.17. Consider a two-factor experiment with $m = 3$ and $s = 4$ and $r = 2$ replicants per cell. Data from such an experiment has been collected and provided in Table 4.6.

The sum of squares and F-statistics are computed in Table 4.7. We observe that both F_τ and F_β exceed their respective critical values. However, the F-statistic for the interaction term does not. We may therefore reject the null hypotheses H_0^τ and H_0^β, concluding that each factor is significant. However, we fail to reject the null hypothesis H_0^ι, so we cannot conclude that there are any interaction effects.

The code used to simulate this example and compute the test statistics is given in Code Block 4.7. $\triangleright$

	$j = 1$	$j = 2$	$j = 3$	$j = 4$
$i = 1$	72, 96	100, 101	90, 86	127, 101
$i = 2$	70, 90	82, 106	89, 100	114, 109
$i = 3$	94, 97	116, 110	139, 112	106, 132

Table 4.6: Data for Example 4.17.

	SSQ	df	MSE	F	F_{crit}
SSB_τ	1632.25	2	816.13	5.099	3.89
SSB_β	2430.46	3	810.15	5.062	3.49
SSI	664.42	6	110.74	0.692	3.00
SSW	1920.5	12	160.04		
SST	6647.63	23			

Table 4.7: Sum of squares and F-statistics for Example 4.17.

4.4.3 Analysis: Single Factor Experiment with Random Block Design

We took some pain to derive the results in the previous section for the general case of a varying number of replicants r_{ij} per cell. This added complexity was unnecessary for the results for two-factor experiments; however, it becomes relevant when applying those results to a single-factor randomized block design, for which r_{ij} is independent of i but not of j. Applying those results to this context, we obtain the following.

Theorem 4.9. *In a single-factor randomized block experiment, the test statistics F_τ and F_ι, as defined in Theorem 4.8, are distributed as*

$$F_\tau \sim \mathrm{F}_{m-1,n-ms}$$
$$F_\iota \sim \mathrm{F}_{(m-1)(s-1),n-ms},$$

under the null hypotheses H_0^τ and H_0^ι, respectively.

Note 4.9. The test statistic F_β is not valid for a single-factor randomized block design, as the number of replicants depends on the stratum: $r_{ij} = r_j$. Therefore, the conditions of Lemma 4.4 are not satisfied. The only exception to this rule is if there are an equal number of units per stratum. This is, however, typically not the case, as the nature of a nuisance factor is there is no control over its various counts.   ▷

Note 4.10. For a single-factor randomized block experiment, there should be no interaction effect. This is because, for each stratum, units are distributed randomly and in equal quantity to the various treatment groups. We can still perform the interaction F-test as a sanity check that our experiment is setup correctly. If we reject the null hypothesis H_0^ι, there could potentially have been an issue with our randomization methodology.   ▷

```
 1   m = 3
 2   s = 4
 3   r = 2
 4   mu = 100
 5   alpha = 0.05
 6   tau = np.array([-10, 0, 10]).reshape((3,1))
 7   beta = np.array([-10, 0, 0, 10]).reshape((1,4))
 8   sigma = 10
 9
10   Y1 = np.random.normal(loc= mu+tau+beta, scale=sigma).astype(int)
11   Y1 = np.random.normal(loc= mu+tau+beta, scale=sigma).astype(int)
12
13   Y = np.array([Y1.T, Y2.T]).T
14
15   SST = np.sum( (Y - Y.mean())**2 )
16   SSB1 = s * r * np.sum( (Y.mean(axis=(1,2)) - Y.mean())**2 )
17   SSB2 = m * r * np.sum( (Y.mean(axis=(0,2)) - Y.mean())**2 )
18   SSI = r * np.sum( (Y.mean(axis=2) -
         Y.mean(axis=(1,2)).reshape((3,1)) - Y.mean(axis=(0,2)) +
         Y.mean())**2 )
19   SSW = np.sum( (Y - Y.mean(axis=2).reshape((3,4,1)))**2 )
20
21   MSB1 = SSB1 / (m-1)
22   MSB2 = SSB2 / (s-1)
23   MSI  = SSI / ((m-1)*(s-1))
24   MSW  = SSW / (m*s*(r-1))
25
26   F1 = MSB1 / MSW
27   F2 = MSB2 / MSW
28   FI = MSI / MSW
29
30   F1_crit = scipy.stats.f.isf(alpha, m-1, m*s*(r-1))
31   F2_crit = scipy.stats.f.isf(alpha, s-1, m*s*(r-1))
32   FI_crit = scipy.stats.f.isf(alpha, (m-1)*(s-1), m*s*(r-1))
```

Code Block 4.7: Simulation and calculation for Example 4.17

4.5 Bandits

In traditional statistical experiments, the significance and power are prescribed, a sample size is calculated, and the experiment is run until completion; i.e., it is not stopped prematurely to "look" at the data, as doing
so could lead to inflated false positive rates (see Reinhart [2015] for a nice
discussion on this). In traditional settings—predominantly clinical trials for
drugs and therapies—this is absolutely necessary to maintain the rigor and
validity of the results. But what if you're not to the standards of clinical

trials but are instead concerned with which font looks best on your website, or if you're running a billion-dollar ad-tech startup? Should your approach be any different?

The idea is simple: experiments are expensive. Each experimental unit comes at a cost. Why not use half-baked results, once a clear leader has emerged? Can we get to the answer more quickly instead of burning money while we are waiting to achieve statistical significance? Put more gently: can we devise an algorithm that can better optimize our *true* objective of finding the optimal treatment using minimal time and expense?

To answer this, we turn next to a problem known as the *multi-armed bandit problem* (or simply *bandits*, for short), first introduced by Robbins [1952]. Essentially, a bandit algorithm seeks to maximize an objective function (total reward or payout) by allocating a fixed total amount of resources across a set of treatment groups, where the strategy (think: randomization mechanism) is allowed to adapt based on experience. The name derives from the casinos: a *single-armed bandit* is a slang term for a slot machine, whereas a multi-armed bandit represents a row of slot machines. Conceptually, the problem is stated as follows: a gambler with a finite number of coins is faced with a row of slot machines, each with an unknown reward distribution (and, hence, an unknown expected payout). Each round, the gambler must choose which machine to feed his next quarter into. The gambler has perfect memory and is allowed to adapt as he gains experience. What strategy should the gambler choose to maximize his total expected reward? Formally, we have the following.

Definition 4.18. *A multi-armed bandit problem consists of a set of real distributions* $\mathcal{B} = \{f_1, \ldots, f_m\}$, *with finite expected values* $\mu_1, \ldots, \mu_m$, *a positive integer* $n \in \mathbb{Z}^+$, *and an agent, who is tasked with selecting one of the arms* $\mathcal{A} = \{1, \ldots, m\}$ *sequentially for a total of* n *rounds.*

A strategy or a policy *is a method for selecting the arms based on the cumulative history; i.e., if we let* A_t *and* R_t *represent the action (choice) and reward for round* t, *respectively, so that* $R_t \sim f_{A_t}$, *we may define a policy* π *as a probability distribution over the arms* $\mathcal{A}$ *that depends on the history; i.e.,*

$$A_t \sim \pi_t = \pi_t \left(\{A_j, R_j\}_{j=1}^{t-1} \right).$$

The goal is then to maximize the total expected reward

$$R = \sum_{t=1}^{n} \mathbb{E}\left[R_t\right].$$

A binomial bandit problem is one in which the m *distributions are over the set* $\{0, 1\}$, *such that the reward is paid out with unknown probability* $p_1, \ldots, p_m$.

Bandit algorithms are related to Markov decision processes and reinforcement learning, a topic to which we will return in Part III. See Kochenderfer [2015] and Sutton and Barto [2018] for more details. Bandit algorithms get to the heart of a common dilemma in reinforcement learning known as the *exploration–exploitation tradeoff*, which balances the extent to which we *explore* or learn about our environment versus the extent to which we *exploit* what we have already learned. Traditional experiments do not address this tradeoff at all: one explores until one achieves statistical significance.

One common approach to the bandit problem is with so-called *greedy algorithms*.

Definition 4.19. *A* greedy algorithm *is one that always selects the best option, based on the available data, except for a reserved set of cases in which one uniformly selects a random action.*

The epsilon-greedy algorithm *selects the best-performing option at each point in time with a probability* $(1 - \epsilon)$, *otherwise, with probability* ϵ, *it selects an action at random. A typical value is* $\epsilon = 0.10$.

The epsilon-decreasing algorithm *is an epsilon-greedy algorithm with the modification that the value of* ϵ *decays with each choice.*

An epsilon-first algorithm *selects an action at random for the first* ϵn *rounds, and selects the best performer for the remaining* $(1 - \epsilon)n$ *rounds.*

In order to implement a greedy algorithm, we must naturally track an estimate for the value of each action as it evolves over time. For this purpose, we may define the following *action-value function*:

$$Q_t(a) = \mathbb{E}[R_t | A_t = a, \{(a_j, r_j)\}_{j=1}^{t-1}] = \frac{1}{N_t(a)} \sum_{j=1}^{t-1} R_j \mathbb{I}[A_j = a], \qquad (4.45)$$

where $N_t(a)$ is the number of times action a was selected prior to round t:

$$N_t(a) = \sum_{i=j}^{t-1} \mathbb{I}[A_j = a].$$

In other words, the action-value function $Q_t(a)$ is simply, for each $a \in \mathcal{A}$, the average reward realized up to the given point in time t.

Let $\mathcal{A}_t^*$ be defined as

$$\mathcal{A}_t^* = \arg\max_{a \in \mathcal{A}} Q_t(a),$$

the set of actions that has produced the optimal reward, up to the current point in time. The ϵ-greedy policy may therefore be represented as

$$\pi_t(a) = \begin{cases} \dfrac{1 - \epsilon}{|\mathcal{A}_t^*|} & \text{for } a \in \mathcal{A}_t^*, \\ \dfrac{\epsilon}{m - |\mathcal{A}_t^*|} & \text{for } a \notin \mathcal{A}_t^* \end{cases}.$$

Typically, $\mathcal{A}_t^*$ will consist of a single action, although the above accounts for the case when multiple actions are tied in first place.

A variation of this, known as *upper-confidence bounds (UCB) algorithms*, replaces the greedy actions $\mathcal{A}_t^*$ with

$$\mathcal{U}_t^* = \arg\max_{a \in \mathcal{A}} \left[Q_t(a) + \lambda \operatorname{se}(Q_t(a)) \right],$$

where $\lambda > 0$ is a constant (e.g., $\lambda = 2$) and $\operatorname{se}(Q_t(a))$ is the standard error of our estimate $Q_t(a)$. The policy distribution for a UCB method is the same as before, except with $\epsilon = 0$. The idea behind UCB methods is instead of selecting a non-optimal action uniformly, we select each action relative to the *potential* for that action to be optimal. This resolves the exploration–exploitation tradeoff in a natural way, as we are always selecting an optimal action relative to a confidence bound for our estimate, as opposed to the raw estimate itself.

An alternative approach to using upper-confidence bounds is simply to select the action with a probability that depends on our estimation of each action's value. A common method is the *softmax* approach, which normalizes the action-value estimates into a proper probability mass function

$$\pi_t(a) = \frac{e^{Q_t(a)/\tau}}{\sum_{i=1}^{m} e^{Q_t(i)/\tau}}.$$

The parameter $\tau > 0$ is referred to as the *temperature*: a high temperate causes the probability distribution to be closer to uniform, whereas a low temperature creates a greater difference between the various actions, with the greedy action resulting in the limit as $\tau \to 0$.

We conclude with an in-depth discussion of a modern advertising problem. For this example, a *creative* is simply a particular ad, which could take the form of a text string, an image, or a video. An ad network is tasked with showing various creatives to its audience, the placement being on websites or on mobile apps within its network. An instance of serving an add to a user is called an *impression*. A *click* occurs if the user decides to click on the ad. Cost is measured two ways: the *cost-per-mille*, or CPM, is the cost per 1,000 impressions and the *cost-per-click*, or CPC, is the cost per click. The *click-through rate*, or CTR is the ratio of clicks per impression. Many ad networks operate on a CPC model, where they charge advertisers a fixed cost per click, and profit based on the arbitrage between the cost per click and the CPM. We can think of the clicks as a binary reward for each impression, as a cost is incurred for each impression served.

Example 4.18. An ad network has a set of five creatives for a particular advertiser, and wants to run an experiment to determine whether or not there is a difference among the creatives and, if so, which creative performs best. We assume that we would like to detect the following effect:

$$p = \langle 0.008, 0.009, 0.01, 0.011, 0.012 \rangle,$$

where p_i represents the expected (true) click-through rate of the ith creative. Our goal is to devise a test with a 5% significance and a 95% power; i.e., a test that will detect a difference between the click-through rates 95% of the time, when the difference is actually present.

An immediate problem arises: if we choose the impression as our experimental unit, the observed responses Y_{ij} will be Bernoulli random variables, which will invalidate our model given by Equation (4.21). Instead, let us take individual milles, or sets of 1,000 impression, as our experimental unit. If each unit represents 1,000 impressions, then the response variable Y_{ij} will constitute a binomial random variable

$$Y_{ij} \sim \mathrm{Binom}(1000, p_i),$$

which is approximately normal with expected value $\mathbb{E}[Y_{ij}] = 1000p_i$ and variance $\mathbb{V}(Y_{ij}) = 1000p_i(1 - p_i) \approx 10$, since $p_i \approx 0.01$, for each i. We can therefore define $\mu = 10$,

$$\tau = \langle -2, -1, 0, 1, 2 \rangle,$$

and $\sigma^2 = 10$. It can be shown that the required sample size to achieve a 95% power test with the above minimum detectable effect is $r = 20$ (see Exercise 4.9). We therefore require a total of $n = rm = 100$ experimental units, which is equivalent to a total of 100,000 ad impressions. (For perspective, if impressions cost a CPM of \$10, then the experiment will cost \$1,000.)

Next, we can write a quick simulation to verify that the hypothesis of no effect is indeed rejected 95% of the time. Moreover, we can capture the probability distribution of which creative ends up in the lead at the end of the experiment. See Code Block 4.8. The simulation resulted in a rejected null hypothesis 95.6% of the time, which matches our expectation. However, the winning lot, after 100 simulated milles, had the following distribution

$$\langle 0, 0, 1.6\%, 16.2\%, 82.2\% \rangle.$$

Thus, our experiment was able to detect an effect 95% of the time, and when it did detect an effect, it correctly identified the best ad 82.2% of the time.   ▷

Next, we turn the experiment described in Example 4.18 into a simulation and compare the results of the experimental approach with various bandit algorithms.

Example 4.19. Compare the results of the randomized experiment described in Example 4.18 with the epsilon-greedy, UCB, and softmax algorithms.

We begin by defining our reward function and four policies in Code Block 4.9. Based on our analysis in Example 4.18, we know that an experiment would run for 100 lots of one mille each. We therefore construct our

```
1   n_sims = 10000
2   alpha = 0.05
3   r, m = 20, 5
4   n = r * m
5   lot_size = 1000
6   mu = 0.01
7   tau = np.array([-0.002, -0.001, 0, +0.001, +0.002]).reshape((m, 1))
8   F_crit = scipy.stats.f.isf(alpha, m-1, n - m) #2.467
9
10  count_wins = np.zeros(5)
11  rejections = 0
12  for i in range(n_sims):
13      Y = np.random.binomial(n=lot_size, p=mu+tau, size=(m, r))
14
15      SSB = np.sum( (Y.mean(axis=1) - Y.mean())**2 ) * r
16      SSW = np.sum( (Y - Y.mean(axis=1).reshape((m,1)))**2 )
17      SST = np.sum( (Y - Y.mean())**2 )
18      MSB = SSB / (m-1)
19      MSW = SSW / (n-m)
20      F = MSB / MSW
21      if F > F_crit:
22          rejections += 1
23          best = np.argmax(Y.sum(axis=1))
24          count_wins[best] += 1
25
26  print(rejections / n_sims) #95.6
27  print(count_wins / rejections) #0, 0, 0.016, 0.162, 0.822
```

Code Block 4.8: Simulation of a single-factor binomial experiment; Example 4.18

experiment as an epsilon-first policy, where the first 100 lots are selected at randomly, and the greedy action is selected for all rounds following the initial 100. (Note that instead of using a pure epsilon-greedy policy, we added a decay parameter, so that the amount of exploration diminishes over time.)

We can then turn these policies into a simulation, as shown in Code Block 4.10. We graph the average reward as a function of time in Figure 4.1. Note that we did perform some quick parameter tuning to obtain this particular set of parameters. Though the epsilon-greedy starts with a significant (50%) amount of exploration, the decay parameter balances this out, so that after 100 rounds, there is only about 18% exploration.

In this simulation, the upper-confidence bound algorithm is the clear winner; however, the epsilon-first strategy, which is based on a carrying out a rigorous experiment and then exploiting it results, does outperform even

```
1   lot_size = 1000
2   mu = 0.01
3   tau = np.array([-0.002, -0.001, 0, +0.001, +0.002]).reshape((m, 1))
4   def get_reward(a):
5       return np.random.binomial(n=lot_size, p=mu+tau[a])[0]
6
7   def eps_first(history, size_explore=100, **kwargs):
8       if history[1, :].sum() < size_explore:
9           return np.random.randint(5)
10      else:
11          b = history[0, :] / (history[1, :]+1)
12          return np.random.choice(np.flatnonzero(b == b.max()))
13
14  def eps_greedy(history, epsilon=0.20, decay=1, **kwargs):
15      if np.random.random() < epsilon*decay**(history[1,:].sum()):
16          return np.random.randint(5)
17      else:
18          b = history[0, :] / (history[1, :]+1)
19          return np.random.choice(np.flatnonzero(b == b.max()))
20
21  def ucb(history, lambda_=3, var=10, **kwargs):
22      b = history[0, :] / (history[1, :]+1) + lambda_ * np.sqrt(var /
                (history[1, :]+1) )
23      return np.random.choice(np.flatnonzero(b == b.max()))
24
25  def softmax(history, tau=1, **kwargs):
26      b = np.exp( history[0, :] / (history[1, :]+1) / tau )
27      b /= b.sum()
28      return np.random.choice(range(5), p=b)
```

Code Block 4.9: Reward function and four policies for Example 4.19

our tuned epsilon-greedy strategy in the long run. However, it should also be noted that the possible payouts are extremely similar in this example, which was, after all, based on our minimal detectable effect. It therefore makes sense that the bandit algorithms should do better based than the experimentation if there is a wider discrepancy between the results. For example, if some treatments are obviously bad and some are obviously winners, a good deal of cost can be saved by recognizing this early as opposed to waiting for the experiment to finish.

To visualize this, we reran the same set of simulations (with the same set of parameters) for a new "true" reward vector

$$p = \langle 0.01\%, 0.2\%, 0.5\%, 1.5\%, 3\% \rangle.$$

We left all of the parameters unchanged. The results are shown in Figure 4.2. Here, the upper-confidence band algorithm wins again, followed by

```
1   total_rounds = 500
2   n_sims = 1000
3   kwargs = {'epsilon': 0.50,
4              'size_explore': 100,
5              'decay': 0.99,
6              'tau': 2,
7              'lambda_': 4}
8
9   i = 0
10  for get_next_action in [eps_first, eps_greedy, ucb, softmax]:
11      color = default_colors[i]
12      i += 1
13      X = np.zeros((n_sims, total_rounds))
14      simulated_rewards = np.zeros(n_sims)
15      for s in range(n_sims):
16
17          history = np.zeros((2,5))
18          for t in range(total_rounds):
19              a = get_next_action(history, **kwargs)
20              r = get_reward(a)
21              history[0, a] += r
22              history[1, a] += 1
23              X[s, t] = history[0, :].sum() / history[1, :].sum()
24          simulated_rewards[s] = history[0, :].sum()
25      print(simulated_rewards.mean()/total_rounds,
              (simulated_rewards/total_rounds).std())
26
27      plt.plot(X.mean(axis=0), color=color)
```

Code Block 4.10: Simulation for Example 4.19

epsilon-greedy (with decay), and third by our epsilon-first (controlled experiment phase for exploration followed by exploitation). For this example, we tuned the parameter for the UCB, but left the other parameters the same. This leaves potential for softmax and epsilon-greedy to be further optimized. The fact that softmax levels out half way between the two best payouts (15 and 30) indicates that the temperature parameter need to be tuned to allow for more early exploration, in order to better distinguish between the top two choices.   ▷

The best approach depends largely on the problem at hand, including the particular cost and reward structure, the time horizon (e.g., can we carry out our experiment on 100 lots in five minutes or five months), as well as the ability to tune the parameters for each algorithm to produce an optimal result. Variance in the final outcomes should also be taken into account: one algorithm might on average produce superior results, but if

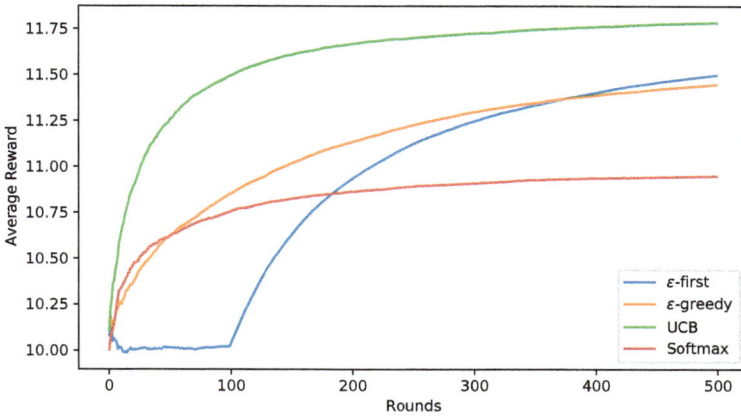

Fig. 4.1: Four bandit algorithms; Example 4.19.

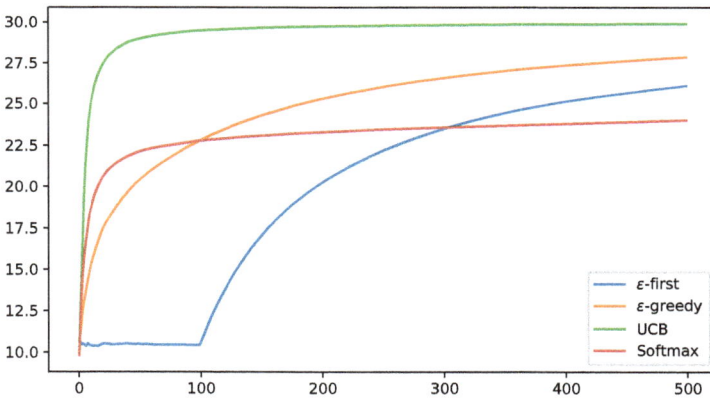

Fig. 4.2: Four bandit algorithms; Example 4.19.

those results have high variance, a more stable algorithm might be better suited for a particular context. More advanced versions of bandit algorithms include Bayesian bandits and contextual bandits, which both are outside of our present scope. They can also be generalized to non-static problems, where the expected reward varies in time or based on context. In addition, Bandit algorithms also form a solid gateway to reinforcement learning. For further reading, see See Kochenderfer [2015] and Sutton and Barto [2018].

Problems

4.1. Write a simulation for Examples 4.1 and 4.2 in order to numerically determine the probability of rejecting the null hypothesis for each of four cases: with and without randomization; no effect and an effect size of 10. As in the examples, assume that

$$R_{0i} \sim N(90 - 2i, 25),$$

and that $R_{1i} = R_{0i} + \delta$, where δ is the effect size (either 0 or 10).

	$\delta = 0$	$\delta = 10$
no randomization		
randomization		

Discuss your findings. Based on the results, estimate the significance and power of the test?

4.2. Suppose an experiment is conducted with two treatment groups and two strata. Suppose that each treatment group contains r replicants, and that $b < r$ units are in stratum 1. Furthermore, suppose the response depends only on the stratum as follows: if a unit is in stratum 0, its response will be 1; otherwise its response will be a value $\lambda > 1$. Now suppose that all b units of stratum 1 are assigned to the treatment group 1. (Oh no!) Show that the T-statistic is given by

$$T = \sqrt{\frac{f(r-1)}{1-f}},$$

where $f = b/r$ is the ratio of stratum 1 units in the treatment group 1. This has the peculiar result that the test statistic depends on the total sample size and the relative coverage of stratum 1 within the treatment group, and not on the size of the parameter λ. Thus, we are just as likely to determine there is a statistically significant difference between the results

Control	1	1	1	1	1
Treatment	1.001	1.001	1.001	1	1

as we are between the results

Control	1	1	1	1	1
Treatment	1001	1001	1001	1	1

(For both cases, the T-statistic is $T = \sqrt{6} \approx 2.449$.)

4.3. Run a simulation of Example 4.5 in order to estimate the fraction of times the hypothesis of no effect is falsely rejected. (It should be close to 5%.)

4.4. Show that Equation (4.18) is properly normalized; i.e., show that

$$\sum_{i=0}^{m-1} \mathbb{P}(Z = i|C_{ij}, x = j) = 1.$$

Bonus: Show that for the first unit in each stratum, that

$$\mathbb{P}(Z = i|C_{ij}, x = j) = \frac{1}{m}.$$

4.5. Perform an F-test for the data in Example 4.14 using a significance level $\alpha = 0.05$. Show that the ANOVA null hypothesis is rejected. Bonus: Assuming $\mu = 100$, $\tau = \langle -20, -10, 0, 10, 20 \rangle$, and variance $\sigma^2 = 400$, determine the power of the test (using $m = 5$ and $r = 6$).

4.6. Show that the cross terms cancel in completing the proof to Theorem 4.7.

4.7. Dr. Robinson works for a lab, where she performs four single-factor randomized block experiments each day of the week, Monday through Friday. After a few months on the job, she notices that she is averaging about one significantly different interaction effect each week. Because of the randomized block design, however, there should not be any interaction effect. What is wrong with her experiment?

4.8. Run a simulation based on Code Block 4.7. Approximately what is the power, for the given values of τ and β, of the F_τ and F_β tests?

4.9. Consider a five-level, single factor experiment with $\mu = 10$, $\sigma^2 = 10$, and

$$\tau = \langle -2, -1, 0, 1, 2 \rangle.$$

Determine the sample size r, for each group, required to achieve a power of 80%, 90%, and 95%.

Estimator 2: Judgement Day

In this chapter, we cover a variety of commonly used methods of statistical estimation and related optimization techniques, including maximum likelihood estimation, simple linear regression, and stochastic gradient descent. We conclude with a brief introduction to regression and classification problems and associated forms of validation.

5.1 Maximum Likelihood Estimation

We begin by considering a technique for estimating the parameter of a distribution, given a sample of data arising from the distribution.

5.1.1 Likelihood and Log Likelihood

Given an IID sample $X_1, \ldots, X_n$ from a PDF (or PMF) $f(x; \theta)$, with parameter θ, recall that the joint distribution is given by

$$f_{X_1, \ldots, X_n}(x_1, \ldots, x_n; \theta) = \prod_{i=1}^{n} f(x_i; \theta).$$

For a given value of parameter θ, the joint distribution tells us how likely a particular set of data is. Conversely, if our set of data is actually observed and therefore fixed, and our parameter θ is unknown, we can make relative assertions: if the true value of the unknown parameter is θ_1 as opposed to θ_2, the data we observed would be more likely to occur. This motivates the following definition.

Definition 5.1. *Given an observed, IID sample $X_1, \ldots, X_n \sim f(x; \theta)$ from a distribution with an unknown (possibly vector) parameter θ, the* likelihood function *is the function of θ defined by*

$$\mathcal{L}(\theta) = \prod_{i=1}^{n} f(x_i; \theta). \tag{5.1}$$

It is often useful to consider the natural logarithm of this quantity, which is referred to as the log-likelihood function *and defined by the relation*

$$\ell(\theta) = \log(\mathcal{L}(\theta)) = \sum_{i=1}^{n} \log f(x_i; \theta). \tag{5.2}$$

Note 5.1. The likelihood function is equivalent to the joint distribution with the exception that it is viewed as a function of the parameter θ for a fixed set of data, as opposed to a function of the data, for a fixed parameter. ▷

Note 5.2. If $\Theta \subset \mathbb{R}^k$ is the set of possible values of the parameter θ^1, then the likelihood function is a mapping $\mathcal{L} : \Theta \to \mathbb{R}_*$ into the set of nonnegative real numbers. In particular, the likelihood function is *not* a density, i.e., it is generally not true that $\int_\Theta \mathcal{L}(\theta) \, d\theta = 1$. ▷

Definition 5.2. *Given an* IID *sample* $X_1, \ldots, X_n \sim f(x; \theta)$ *from a distribution with unknown parameter* θ, *the* maximum likelihood estimate (MLE) *for* θ, *denoted* $\hat{\theta}_n$, *is the value of* θ *that maximizes the likelihood function; i.e.,*

$$\hat{\theta}_n = \arg \max_{\theta \in \Theta} \mathcal{L}(\theta). \tag{5.3}$$

Proposition 5.1. *The* MLE *for* θ, *as given in Definition 5.2, is equivalent to*

$$\hat{\theta}_n = \arg \max_{\theta \in \Theta} \ell(\theta). \tag{5.4}$$

Proof. Since the logarithm is a monotonically increasing function, the maximum value $\hat{\theta}_n$ of $\mathcal{L}(\theta)$ is also the maximum value of $\log(\mathcal{L}(\theta)) = \ell(\theta)$. □

Example 5.1. Let $X_1, \ldots, X_n \sim \text{Bern}(p)$ be IID Bernoulli random variables with PMF $f(x; p) = p^x (1 - p)^{1-x}$ for $x = 0, 1$. The likelihood function is given by

$$\mathcal{L}(p) = \prod_{i=1}^{n} f(x_i; p) = \prod_{i=1}^{n} p^{x_i} (1 - p)^{1-x_i} = p^k (1 - p)^{n-k},$$

where $k = \sum_{i=1}^{n} x_i$ represents the number of successes. The log-likelihood is therefore given by

$$\ell(p) = k \log(p) + (n - k) \log(1 - p).$$

By setting the derivative $\ell'(p) = 0$ and solving, one finds the MLE as $\hat{p}_n = k/n$. ▷

---

[1] Here, we are considering the general case where $\theta \in \mathbb{R}^k$, though we will primarily be focused on the case of a single scalar parameter $k = 1$.

Example 5.2. Let $X_1, \ldots, X_n \sim \mathrm{Exp}(\beta)$ be IID exponential random variables with PDF $f(x; \beta) = (1/\beta)e^{-x/\beta}$. The likelihood function is given by

$$\mathcal{L}(\beta) = \prod_{i=1}^{n} f(x_i; \beta) = \frac{1}{\beta^n} e^{-n\bar{x}/\beta},$$

where $\bar{x} = (1/n)\sum_{i=1}^{n} x_i$ is the sample mean. The log-likelihood is therefore

$$\ell(\beta) = -\frac{n\bar{x}}{\beta} - n\log(\beta).$$

The first derivative

$$\frac{d\ell(\beta)}{d\beta} = \frac{n\bar{x}}{\beta^2} - \frac{n}{\beta}$$

can be set equal to zero and solved for β to obtain the MLE $\hat{\beta} = \bar{x}$. The second derivative,

$$\frac{d^2\ell(\beta)}{d\beta^2} = -\frac{2n\bar{x}}{\beta^3} + \frac{n}{\beta^2},$$

when evaluated at the critical point $\hat{\beta} = \bar{x}$, is given by $\ell''(\hat{\beta}) = -n/\bar{x}^2 < 0$, proving that our critical point is indeed a maximum.   ▷

5.1.2 Score Statistic and Fisher Information

In studying maximum likelihood estimation and, as we shall see later, generalized linear models, it is useful to make several definitions. For additional details, see Dobson and Barnett [2018] and Dunn and Smyth [2018].

Definition 5.3. *Given an IID sample $X_1, \ldots, X_n \sim f(x; \theta)$ from a distribution with parameter $\theta \in \mathbb{R}^p$ and the data's associated log-likelihood $\ell(\theta)$, the* score function $U_n \in \mathbb{R}^p \to \mathbb{R}^{p*}$ *(see note[2]) is defined as the derivative*

$$U_n(\theta) = \frac{\partial \ell(\theta)}{\partial \theta}. \tag{5.5}$$

The score function U_n is sometimes referred to as the score statistic, *as it is also a function of the data.*

The equation obtained by setting the score function to zero, $U_n(\theta) = 0$, is known as the *score equation*; its solution $\hat{\theta}_n$ is the MLE estimate for the parameter θ. (We will assume that the log-likelihood functions we encounter throughout this text are unimodal and continuously differentiable, so that this is always a *maximum* solution.)

---

[2] We use the star $\mathbb{R}^{p*}$ to remind us that the quantity is a *covector*, which is represented as a row vector, not a column vector.

Theorem 5.1. *Let* $X_1, \ldots, X_n \sim f(x; \theta_0)$ *be* IID *from a distribution with true parameter* θ_0, *and let* $U_n(\theta)$ *be the data's associated score function. Then*

$$\mathbb{E}[U_n(\theta_0)] = 0; \tag{5.6}$$

i.e., when evaluated at the true parameter value θ_0, *the expected value of the score function is zero. The expectation is taken with respect to the distribution that generated the data, similar to Note 1.5.*

Proof. The score function from Equation (5.5) is equivalent to

$$U_n(\theta) = \sum_{i=1}^{n} \frac{1}{f(x_i; \theta)} \frac{\partial f(x_i; \theta)}{\partial \theta}. \tag{5.7}$$

Note that this depends on the observed data $X_1, \ldots, X_n$ and the family of distributions $f(x; \theta)$, but not the true value of $\theta = \theta_0$ that actually generated the data. The joint distribution of our sample is given by

$$f_{X_1, \ldots, X_n}(x_1, \ldots, x_n; \theta_0) = \prod_{j=1}^{n} f(x_j; \theta_0).$$

Therfore, the expected value of the score function is given, as a function of the parameter θ, by the relation

$$\mathbb{E}[U_n(\theta)] = \underbrace{\int_{-\infty}^{\infty} \cdots \int_{-\infty}^{\infty}}_{n \text{ times}} \sum_{i=1}^{n} \prod_{j=1}^{n} \frac{f(x_j; \theta_0)}{f(x_i; \theta)} \frac{\partial f(x_i; \theta)}{\partial \theta} \, dx_1 \cdots dx_n$$

$$= \sum_{i=1}^{n} \left(\int_{-\infty}^{\infty} \cdots \int_{-\infty}^{\infty} \prod_{j=1}^{n} \frac{f(x_j; \theta_0)}{f(x_i; \theta)} \frac{\partial f(x_i; \theta)}{\partial \theta} \, dx_1 \cdots dx_n \right);$$

that is, we are taking the expectation with respect to all possible realizations of the data from its true parameter θ_0. Now, if we evaluate this expression at the true value $\theta = \theta_0$, the multiple integral becomes separable, and we have

$$\mathbb{E}[U_n(\theta_0)] = \left(\int_{-\infty}^{\infty} \frac{\partial f(x_i; \theta_0)}{\partial \theta} \, dx_i \right) \cdot \prod_{\substack{j=1 \\ j \neq i}}^{n} \left(\int_{-\infty}^{\infty} f(x_j; \theta_0) \, dx_j \right).$$

However, the $(n-1)$ integrals of $f(x_j; \theta_0)$ are equal to unity, since a PDF must be normalized; similarly the lead factor,

$$\mathbb{E}[U_n(\theta_0)] = \frac{\partial}{\partial \theta} \int_{-\infty}^{\infty} f(x_i; \theta_0) \, dx_i = \frac{\partial 1}{\partial \theta} = 0,$$

vanishes. This completes the result.   □

Corollary 5.1. *The variance–covariance matrix of the score function is given by*

$$\mathbb{V}(U_n(\theta_0)) = \mathbb{E}[U_n(\theta_0)^T U_n(\theta_0)], \tag{5.8}$$

when evaluated at the true value $\theta = \theta_0$.

Proof. Combining Proposition 1.7 and Theorem 5.1 yields the result. (Recall that $U_n(\theta)$ is a covector.) □

Definition 5.4. *Consider an* IID *sample $X_1, \ldots, X_n \sim f(x; \theta_0)$ and its associated score function $U_n(\theta)$. The* observed information *is the matrix*

$$\mathcal{J}_n(\theta) = -\frac{\partial U_n(\theta)}{\partial \theta^T} = -\frac{\partial^2 \ell(\theta)}{\partial \theta^T \partial \theta}. \tag{5.9}$$

Similarly, the expected information *or* Fisher information *is the matrix*

$$\mathcal{I}_n(\theta) = \mathbb{E}[\mathcal{J}_n(\theta)]. \tag{5.10}$$

The difference between Equations (5.9) and (5.10) is that the observed information is a function of the data, i.e., a statistic, whereas the Fisher information is the expected value over all possible random samples from $f(x; \theta)$. (Both are a function of the parameter θ of the model.)

Note 5.3. The ijth component of the observed information is given by

$$\mathcal{J}_n(\theta)_{ij} = -\frac{\partial^2 \ell(\theta)}{\partial \theta_i \partial \theta_j}.$$

This is because the derivative $\partial/\partial\theta$ creates columns[3], whereas $\partial/\partial\theta^T$ creates rows. ▷

Note 5.4. Due to the linearity of the log-likelihood and derivative operators, it follows that

$$\mathcal{I}_n(\theta) = n\mathcal{I}(\theta), \tag{5.11}$$

where we write $\mathcal{I}(\theta)$ as a short-hand for $\mathcal{I}_1(\theta)$. ▷

Theorem 5.2. *Given an* IID *sample $X_1, \ldots, X_n \sim f(x; \theta_0)$, the variance–covariance matrix of the score function is given by*

$$\mathbb{V}(U_n(\theta_0)) = \mathcal{I}_n(\theta_0), \tag{5.12}$$

when evaluated at the true parameter $\theta = \theta_0$.

---

[3] So that $U_n(\theta) = \partial\ell/\partial\theta$ is a row vector. Also, this is consistent with the familiar case of $f : \mathbb{R}^n \to \mathbb{R}^n$, with $y = f(x)$, then $(\partial y/\partial x)_{ij} = \partial y_i/\partial x_j$.

Proof. Given Corollary 5.1, the result follows as long as we can prove that

$$\mathbb{E}\left[\frac{\partial^2 \ell}{\partial \theta^T \partial \theta}\bigg|_{\theta=\theta_0}\right] = -\mathbb{E}\left[\frac{\partial \ell}{\partial \theta^T}\frac{\partial \ell}{\partial \theta}\bigg|_{\theta=\theta_0}\right]. \tag{5.13}$$

Starting with the left-hand side, we may differentiate Equation (5.7) to find

$$\frac{\partial^2 \ell}{\partial \theta^T \partial \theta} = \sum_{i=1}^{n}\left[\frac{1}{f(x_i;\theta)}\frac{\partial^2 f(x_i;\theta)}{\partial \theta^T \partial \theta} - \frac{1}{f(x_i;\theta)^2}\frac{\partial f(x_i;\theta)}{\partial \theta^T}\frac{\partial f(x_i;\theta)}{\partial \theta}\right].$$

Now, when evaluated at the true parameter value $\theta = \theta_0$, the first term vanishes; i.e., we have

$$\mathbb{E}\left[\sum_{i=1}^{n}\frac{1}{f(x_i;\theta_0)}\frac{\partial^2 f(x_i;\theta_0)}{\partial \theta^T \partial \theta}\right] = 0;$$

for the same reason that $\mathbb{E}[U_n(\theta_0)] = 0$ in the proof of Theorem 5.1. We are thus left with

$$\mathbb{E}\left[\frac{\partial^2 \ell}{\partial \theta^T \partial \theta}\bigg|_{\theta=\theta_0}\right] = -\mathbb{E}\left[\sum_{i=1}^{n}\frac{1}{f(x_i;\theta_0)^2}\frac{\partial f(x_i;\theta_0)}{\partial \theta^T}\frac{\partial f(x_i;\theta_0)}{\partial \theta}\right],$$

which differs from the right-hand side of Equation (5.13) only by the absence of the cross terms

$$\mathbb{E}\left[\sum_{i=1}^{n}\sum_{\substack{j=1\\j\neq i}}^{n}\frac{1}{f(x_i;\theta_0)}\frac{\partial f(x_i;\theta_0)}{\partial \theta^T}\frac{1}{f(x_j;\theta_0)}\frac{\partial f(x_j;\theta_0)}{\partial \theta}\right].$$

However, the ijth term,

$$\mathbb{E}\left[\frac{1}{f(x_i;\theta_0)}\frac{\partial f(x_i;\theta)}{\partial \theta^T}\frac{1}{f(x_j;\theta_0)}\frac{\partial f(x_j;\theta)}{\partial \theta}\right],$$

is equivalent to

$$\left(\int_{-\infty}^{\infty}\frac{\partial f(x_i;\theta_0)}{\partial \theta}\,dx_i\right)\left(\int_{-\infty}^{\infty}\frac{\partial f(x_j;\theta_0)}{\partial \theta}\,dx_j\right)\prod_{\substack{k=1\\k\notin\{i,j\}}}^{n}\left(\int_{-\infty}^{\infty}f(x_k;\theta_0)\,dx_k\right) = 0$$

whenever $i \neq j$. (This follows similar logic as the proof to Theorem 5.1.) The result follows.   □

5.1.3 The Method of Scoring

We are concerned with finding a solution to the score equation $U_n(\theta) = 0$. One such approach uses a modification of the Newton–Raphson method, often simply referred to as Newton's method. This method is based on the following approximation. A more detailed treatment of this and related methods is discussed in Chong and Zak [2008].

Theorem 5.3. *Let $f : \mathbb{R}^n \to \mathbb{R}$ be twice continuously differentiable with invertible Hessian matrix at a given point $x_0 \in \mathbb{R}^n$. Then the global extremum of Taylor's quadratic approximation of f at x_0 is given by*

$$x^* = x_0 - D^2 f(x_0)^{-1} \cdot \nabla f(x_0), \tag{5.14}$$

where

$$\nabla f(x) = \frac{\partial f}{\partial x^T} \quad and \quad D^2 f(x) = \frac{\partial f}{\partial x^T \partial x}$$

are the gradient *and* Hessian *of the function f, respectively. Moreover, if the Hessian is positive definite, the point x^* is the global minimum, and if the Hessian is negative definite, it is the global maximum.*

Proof. The second-order Taylor approximation for f at x_0 is given by

$$f(x) \approx q(x) = f(x_0) + (x - x_0)^T \cdot \nabla f(x_0) + \frac{1}{2}(x - x_0)^T \cdot D^2 f(x_0) \cdot (x - x_0).$$

As the function $q : \mathbb{R}^n \to \mathbb{R}$ is quadratic, its extremum is straightforward to find: the gradient of q,

$$\nabla q(x) = \nabla f(x_0) + D^2 f(x_0) \cdot (x - x_0),$$

can be set equal to zero to obtain the result. $\qquad\square$

The *Newton–Raphson method*, usually referred to simply as *Newton's method*, is an iterative approach that leverages Theorem 5.3 to approximate a critical point of a multivariate function given a nearby starting point, as shown in Algorithm 5.1.

Input : Function $f : \mathbb{R}^n \to \mathbb{R}$; starting value $x^0 \in \mathbb{R}^n$; stopping
threshold h; max iterations T.
Output: Approximation of critical value $x^* \in \mathbb{R}^n$; message.
1 Set $k = 0$;
2 Set $\Delta = 1$;
3 Set message = Null;
4 while $k < T$ and $\Delta > h$ do
5 | Compute $x^{k+1} = x^k - D^2 f(x^k)^{-1} \cdot \nabla f(x^k)$;
6 | Set $\Delta = ||\nabla f(x^k)||_1$;
7 | Set $k = k + 1$;
8 end
9 Set $x^* = x^k$;
10 if $\Delta > h$ then
11 | Set message = Failed to Converge;
12 end

Algorithm 5.1: Newton–Raphson method.

We next apply this approach to the case where the function $f : \mathbb{R}^n \to \mathbb{R}$ is the log-likelihood function associated with an IID sample $X_1, \ldots, X_n \sim f(x; \theta_0)$. In this context, line 5 of Algorithm 5.1 becomes

$$\theta^{k+1} = \theta^k + \mathcal{J}_n(\theta^k)^{-1} \cdot U_n(\theta^k)^T.$$

In practice, however, the observed information $\mathcal{J}_n(\theta)$ is often difficult to compute. A modification on the preceding equation, for which the (negative) Hessian $\mathcal{J}_n(\theta^k)$ is replaced with its expected value, yields the *method of scoring* or *Fisher scoring*:

$$\theta^{k+1} = \theta^k + \mathcal{I}_n(\theta^k)^{-1} \cdot U_n(\theta^k)^T. \tag{5.15}$$

5.1.4 Asymptotic Properties of the MLE

Before we discuss the asymptotic behavior of the MLE, we first state several important properties.

Proposition 5.2 (Invariance Property). *Given an* IID *sample $X_1, \ldots, X_n \sim f(x; \theta_0)$, and let $\hat{\theta}_n$ be the* MLE *of the parameter θ. If $g : \mathbb{R} \to \mathbb{R}$ is one-to-one on the domain of the* PDF *f, then the* MLE *of the transformed parameter $\tau = g(\theta)$ is given by $\hat{\tau}_n = g(\hat{\theta}_n)$.*

Proof. Let $\mathcal{L}'(\tau) = \mathcal{L}(g^{-1}(\tau))$ represent the likelihood function expressed as a function of the new parameter τ, and define $\hat{\tau}_n = g(\hat{\theta}_n)$. Then, for any τ, we have $\mathcal{L}'(\tau) = \mathcal{L}(g^{-1}(\tau)) = \mathcal{L}(\theta) \leq \mathcal{L}(\hat{\theta}_n) = \mathcal{L}'(\hat{\tau}_n)$. This completes the proof. □

Proposition 5.3 (Consistency Property). *Given an infinite sequence of* IID *random variables* $X_1, X_2, \ldots \sim f(x; \theta_0)$, *define* $\hat{\theta}_n$ *to be the* MLE *obtained by using the first n samples. Then* $\hat{\theta}_n \to \theta_0$ *in probability as* $n \to \infty$[4].

For a proof, see Wasserman [2004]. Essentially, Proposition 5.3 means that the MLE is asymptotically unbiased. We are now ready to state our main theorem.

Theorem 5.4. *Given an infinite sequence of* IID *random variables* $\{X_i\}_{i=1}^{\infty} = X_1, X_2, \ldots \sim f(x; \theta_0)$, *generated by a distribution with parameter* $\theta_0 \in \mathbb{R}^k$, *and the associated sequence* $\{\hat{\theta}_n\}_{n=1}^{\infty}$ *of estimators, where* $\hat{\theta}_n$ *is defined as the the* MLE *of the sample* $X_1, \ldots, X_n$ *consisting of the first n terms of the sequence* $\{X_i\}_{i=1}^{\infty}$, *the sequence of random vectors* $\{W_n\}_{n=1}^{\infty}$ *defined by*

$$W_n = \mathcal{I}_n(\theta_0)^{1/2}(\hat{\theta}_n - \theta_0) \tag{5.16}$$

converges in distribution to the standard normal random vector $Z \sim$ $N(0_k, I_k)$; *i.e.,* $W_n \to Z$ *in distribution as* $n \to \infty$.

Proof. Consider the first-order Taylor expansion of the score statistic about $\theta = \theta_0$. We have

$$U_n(\theta) = U_n(\theta_0) - (\theta - \theta_0)^T \mathcal{J}_n(\theta_0) + O(||\theta - \theta_0||^2).$$

Now, by definition, $U_n(\hat{\theta}_n) = 0$, when evaluated at the MLE $\hat{\theta}_n$. We therefore have, to first order,

$$\sqrt{n}(\hat{\theta}_n - \theta_0) = \underbrace{n\mathcal{J}_n(\theta_0)^{-1}}_{B_n^{-1}} \underbrace{U_n(\theta_0)^T/\sqrt{n}}_{T_n}. \tag{5.17}$$

(Recall that $\mathcal{J}_n(\theta)$ is symmetric.) We have defined the statistics $B_n = \mathcal{J}_n(\theta_0)/n$ and $T_n = U_n(\theta_0)^T/\sqrt{n}$, as shown in preceding equation, for convenient usage momentarily.

Let us next define an infinite sequence of IID statistics

$$U^i(\theta) = \frac{\partial \log f(X_i; \theta)}{\partial \theta} \quad \text{and} \quad J^i(\theta) = \frac{\partial^2 f(X_i; \theta)}{\partial \theta^T \partial \theta},$$

for $i = 1, 2, \ldots$, so that

$$U_n(\theta) = \sum_{i=1}^{n} U^i(\theta) \quad \text{and} \quad \mathcal{J}_n(\theta) = \sum_{i=1}^{n} J^i(\theta).$$

The statistic B_n, defined above, is simply the sample mean of the first n IID random matrices $\{J^i\}_{i=1}^{n}$; i.e., $B_n = \overline{J}_n$. However, from Definition 5.4,

---

[4] We refer to any estimator that is asymptotically unbiased, in the sense that $\hat{\theta} \to \theta_0$ in probability, as a *consistent estimator*.

we have $\mathbb{E}[J^i(\theta)] = \mathcal{I}(\theta)$, which is, in particular, also true for $\theta = \theta_0$. By the law of large numbers (Theorem 1.3), therefore, we have $B_n(\theta_0) \to \mathcal{I}(\theta_0)$ in probability as $n \to \infty$.

Next, let us consider the statistic $T_n = U_n(\theta_0)^T/\sqrt{n} = \sqrt{n}\,\overline{U}_n(\theta_0)^T$, where the sample mean of the sequence $\{U^i\}_{i=1}^n$ is defined as $\overline{U}_n(\theta) = (1/n)U_n(\theta)$. Now, from Theorems 5.1 and 5.2, we have $\mathbb{E}[U^i(\theta_0)] = 0$ and $\mathbb{V}(U^i(\theta_0)) = \mathcal{I}(\theta)$, respectively. From the central limit theorem (Theorem 2.11), the statistic $T_n(\theta_0)$ therefore converges in distribution to $\mathrm{N}(0_k, \mathcal{I}(\theta_0))$.

By applying Slutsky's theorem (Theorem 1.4) to Equation (5.17), we therefore have that

$$\sqrt{n}(\hat{\theta}_n - \theta_0) \to \mathcal{I}(\theta_0)^{-1}\mathrm{N}(0_k, \mathcal{I}(\theta_0)) = \mathrm{N}(0_k, \mathcal{I}(\theta_0)^{-1}),$$

in distribution as $n \to \infty$, where the equality holds due to a multivariate generalization of Lemma 2.1.

Applying Proposition 2.34, we may rewrite this as $\sqrt{n}\mathcal{I}^{1/2}(\theta_0)(\hat{\theta}_n - \theta_0) \to \mathrm{N}(0_k, I_k)$. Finally, recognizing that $\mathcal{I}_n(\theta) = n\mathcal{I}(\theta)^5$, we obtain our result.   □

Corollary 5.2. *Let $X_1, \ldots, X_n \sim f(x; \theta_0)$ be an IID sample, as before, and consider the MLE $\hat{\theta}_n$. Then, for large n, we have the approximation*

$$\hat{\theta}_n \approx \mathrm{N}(\theta_0, \mathcal{I}_n(\theta_0)^{-1}); \qquad (5.18)$$

i.e., $\mathbb{V}(\hat{\theta}_n) \approx \mathcal{I}_n(\theta_0)^{-1} = \mathcal{I}(\theta_0)^{-1}/n$ *and* $\mathrm{se}(\hat{\theta}_n) \approx \mathcal{I}(\theta_0)^{-1/2}/\sqrt{n}$.

In particular, note that Equation (5.16) is equivalent to the statement that

$$\mathrm{se}(\hat{\theta}_n)^{-1}(\hat{\theta}_n - \theta_0) \to \mathrm{N}(0_k, I_k)$$

in distribution as $n \to \infty$. We can approximate this relation one step further, as stated in the following corollary.

Corollary 5.3. *Consider $\hat{\theta}_n$ and $\mathrm{se}(\hat{\theta}_n)$, as given in Corollary 5.2, and let the* approximate standard error *of the MLE be given by* $\hat{\mathrm{se}}(\hat{\theta})_n = \mathcal{I}(\hat{\theta}_n)^{-1/2}/\sqrt{n}$. *Then the sequence of random vectors*

$$W_n = \mathcal{I}_n(\hat{\theta}_n)^{1/2}(\hat{\theta}_n - \theta_0) = \hat{\mathrm{se}}(\hat{\theta}_n)^{-1}(\hat{\theta}_n - \theta_0) \qquad (5.19)$$

converges to $\mathrm{N}(0_k, I_k)$ in distribution as $n \to \infty$.

This follows from a further application of Slutsky's theorem along with the fact that, assuming $\mathcal{I}(\theta)$ is continuous, $\mathcal{I}(\hat{\theta}_n) \to \mathcal{I}(\theta_0)$ in probability as $n \to \infty$. We leave the details to Casella and Berger [2002] and Wasserman [2004].

The upshot, however, is that, for large n, the MLE can be approximated by

$$\hat{\theta}_n \approx \mathrm{N}(\theta_0, \mathcal{I}_n(\hat{\theta}_n)^{-1}) = \mathrm{N}(\theta_0, \mathcal{I}(\hat{\theta}_n)^{-1}/n). \qquad (5.20)$$

---

[5] See Note 5.4.

Note 5.5. Recall from the definition of the variance–covariance matrix that the diagonal elements of $\mathcal{I}_n(\hat{\theta}_n)^{-1}$ approximate the variance of each individual parameter within the vector $\hat{\theta}_n = (\hat{\theta}_n^1, \ldots, \hat{\theta}_n^n)$, i.e.,

$$\mathbb{V}(\hat{\theta}_n^i) \approx \left[\mathcal{I}_n(\hat{\theta}_n)^{-1}\right]_{ii},$$

whereas the covariances are given by the off-diagonal elements

$$\mathrm{COV}(\hat{\theta}_n^i, \hat{\theta}_n^j) \approx \left[\mathcal{I}_n(\hat{\theta}_n)^{-1}\right]_{ij}.$$

In particular, if the off-diagonal elements of $\mathcal{I}(\hat{\theta}_n)$ are zero, the estimates of the individual components of $\hat{\theta}_n$ are independent.   ▷

5.1.5 Example: Gamma Random Variable

We conclude the section with an in-depth analysis of a particular example: IID gamma random variables.

Example 5.3. Let $X_1, \ldots, X_n \sim \mathrm{Gamma}(\alpha, \beta)$ be an IID gamma random variables. Determine the log-likelihood, score functions, and Fisher information.

We can use the PDF for a gamma random variable, as given in Equation (2.52), to construct the log-likelihood for the parameter $\theta = (\alpha, \beta) \in \mathbb{R}_+^2$, thereby obtaining

$$\ell(\alpha, \beta) = \sum_{i=1}^{n} \left[(\alpha - 1)\log(x_i) - \frac{x_i}{\beta}\right] - n\log(\Gamma(\alpha)) - \alpha n \log(\beta). \quad (5.21)$$

The score function is given by

$$U_n(\alpha, \beta)^T = \frac{\partial \ell(\alpha, \beta)}{\partial(\alpha, \beta)^T} = \begin{bmatrix} \sum_{i=1}^{n} \log(x_i) - n\psi(\alpha) - n\log(\beta) \\ \frac{\sum_{i=1}^{n} x_i}{\beta^2} - \frac{\alpha n}{\beta} \end{bmatrix}, \quad (5.22)$$

where we define the *digamma function*[6] $\psi(z)$ by the relation

$$\psi(z) = \frac{d\log(\Gamma(z))}{dz} = \frac{\Gamma'(z)}{\Gamma(z)}.$$

Similarly, the observed information is given by

---

[6] The digamma function is available in Python by using `scipy.special.psi(z)`. Its nth derivative is available under `scipy.special.polygamma(n, z)`.

$$\mathcal{J}_n(\alpha, \beta) = -\frac{\partial \ell^2(\theta)}{\partial \theta^T \partial \theta} = \begin{bmatrix} n\psi'(\alpha) & n/\beta \\ n/\beta & \frac{2\sum_{i=1}^{n} x_i}{\beta^3} - \frac{\alpha n}{\beta^2} \end{bmatrix}.$$

Recall from Equation (2.53) that the expected value of a gamma random variable $X_i \sim \text{Gamma}(\alpha, \beta)$ is given by $\mathbb{E}[X_i] = \alpha\beta$. Therefore, the Fisher information is given by

$$\mathcal{I}_n(\alpha, \beta) = \mathbb{E}\left[\mathcal{J}_n(\alpha, \beta)\right] = \begin{bmatrix} n\psi'(\alpha) & n/\beta \\ n/\beta & \alpha n/\beta^2 \end{bmatrix} = n\mathcal{I}(\alpha, \beta),$$

where $\mathcal{I}(\theta)$ is the Fisher information of a single observation:

$$\mathcal{I}(\alpha, \beta) = \begin{bmatrix} \psi'(\alpha) & 1/\beta \\ 1/\beta & \alpha/\beta^2 \end{bmatrix}. \tag{5.23}$$

This matrix is invertible since $\alpha\psi'(\alpha) \neq 1$, and its inverse is given by

$$\mathcal{I}(\alpha, \beta)^{-1} = \frac{\beta^2}{\alpha\psi'(\alpha) - 1} \begin{bmatrix} \alpha/\beta^2 & -1/\beta \\ -1/\beta & \psi'(\alpha) \end{bmatrix}. \tag{5.24}$$

The MLE estimates for $\theta = (\alpha, \beta)$ are obtained by solving the score equation, which is simply Equation (5.22) set to zero. Moreover, for large n, the variance of the estimates is given by the diagonal elements of Equation (5.24) divided by n, or

$$\mathbb{V}(\hat{\alpha}_n) \approx \frac{\alpha}{n(\alpha\psi'(\alpha) - 1)} \quad \text{and} \quad \mathbb{V}(\hat{\beta}_n) \approx \frac{\beta^2\psi'(\alpha)}{n(\alpha\psi'(\alpha) - 1)}.$$

Since the off-diagonal elements are non-zero, the estimates are not independent. In fact, $\text{COV}(\hat{\alpha}_n, \hat{\beta}_n) \approx -\beta/(\alpha\psi'(\alpha) - 1)/n$. ▷

Next, we will continue this example, showing how the score function and FIsher information may be computed in Python.

Example 5.4. (Continued from Example 5.3.) Use Python to generate a random sample $X_1, \ldots, X_{100} \sim \text{Gamma}(\alpha_0, \beta_0)$, where $\alpha_0 = 2$ and $\beta_0 = 3$. Define Python functions for the score function and Fisher information. Then compute the expected variance of the MLE $\hat{\theta}_n = (\hat{\alpha}_n, \hat{\beta}_n)$.

To begin, let us generate a random sample of 100 IID gamma random variables, as shown in Code Block 5.1. This should be old hat, as we are using the `numpy.random` library to generate our random sample and the `scipy.stats` library to evaluate our PDF for an array of values. We can further plot a histogram of our sample concurrently with the PDF of Gamma(2,3), which is shown in Figure 5.1.

Next, let us define two Python functions to represent our score function, given by Equation (5.22), and the Fisher information, as given by Equation (5.23). This is done in Code Block 5.2. Notice the usage of the built-in

```
1   # Generate random sample from Gamma(2, 3)
2   alpha_0 = 2
3   beta_0 = 3
4   n_samples = 100
5   samples = np.random.gamma(alpha_0, scale=beta_0, size=n_samples)
6
7   # Compute PDF
8   x = np.linspace(0, 10)
9   y = scipy.stats.gamma.pdf(x, alpha_0, scale=beta_0)
10
11  # Plot PDF along with histogram of Sample.
12  plt.plot(x,y)
13  plt.hist(samples, bins=20, normed=True)
```

Code Block 5.1: Generate Random Sample of IID Gamma Random Variables

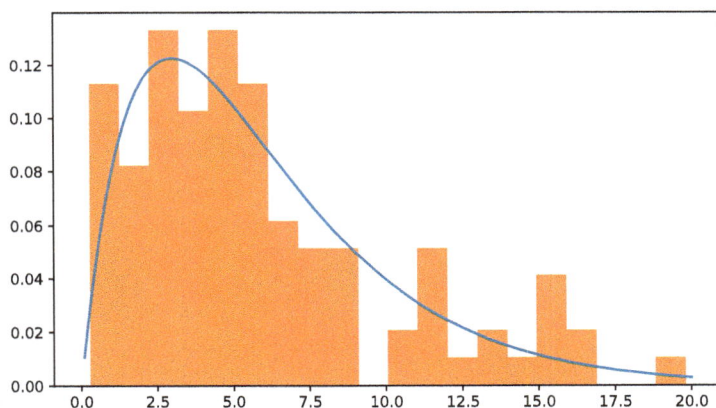

Fig. 5.1: PDF and histogram of random sample from Gamma(2,3).

functions `scipy.special.psi(z)` and `scipy.special.polygamma(n, z)` to evaluate the digamma function $\psi(z)$ and its derivatives. Furthermore, we construct the `FisherInfo` function to accept an optional argument `inverse`, which, when set to `True`, will return the inverse of the Fisher information matrix.

We can use the `FisherInfo` function to approximate the Fisher information at (α_0, β_0).

$$\mathcal{I}(2,3) \approx \begin{bmatrix} 0.6449 & 1/3 \\ 1/3 & 2/9 \end{bmatrix} \quad \text{and} \quad \mathcal{I}(2,3)^{-1} \approx \begin{bmatrix} 6.8997 & -10.3495 \\ -10.3495 & 20.0243 \end{bmatrix}.$$

```
1   def score(alpha, beta, samples):
2       """ Compute the score statistic
3       for an IID Gamma sample with parameters alpha, beta.
4       """
5       n = len(samples)
6
7       U_alpha = np.sum( np.log(samples) ) - n*scipy.special.psi(alpha)
            - n*np.log(beta)
8       U_beta = np.sum(samples) / beta**2 - alpha*n / beta
9
10      return np.array([U_alpha, U_beta]).reshape((2,1))
11
12  def FisherInfo(alpha, beta, n=1, inverse=False):
13      """ Returns the 2x2 Fisher Information for a Gamma random
            variable.
14      Set optional argument n=1 for sample size.
15      Set inverse=True to return the inverse of the Fisher Information.
16      """
17
18      I = n * np.array([ [scipy.special.polygamma(1, alpha), 1/beta],
            [1/beta, alpha / beta**2] ])
19
20      return np.linalg.inv( I ) if inverse else I
```

Code Block 5.2: Score Function and Fisher Information for a Gamma sample

From Corollary 5.2, the variance–covariance matrix for the MLE is therefore approximated by

$$\mathbb{V}((\hat{\alpha}_n, \hat{\beta}_n)) \approx \frac{1}{n} \begin{bmatrix} 6.8997 & -10.3495 \\ -10.3495 & 20.0243 \end{bmatrix}.$$

It is also interesting to note that one can use the `scipy.linalg.sqrtm` function to compute the *matrix square root* of the variance, obtaining

$$\mathbb{V}((\hat{\alpha}_n, \hat{\beta}_n))^{1/2} \approx \frac{1}{\sqrt{n}} \begin{bmatrix} 2.0214 & -1.6774 \\ -1.6774 & 4.1486 \end{bmatrix}.$$

$\triangleright$

In our next example, we use Python to compute the MLE estimates for the random sample generated in Example 5.3.

Example 5.5. (Continued from Example 5.4.) Use the score and information functions from Example 5.4 to compute the MLE for the sample data generated in Example 5.3.

```
scipy.optimize.root(lambda x: score(x[0], x[1], samples), [1, 1])
### MLE Estimate: alpha_hat = 1.73779384 beta_hat = 3.37642878

def FisherScoring(samples, score, FisherInfo, x0=[2,3],
    max_iter=1000, tol=1e-6):

    k = 0
    Delta = 1
    theta = np.array(x0)
    n = len(samples)

    while k < max_iter and Delta > tol:
        theta = theta + FisherInfo(theta[0], theta[1], n=n,
            inverse=True) @ score(theta[0], theta[1], samples)
        Delta = np.linalg.norm(score(theta[0], theta[1], samples))
        k += 1
        print(k, theta)

    if Delta > tol:
        message = 'Failed to converge in {n_iter}
            iterations.'.format(n_iter=k)
    else:
        message = 'Success.'

    return theta, message

beta = samples.var() / samples.mean()
alpha = samples.mean() / beta
FisherScoring(samples, score, FisherInfo, x0=[alpha,beta])
```

Code Block 5.3: Determining the MLE in Python

The code is shown in Code Block 5.3. First, we compute the MLE using the built-in root finder `scipy.optimize.root`, as shown on line 1. Note the use of a "lambda function," which is simply a way of defining a function in place[7]. For our particular sample, we obtain the estimate

$$\hat{\alpha}_{100} = 1.73779384 \qquad \text{and} \qquad \hat{\beta}_{100} = 3.37642878.$$

We then proceed to implement the method of scoring in lines 4–22. We found, however, that unless we start with values very close to the true values, this method does not converge. To facilitate this, we generate an initial "guess" for the true values of the parameters by setting Equations (2.53)

---

[7] The quantity `lambda x: score(x[0], x[1], samples)` represents a function that takes in an argument x and outputs the value `score(x[0], x[1], samples)`.

and (2.54) equal to the sample mean and variance and solving for α and β. This is accomplished in lines 24–26, and it produces the same estimate as line 1.

The upshot is that the built-in `scipy.optimize` routines likely do a better job at determining the MLE of a dataset than a simple implementation of the method of scoring. Note that the function **FisherScoring** takes as *inputs* the methods **score** and **FisherInfo**, defined in Code Block 5.2.
▷

Example 5.6. Using the variance–covariance matrix obtained in Example 5.4, plot the $z = 1$ and $z = 2$ Mahalanobis contours using Code Block 2.23. Then repeat Examples 5.4 and 5.5 one thousand times, using different random samples, and plot the MLE for each. What fraction of estimates are within a Mahalanobis distance of 1 or 2?

The code to accomplish this is shown in Code Block 5.4. The result in shown in Figure 5.2 (left). Moreover, we see that 39.5% of our samples had a Mahalanobis distance $z = d_\Sigma(\hat{\theta}_{100}, \theta_0)$ less than one and 83.4% had $z < 2$, agreeing nicely with the expected result from Table 2.1 for the case $k = 2$.

```
1   alpha_0 = 2
2   beta_0 = 3
3   n_samples = 100
4   n_simus = 1000
5   mu = np.array([alpha_0, beta_0])
6   Sigma = FisherInfo(alpha_0, beta_0, n=n_samples, inverse=True)
7   z_scores = np.zeros(n_simus)
8
9   for i in range(n_simus):
10      samples = np.random.gamma(alpha_0, scale=beta_0, size=n_samples)
11      output = scipy.optimize.root(lambda x: score(x[0], x[1],
            samples), [2, 2])
12      theta_hat = output['x']
13      plt.plot(theta_hat[0], theta_hat[1], '.', color='#0099ff')
14      z_scores[i] = mahalanobis(theta_hat, mu, Sigma)
15
16  x, y = ellipse(mu, Sigma, z=1)
17  plt.plot(x, y, 'r', linewidth=2)
18  x, y = ellipse(mu, Sigma, z=2)
19  plt.plot(x, y, 'r', linewidth=2)
20
21  print( 'z < 1', np.sum( z_scores < 1 ) / n_simus )
22  print( 'z < 2', np.sum( z_scores < 2 ) / n_simus )
```

Code Block 5.4: Code to run 1 000 simulations and plot the resulting maximum likelihood estimates.

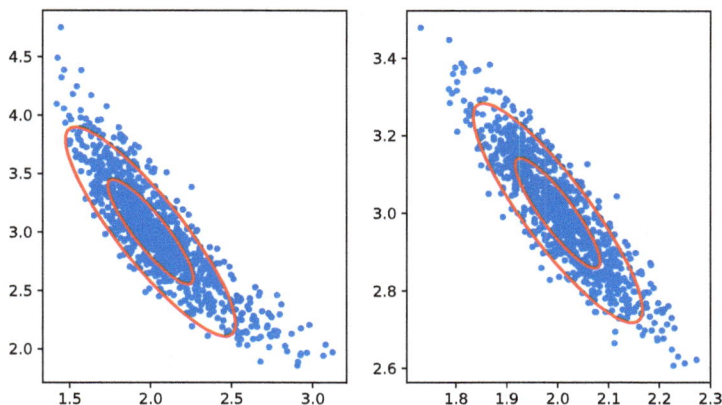

Fig. 5.2: 1,000 simulations with sample size 100 (left) and 1,000 (right).

We can repeat this game, changes `n_samples=1000` on line 3, to generate Figure 5.2 (right). (Note the different scale of Figure 5.2 (left) and (right).) The 1,000 computed MLE estimates behave much more closely to our theoretical normal distribution. However, we still observe 36.8% of simulations having $z < 1$ and 86.8% having $z < 2$. (An even closer agreement to Table 2.1, as expected.)

▷

5.2 Gradient Descent Method

The *Method of Gradient Descent* is an algorithm that locates a local minimum of an *objective function*. It is an optimization technique that can not only be used for solving maximum likelihood problems, but can also be applied to many problems in machine learning, as we will see later in this text. For more details on gradient descent and other optimization techniques, see Chong and Zak [2008].

5.2.1 Basic Gradient Descent

Classic gradient descent is an iterative technique used to solve the optimization problem

$$\min f(x),$$

for a given $f : \mathbb{R}^n \to \mathbb{R}$. Gradient descent works by taking a number of small, successive "steps" in the direction opposite of the objective function's gradient. An outline of this algorithm is provided in Algorithm 5.2.

Data: A given function $f : \mathbb{R}^n \to \mathbb{R}$ and its gradient
$Df : \mathbb{R}^n \to \mathbb{R}^n$; starting point $x_0 \in \mathbb{R}^n$, step size h, error
tolerance `tol`.

Result: Approximate local minimum x^*.

1 $i = 0$;
2 **while** $||Df(x_i)|| > $ `tol` **do**
3 $\quad$ Compute $Df(x_i)$;
4 $\quad$ (Optional) Let $h = \arg\min_{h \geq 0} f(x_i - hDf(x_i))$;
5 $\quad$ Compute $x_{i+1} = x_i - h \cdot Df(x_i)$;
6 $\quad$ $i = i + 1$
7 **end**
8 Output x_i, $f(x_i)$.

Algorithm 5.2: Gradient Descent Algorithm. (With Line 4, Steepest Descent).

The task of tuning the step size is a delicate one, as a smaller step size leads to a greater number of steps, and more computational tasks before convergence, whereas a larger step size may lead to overshooting, resulting in a more zig-zagged path. Some variations of the algorithm allow for variable step sizes. In particular, when line 4 of Algorithm 5.2 is deployed, the algorithm is referred to as *Method of Steepest Descent*. In this version of gradient descent, the task of selecting the step size at each step is tantamount to solving a one-dimensional line-search problem. Details of such an implementation can be found in Chong and Zak [2008].

It goes without saying that the method, in some cases, might fail to converge, or might converge to a local minimum that is not the global minimum. As with many optimization algorithms, success sometimes hinges on finding a suitable starting point x_0.

Example 5.7. Use gradient descent to determine the minimum of

$$f(x, y) = \frac{x^2}{5} + y^2.$$

A contour plot of this quadratic can be produced, as shown in Code Block 5.5. The contours are shown in Figure 5.3.   Gradient descent is implemented in Code Block 5.6.

Gradient descent, following Code Block 5.6, was run three times, with step sizes $h = 0.1$, 0.9, and 0.99. The results are shown in Figure 5.3. For a step size of $h = 0.1$, gradient descent follows the path of steepest descent, but it takes 311 steps to converge. (A step size of $h = 0.01$ would require 3,163 steps to converge.) A step size of $h = 0.9$ does much better, converging in only 60 steps. However, since the individual steps are larger, the algorithm results in a more zig-zagged path. Increasing the step size to $h = 0.99$, the algorithm converges in 659 steps. A step size of $h = 1$, the algorithm fails to converge. (At least, it does not converge within 10,000 steps.)

```
delta = 0.025
x = np.arange(-9.0, 9.0, delta)
y = np.arange(-4.0, 4.0, delta)
X, Y = np.meshgrid(x, y)
Z = X**2/5 + Y**2

fig, ax = plt.subplots(figsize=(8, 9/2))
CS = ax.contour(X, Y, Z)
ax.clabel(CS, inline=1, fontsize=10)
ax.set_title('Gradient Descent Example')
ax.set_aspect('equal')
```

Code Block 5.5: Contours for Example 5.7

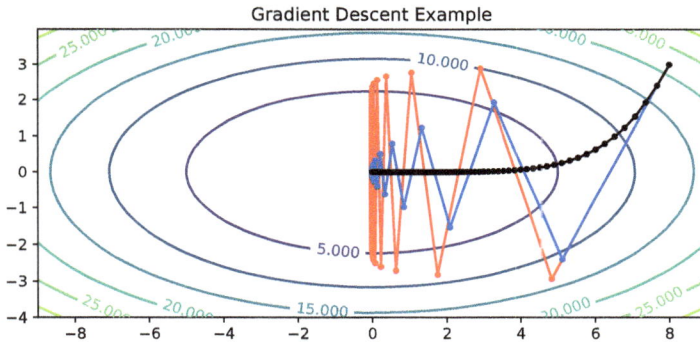

Fig. 5.3: Gradient Descent: $h = 0.1$ (black), $h = 0.9$ (blue), $h = 0.99$ (red).

This example illustrates the trade off between small and large step sizes. The number of steps required to converge is shown as a function of step size in Table 5.1. We observe that having a step size that is too small results in

step size	0.01	0.1	0.4	0.5	0.7	0.8	0.9	0.99
steps	3,163	311	73	57	39	33	60	659

Table 5.1: Converge time (number of steps) as a function of step size

an increased time to convergence, even though the path that is followed is a smooth one. Here, the converge time is a result of taking extremely small

```
1   def f(x):
2       return x[0]**2 / 5 + x[1]**2
3
4   def Df(x):
5       A = np.zeros(2)
6       A[0] = 2 * x[0] / 5
7       A[1] = 2 * x[1]
8       return A
9
10  x = np.array([8, 3]) # Initialize
11  h = 0.1
12  tol = 1e-5
13  max_steps = 10000
14  go = True
15  converged = False
16  X = x.reshape((2,1))
17
18  i = 0
19  while go: # Implement Gradient Descent
20      x = x - h * Df(x)
21      X = np.concatenate((X, x.reshape((2,1))), axis=1)
22      i += 1
23      if np.linalg.norm(Df(x)) < tol:
24          go = False
25          converged = True
26      if i > max_steps:
27          go = False
28  if not converged: # Warning
29      print ('Failed to converge')
30
31  for i in range(X.shape[1]): # Plot
32      plt.plot(X[0,i], X[1,i], 'k.')
33  plt.plot(X[0,:], X[1,:], 'k')
```

Code Block 5.6: Gradient Descent Algorithm for Example 5.7

steps. On the other hand, if the step size is too large, the path can end in a zig-zag path that bounces around before finally converging, or it might fail to converge altogether.   ▷

5.2.2 Stochastic Gradient Descent

Gradient descent is a commonly used optimization algorithm in statistical and machine learning. In this context, however, the objective function typically takes the form of a sum, as we saw with the log-likelihood function given by Equation (5.2). Thus, let us consider an objective function of the form

$$J(\theta) = \frac{1}{n} \sum_{i=1}^{n} f(\theta; x_i), \tag{5.25}$$

where the set $\mathcal{X} = \{x_1, \ldots, x_n\}$ constitutes a set of data. We will refer to the function f as the *kernel* of the objective function J, as it is used, along with the data $\mathcal{X}$, to generate the objective function J. In machine learning, the data are sometimes referred to as *training samples*.

There are three main approaches for applying gradient descent to an objective function of the form Equation (5.25), as outlined in the following definition.

Definition 5.5. *Given a set of data $\mathcal{X}$ and an optimization kernel $f : \mathbb{R}^n \times \mathbb{R} \to \mathbb{R}$, the gradient descent algorithm, given by Algorithm 5.2, may be applied to the objective function Equation (5.25) with any of the following modifications:*

1. (Batch gradient descent) *Algorithm 5.2 is implemented as written; i.e., the gradient for each step is computed using the full data set.*
2. (Mini-batch gradient descent) *A positive integer k, known as the* batch size, *is chosen (e.g., $k = 32$). For each step of gradient descent, the gradient is computed using k training examples. Once all the data have been used, the data are shuffled and the algorithm continues.*
3. (Stochastic gradient descent) *For each step of gradient descent, the gradient is computed using only a single training example. Once all the data have been used, the data are shuffled and the algorithm continues.*

A training epoch is said to occur each time the algorithm has cycled through all n training samples, regardless as to how they were grouped.

We note that batch and stochastic gradient descent are special cases of mini-batch with $k = n$ and $k = 1$, respectively. Batch gradient descent is exactly what one might expect, as the full gradient

$$DJ = \frac{1}{n} \sum_{i=1}^{n} \frac{\partial f(\theta; x_i)}{\partial \theta}$$

is computed with each step of the process. Thus, each step of gradient descent represents one training epoch. Batch gradient descent has the disadvantage that the gradient of the kernel must be calculated for each datum on each step of gradient descent. This can easily become computationally prohibitive, especially when dealing with large data sets.

On the opposite end of the spectrum is stochastic gradient descent, where the gradient is replaced with the gradient of a single datum,

$$DJ = \frac{\partial f(\theta; x_j)}{\partial \theta}.$$

The datum x_j used for the ith step is computed as $j = i \mod n$, and the ordering of the data is shuffled each time a new training epoch is reached. In this case, a training epoch corresponds to n steps of gradient descent. Due to the stochastic nature of each update, stochastic gradient descent is less likely to converge on a local minimum. However, the updates are much more noisy, as each update is determined based on a single training example.

In mini-batch gradient descent, we group the data into $n_b = \text{ceil}(n/k)$ batches[8], $K_0, \ldots, K_{n_b-1}$, where $K_i = \{ki, \ldots, k(i+1) - 1\}$, with the exception that the final batch K_{n_b-1} terminates at n instead of $(kn_b - 1)$. For each step of gradient descent, we then use the partial gradient

$$DJ = \frac{1}{|K_i|} \sum_{j \in K_i} \frac{\partial f(\theta; x_j)}{\partial \theta}, \qquad (5.26)$$

where i is the step number modded by n_b, and where the data are again shuffled at the arrival of each new training epoch. In this case, a training epoch consists of precisely n_b steps of gradient descent. Mini-batch gradient descent alleviates some of the noise of stochastic gradient descent, but yet retains the advantage of the model being updated more frequently than in batch gradient descent.

In Code Blocks 5.7 and 5.8, we create an abstract class to serve as the parent for any objective function of the form of Equation (5.25). Note that this class is a subclass of ABC, which is imported from the abc package (for abstract classes). Overall, the ObjectiveFunction class looks like a typical class, except for the two methods kernel and kernelGrad, which don't seem to do anything, and which are adorned with the decorator @abstractmethod. This decorator ensures that any subclass must overwrite (and actually define) these two methods. In order to use this class, we simply create a subclass that overwrites the methods kernel and kernelGrad. (We will show how this is done in Example 5.8.)

We further utilize Python generators in order to implement all three use cases: batch, mini-batch, and stochastic. This is done within the (private) __batchGenerator method, which returns a generator object. This method is called whenever an object from this class instantiated, and the generator is saved as self.batch_gen. A generator is a function that essentially

---

[8] It would be inappropriate to use the floor function here, as the formula would not work correctly on certain edge cases. For example, consider batch size $k = 20$ and sample sizes $n = 80, 95, 100$. Naturally, the case $n = 80$ should be divided into four batches, whereas both $n = 95$ and $n = 100$ must be divided into five batches. The ceiling function of n/k yields 4, 5, and 5, whereas the floor function yields 4, 4, and 5. The point here is not that the floor function would need to be adjusted by $+1$, but rather that the floor function would group 80 and 95 into the same number of batches, which should not be the case.

```python
from abc import ABC, abstractmethod
class ObjectiveFunction(ABC):

    def __init__(self, data, batch_size=0):
        """
        data (array or list)
        batch_size (int):
            0 for full-batch gradient descent
            1 for stochastic gradient descent
            >1 for mini-batch gradient descent
        """
        assert batch_size >= 0, "Batch size must be non-negative"
        assert batch_size < len(data), "Batch size too large"
        self.data = np.array(data) # In case input is list
        self.batch_size = batch_size
        self.batch_gen = self.__batchGenerator() #generator object

    def __len__(self):
        return len(self.data)

    def __batchGenerator(self):
        cycle_index = 0
        n_batches = int( np.ceil( len(self.data) / self.batch_size )
            )
        while True:
            if self.batch_size == 0:
                yield self.data
            else:
                if cycle_index == n_batches:
                    # Reset Training Epoch
                    np.random.shuffle(self.data)
                    cycle_index = 0

                start = self.batch_size * cycle_index
                stop = self.batch_size * (cycle_index + 1)
                cycle_index += 1
                yield self.data[start:stop]
```

Code Block 5.7: Abstract class for Objective Function. (Continued in Code Block 5.8.)

```python
def fEval(self, theta):
    """ Evaluates Objective Function using full data set """
    return np.sum( self.kernel(theta, self.data) ) / \
        len(self.data)

def grad(self, theta):
    """ Evaluates Gradient Function using current batch """
    data = next(self.batch_gen)
    return np.sum( self.kernelGrad(theta, data), axis=1 ) / \
        len(data)

@abstractmethod
def kernel(self, theta, x):
    pass

@abstractmethod
def kernelGrad(self, theta, x):
    pass
```

Code Block 5.8: Abstract class for Objective Function (continued).

remembers its own local variables, and executes the code until it reaches a `yield` statement. This is similar to a function that has a **return** statement, except that a generator remembers its spot in the code, as well as the values of local variables, and returns to this location the next time it is called. This way, whenever we call upon the **grad** method to obtain the gradient, it computes the gradient using only the next batch of data that is tracked by our generator.

As a final note, most generators terminate after a finite number of calls (think for-loop). However, our use of `while True`, ensures our generator will continue to run regardless of how many times it is called. For the case of (full) batch gradient descent, it simply returns the full data set each time it is called. For mini-batch (and, as a special case, stochastic) gradient descent, it defines a variable `cycle_index`, which tracks where it is within each training epoch. Each time `cycle_index` reaches the value `n_batches`, it shuffles the data set and resets the cycle index back to zero. Each time it is called, it implements Equation (5.26), returning the appropriate subset of data. When `batch_size=1`, it returns the single, next datum, and the training epoch is complete when it has returned the full data set.

The nice thing about this abstract class, is that it does all the heavy lifting for us, in terms of implementing batch, mini-batch, or stochastic gradient descent. Our actual gradient descent code can therefore be method-agnostic, as the method for tracking the batches is implemented internally within the objective function class.

Example 5.8. Next, let us continue our discussion of Example 5.5, the maximum likelihood estimation of a Gamma random variable, by showing how gradient descent can be used to estimate the parameter values. Let us start off by subclassing our `ObjectiveFunction` abstract class from Code Block 5.7, to create a class specific to the Gamma random variable. For this subclass, we only need to specify the definitions of the `kernel` and `kernelGrad` methods, which is shown in Code Block 5.9.

```
class GammaLikelihood(ObjectiveFunction):

    def kernel(self, theta, x):
        """ theta = [alpha, beta]
            x (np.array) len n
        output:
            1 x n array
        """
        return (theta[0]-1)*np.log(x) - x / theta[1] -
            np.log(scipy.special.gamma(theta[0])) -
            theta[0]*np.log(theta[1])

    def kernelGrad(self, theta, x):
        """ theta = [alpha, beta]
            x (np.array) len n
        output:
            2 x n array
        """
        U_alpha = np.log(x) - scipy.special.psi(theta[0]) -
            np.log(theta[1])
        U_beta = x / theta[1]**2 - theta[0] / theta[1]
        return np.array([U_alpha, U_beta])
```

Code Block 5.9: Gamma Likelihood Objective Function

Note that both of these methods return matrices with as many columns as training instances. That is, `kernel` returns a $1 \times n$ array, and `kernelGrad` returns a $2 \times n$ array. The summation of these values is taken care of in the `fEval` and `grad` methods in the parent class.

Now that we have defined our subclass, we can implement gradient descent in a few lines of code, as shown in Code Block 5.10. Note that we can easily switch between (full) batch gradient descent (`batch_size = 0`), mini-batch gradient descent (e.g., `batch_size = 1000`), and stochastic gradient descent (`batch_size = 1`) by changing the input to the constructor on line 7. We also define a list L that tracks the value of the full likelihood function with each step in our gradient descent algorithm; see lines 15 and 20. We note that the true value is $\theta = (2, 3)$, whereas our starting point

```
1   alpha_0 = 2
2   beta_0 = 3
3   n_samples = 1000000
4   samples = np.random.gamma(alpha_0, scale=beta_0, size=n_samples)
5
6   ## Implement Gradient Descent
7   G = GammaLikelihood(samples, batch_size=0) # set batch_size here
8   theta = np.array([3, 2])
9   h = 0.1
10  tol = 1e-3
11  max_steps = 10000
12  go = True
13  converged = False
14  X = theta.reshape((2,1))
15  L = [G.fEval(theta)]
16
17  i = 0
18  while i < max_steps and np.linalg.norm( np.sum(G.kernelGrad(theta,
        G.data), axis=1) ) / len(G) > tol:
19      theta = theta + h * G.grad(theta)
20      L.append(G.fEval(theta))
21      X = np.concatenate((X, theta.reshape((2,1))), axis=1)
22      i += 1
```

Code Block 5.10: Gradient Descent for Example 5.8

is $\theta_0 = (3, 2)$. Also, since we are trying to *maximize* the log-likelihood function, we use a *plus* instead of a *minus* on line 19, so that the algorithm allows us to drift up-hill instead of down-hill.

We ran this code three times, using full batch (batch_size = 0), mini-batch (batch_size = 32), and stochastic (batch_size = 1) gradient descent. The results are shown in Figure 5.4. We observe that the path traced by stochastic gradient descent does quite a bit more "wandering" before it finally converges. The stochastic nature of this path helps prevent the algorithm from converging on a local minimum that is not a global minimum. Using a batch size of 32 dramatically reduces the stochasticity, whereas the full batch gradient descent follows a path directly toward the maximum.

Finally, for each case we plot the log-likelihood as a function of the number of steps, in Figure 5.5. We observe that the full batch gradient descent, as expected, shows a much smoother and monotonic improvement in log-likelihood with each step, whereas the stochastic gradient descent makes numerous steps that result in a worse log-likelihood.

As a final reminder, each step of the full batch gradient descent uses all 1,000,000 pieces of data, and so there are many times more computations done with each individual step, whereas stochastic gradient descent uses

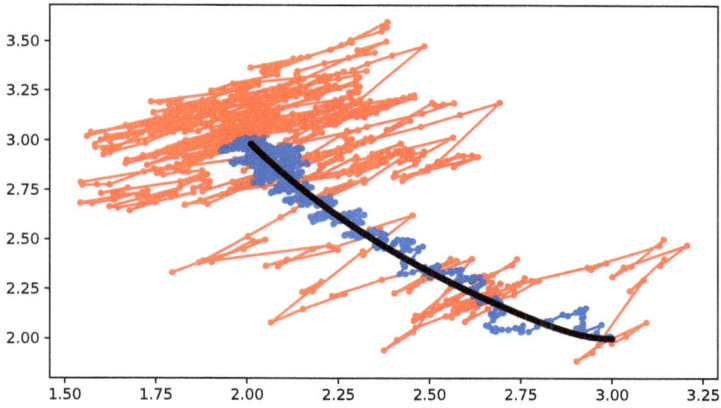

Fig. 5.4: Gradient Descent: full batch (black), mini-batch size 32 (blue), stochastic (red). Curves start at $(3, 2)$ and end at $(2, 3)$.

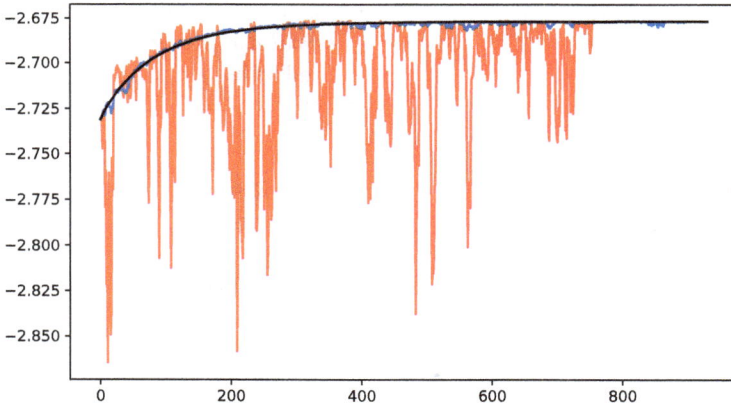

Fig. 5.5: Log-likelihood for three cases: full batch (black), mini-batch size 32 (blue), stochastic (red).

only 1 datum with each step. So even though they converge in approximately the same number of steps (full batch 931, mini-batch 900, stochastic 757), the number of *computations* is dramatically lower. This distinction is only exaggerated within the realms of big data, where a single step of the full batch gradient descent might be cost prohibitive.   ▷

5.3 Censored and Truncated Data

In this section, we consider statistical estimation for incomplete data sets. There are two primary ways that lead to (systematically) incomplete data sets: censorship and truncation. We will go over the basics of these mechanisms. For a more complete review, see Klein and Moeschberger [2003].

5.3.1 Overview

Before diving in, let us provide an overview of the distinction between censorship and truncation. As stated previously, they each deal with incomplete data. They differ, however, in the way in which data are left out.

Definition 5.6. *A set of data* $\mathcal{X} = X_1, \ldots, X_n$ *is said to be* censored *if its values are known only if they fall within an interval* $I = (x_l, x_r)$, *and if the counts of data below or above the interval are also known. Moreover, the data are considered to be* interval-censored *if both* x_l *and* x_r *are finite,* left-censored *if* $x_r = \infty$, *and* right-censored *if* $x_l = -\infty$ *(or if* $x_l = 0$, *depending on context).*

As an example of censored data, consider a small kitchen scale that can weight objects up to 3 kg. Upon weighing seven household objects, we obtain the data $\mathcal{X} = (1.27, 0.65, 3, 2.12, 3, 3, 2.86)$. Three of these data are right-censored at 3, whereas four of these data have observed values. Note that, critically, we know *how many* objects fall outside of our censorship interval $(0, 3)$. We say that these data are right-censored at $X = 3$.

Definition 5.7. *A set of data* $\mathcal{X} = X_1, \ldots, X_n$ *is said to be* truncated *if data are observed only if they fall within an interval* $I = (x_l, x_r)$. *Moreover, the data are considered to be* interval-truncated *if both* x_l *and* x_r *are finite,* left-truncated *if* $x_r = \infty$, *and* right-truncated *if* $x_l = -\infty$ *(or if* $x_l = 0$, *depending on context).*

As an example, consider the height distribution of Royal Marines. Since the Royal marines have a height restriction, only admitting persons at least 64 inches tall, the data will be left-truncated Trussell and Bloom [1979]. Note, critically, that we do not know how many individuals did not meet the height requirement and therefore did not find their way into the data set. As a result, the average height of a Royal Marine will be higher than the average height of the population. We say that the data in this case are left-truncated at $X = 64$.

Interestingly, data can be simultaneously both truncated and censored. For example, consider an auto-insurance policy that has a deductible of $500 and a maximum payout of $50,000. We would say that the set of claim values is left-truncated at 500, because individuals will not file a claim unless the total amount is above their deductible, and right-censored at 50,000, since claims greater than this amount are capped at this value. Note that we do not know how many claims there might have been under $500 (since those claims were never made), but we do know the count of claims above $50,000, as they were each capped at this maximum payout value.

5.3.2 Censored Data

In the case of censored data, we observe the values of data within our interval I, as well as the counts of data that were to the left or right of our interval. We may therefore modify our likelihood function with an appropriate number of factors of the cumulative distribution function, evaluated at the left-censorship point, and the survival function evaluated at the right censorship point. In particular, we have the following:

Proposition 5.4. *Consider a set of* IID *data* $X_1, \ldots, X_n \sim f(x; \theta)$ *that is censored by the interval* $I = (x_l, x_r)$. *Suppose that exactly* p *of the data points are left censored and* q *are right censored. Without loss of generality, we can arrange our data such that the first* $n - p - q$ *points are the values that lie within the interval* I. *Then the likelihood function of our data is given by*

$$\mathcal{L}(\theta) = [1 - S(x_l; \theta)]^p S(x_r; \theta)^q \prod_{i=1}^{n-p-q} f(x_i; \theta), \qquad (5.27)$$

where $S(x; \theta) = \int_x^\infty f(t; \theta) \, dt$ *is the survival function of the* PDF f.

The quantity $[1 - S(x_l; \theta)] = C(x_l; \theta)$ is equivalent to the CDF, which represents the total amount of probability to the left of the point $X = x_l$, whereas the survival function $S(x_r; \theta)$ represents the total probability to the right of x_r. (We express the entire formula in terms of the survival function to follow custom.) We therefore treat each censored data point by adding a factor equal to the probability that the data point would have been censored. Note that Equation (5.27) treats the case of left, right, and interval censored data using a single, compact formula.

Example 5.9. Determine the likelihood function for IID data $T_1, \ldots, T_n \sim \text{Exp}(\beta)$ that are right-censored at the point $T = \tau$. Determine an analytical formula for the MLE $\hat{\beta}$.

We begin by counting the number of data points that are censored, obtaining

$$k = \sum_{i=1}^{n} \mathbb{I}[t_i \geq \tau].$$

Since the survival function at $T = \tau$ is given by $S(\tau; \beta) = e^{-\tau/\beta}$, Equation (5.27) yields the likelihood function

$$\mathcal{L}(\beta) = \frac{e^{-k\tau/\beta}}{\beta^{n-k}} \prod_{i=1}^{n} e^{-t_i/\beta \cdot \mathbb{I}[t_i < \tau]}. \tag{5.28}$$

Note that the product in this equation is equivalent to taking the product only over the $n - k$ uncensored lifetimes. The log-likelihood is therefore given by

$$\ell(\beta) = \frac{-k\tau}{\beta} - (n - k) \log(\beta) - \frac{1}{\beta} \sum_{i=1}^{n} t_i \mathbb{I}[t_i < \tau]. \tag{5.29}$$

Differentiating with respect to the parameter β and equating the result to zero, we find the MLE

$$\hat{\beta} = \frac{1}{n - k} \left[k\tau + \sum_{i=1}^{n} t_i \mathbb{I}[t_i < \tau] \right]. \tag{5.30}$$

This represents the total sum of observed lifetimes (including the k censored lifetimes, which have a current value of k) divided by the total number of uncensored lifetimes. This expression is therefore equivalent to

$$\hat{\beta} = \frac{1}{n - k} \sum_{i=1}^{n} \min(t_i, \tau).$$

Compare this for the MLE of uncensored data, which is given by $\hat{\beta} = (1/n) \sum t_i$. ▷

Example 5.10. A cohort of 1,000 players downloads the game *Angry Penguins* on its release date. Let the random variable T represent the customer lifespan of each player, i.e., the time it takes for each customer to *churn*, and suppose that $T \sim \text{Exp}(\beta)$ is exponentially distributed. Assume that the scale parameter has a true value of $\beta = 6$, measured in months, and that nine months have passed since the game's release. Simulate the lifetime of each user, indicating which users' lifetimes are censored, and then use Equation (5.30) to estimate the value of the parameter β.

We simulate 1,000 user lifetimes from $\text{Exp}(6)$, as shown in Code Block 5.11. From this shining simulation, we estimate $\hat{\beta} = 6.146003$, which is quite close to the true value of $\beta = 6$. Without correcting for the fact that our data are right-censored, our estimate, obtained by `np.sum(np.fmin(9, lifetimes)`

```
1  beta = 6
2  tau = 9
3  n_samples = 1000
4  lifetimes = np.random.exponential(scale=beta, size=n_samples)
5
6  k = sum( lifetimes >= tau) # 237
7  beta_hat = np.sum(np.fmin(9, lifetimes)) / (n_samples - k) #
       6.146003
```

Code Block 5.11: Simulation of user lifetime

/ n_samples, would equal 4.6894, which is definitely incorrect. The estimate using the full set of data, with the censorship removed, is given by
np.sum(lifetimes) / n_samples, which results in an estimate of 6.187.
▷

5.3.3 Truncated Data

In the case of truncated data, we only observe the each datum if it lies
within our interval I, and we receive no information regarding the counts
of data that fell outside of our interval. In this case, we modify our likelihood function by renormalizing the probability density to only consider the
probability within the interval I. In particular, we have the following:

Proposition 5.5. *Consider a set of* IID *data* $X_1, \ldots, X_n \sim f(x; \theta)$ *that
are truncated by the interval* $I = (x_l, x_r)$, *so that* $x_l \leq X_i \leq x_r$, *for all*
$i = 1, \ldots, n$. *Then the likelihood function of our data is given by*

$$\mathcal{L}(\theta) = \frac{1}{[S(x_l) - S(x_r)]^n} \prod_{i=1}^{n} f(x_i; \theta), \qquad (5.31)$$

where $S(x; \theta) = \int_x^\infty f(t; \theta)\, dt$ *is the survival function of the* PDF f.

In Equation (5.31), we have replaced the PDF with the renormalized PDF

$$f(x; \theta) \to \frac{f(x; \theta)}{S(x_l) - S(x_r)},$$

defined on the interval I. For the special case of left-truncated data, Equation (5.31) reduces to

$$\mathcal{L}(\theta) = \frac{1}{S(x_l)^n} \prod_{i=1}^{n} f(x_i; \theta), \quad \text{(left-truncated)}$$

since $S(\infty) \to 0$. (Recall $S(x_l) = \mathbb{P}(X \geq x_l)$.) Similarly, for the case of
right-truncated data, Equation (5.31) reduces to

$$\mathcal{L}(\theta) = \frac{1}{[1 - S(x_r)]^n} \prod_{i=1}^{n} f(x_i; \theta), \quad \text{(right-truncated)}$$

since $S(-\infty) \to 1$. (Recall $1 - S(x_r) = \mathbb{P}(X \leq x_r)$.)

Example 5.11. Determine the likelihood function for IID data $T_1, \ldots, T_n \sim$ Exp(β) that are right-truncated at the point $T = \tau$.

Following Equation (5.31), the likelihood function is given by

$$\mathcal{L}(\beta) = \frac{1}{[1 - e^{-\tau/\beta}]^n \beta^n} \prod_{i=1}^{n} e^{-t_i/\beta}.$$

Here, each of the t_i are observed, i.e., $t_i \leq \tau$, and the number of unobserved observations, for which $T > \tau$, is unknown. The log-likelihood is given by

$$\ell(\beta) = -n \log(1 - e^{-\tau/\beta}) - n \log(\beta) - \frac{1}{\beta} \sum_{i=1}^{n} t_i.$$

By differentiating with respect to β and setting the result equal to zero, we obtain the following transcendental equation,

$$\beta - \frac{\tau}{e^{\tau/\beta} - 1} = \frac{1}{n} \sum_{i=1}^{n} t_i, \tag{5.32}$$

which may be solved for β to yield the MLE.   ▷

Example 5.12. A cohort of 10,000 players downloads the freemium game *Angry Penguins* on its release date. The game developer wants to estimate the number of players who will "convert" into paying users, i.e., the number of players who will make a purchase within the game. Suppose the random variable T represents the time until first purchase of a payer. Suppose that $T \sim$ Exp(β). Use the value $\beta = 6$ to simulate the first-purchase time for a total of 1,000 payers. Then truncate the data at $\tau = 9$, and estimate the value of β from the simulated data.

The code is shown in Code Block 5.12. Upon sampling 1,000 data points, corresponding to the 1,000 payers, on line 4, we then proceed to truncate the data at $\tau = 9$. That is, for the purpose of our simulation, we have two pieces of knowledge that we would not have had in real life: the true value of the parameter $\beta = 6$ and the true total number of payers 1,000.

After truncating the data (line 5), we are left with 791 observed first-purchase times. The story for the data we have simulated is as follows: 10,000 players download the game on its release date. We do not know how many will ultimately "convert" into payers. However, by day 9, we have observed 791 users who have made their first purchase. This is an example of truncated data, not censored data, as we ultimately do not

```
1  beta = 6
2  tau = 9
3  n_samples = 1000
4  conversion_times = np.random.exponential(scale=beta, size=n_samples)
5  data = conversion_times[conversion_times < tau] #791
6  data.mean() # 3.391
7  def f(beta):
8      return beta - tau / (np.exp(tau / beta)-1) - data.mean()
9  scipy.optimize.root(f,5) # 5.8588
```

Code Block 5.12: Simulation of conversion curve

know how many payers we have yet to observe. Since the data are right-truncated and exponentially distributed, we may use Equation (5.32), which is implemented in lines 7–9. Note that the mean first-purchase time of our observed data is 3.391 (line 6), whereas the MLE, when accounting for truncation, is 5.8588, an excellent approximation to the true value of $\beta = 6$.

We should mention that since these simulations are computed using random number generators, the precise values of the estimates will change each time they are run. To decrease the variance of the estimates, use a larger sample size, e.g., we ran the same code using **n_samples=1000000** and obtained an estimate $\hat{\beta} = 5.979$. Repeat simulations with a larger sample size will consistently return estimates that are much closer to the true value of $\beta = 6$.

The difference between this example and Example 5.10 is that in Example 5.10, we know the total count of users, and therefore know the total count of the users who have yet to churn. In this example, while we again know that there are a total of 10,000 players, we do not know how many of those players will ultimately make a purchase. We only know that, after nine days, a total of 791 users have already made their first purchase. However, we do not know how many of the remaining 9,209 *players* will ultimately become *payers*. Therefore, we must rely on truncation and not censorship to obtain our estimate.

With our estimate of $\hat{\beta} = 5.8588$ in hand, we can compute the survival function $S(9; 5.8588) = 0.2152$, which estimates the fraction of total payers who have yet to convert (i.e., who have yet to make their first purchase). We therefore estimate that a total of 1,008 users (792 observed / 0.7845 estimated fraction of observed) will ultimately convert into payers, which is reasonably close to the true value of 1,000, which was used to generate the data. ▷

5.3.4 Hazard Function

The examples we have visited so far can be viewed as industry applications of a field called *survival analysis*. Survival analysis is concerned with longevity of individual items within a collective; in particular, it provides a framework for analyzing the duration until a given event happens. There are many applications in medicine, biology, and engineering, including time-to-failure of a mechanical system as well as more literal applications involving the time-to-death of patients in a clinical study[9]. There are many examples of more traditional applications in Klein and Moeschberger [2003].

One quantity often discussed in survival analysis is the *hazard function*, defined as the probability density of an item failing now, given that it has already lasted this long. Specifically, we can think of this as the following limit

$$h(x) = \lim_{\Delta x \to 0} \frac{\mathbb{P}(x \leq X \leq x + \Delta x | X \geq x)}{\Delta x}.$$

We will, however, define the hazard function in an equivalent but more practical way.

Definition 5.8. *Given a random variable $X \sim f$, its* hazard function *is defined by the ratio*

$$h(x) = \frac{f(x)}{S(x)}, \tag{5.33}$$

where $S(x)$ is the survival function.

Example 5.13. Show that the hazard function of an exponential random variable $X \sim \text{Exp}(\beta)$ is constant.

Recall that

$$f(x; \beta) = \frac{1}{\beta} e^{-x/\beta} \qquad \text{and} \qquad S(x) = e^{-x/\beta}.$$

Taking the ratio, we see that the hazard function is given by

$$h(x; \beta) = \frac{1}{\beta},$$

a constant.   ▷

Example 5.13 shows that an item's probability of failing eminently is always the same, regardless of its age. This is related to the fact that the exponential random variables are *memoryless*.

---

[9] To eschew the macabre, however, we opted for applications dealing with the lifetime of players within a game.

5.3.5 The Pareto Distribution

We next introduce two new distributions that have simple hazard functions. The first of these is a basic power-law distribution that has many applications in various areas, including the distribution of wealth. It is also the distribution that gives rise to the so-called "80-20" rule, which states that 80% of wealth of a society is held by the top 20% of its population, or that 20% of your effort typically produces 80% of your return (in work, sports, etc.). This distribution is, of course, the Pareto distribution, which is defined as follows.

Definition 5.9. *A random variable X is considered a Pareto random variable, denoted $X \sim \text{Pareto}(m, \alpha)$, if it satisfies the Pareto distribution function*

$$f(x; m, \alpha) = \frac{\alpha m^\alpha}{x^{\alpha+1}} \tag{5.34}$$

on the domain $x \in [m, \infty)$. The parameter m is called the scale parameter, and the parameter α is called the shape parameter.

The Pareto distribution has the following properties.

Proposition 5.6. *Let $X \sim \text{Pareto}(m, \alpha)$. Then*

$$S(x) = \left(\frac{m}{x}\right)^\alpha, \tag{5.35}$$

$$\mathbb{E}[X] = \frac{\alpha m}{\alpha - 1} \quad \text{for } \alpha > 1, \tag{5.36}$$

$$\mathbb{V}(X) = \frac{m^2 \alpha}{(\alpha - 1)^2(\alpha - 2)} \quad \text{for } \alpha > 2. \tag{5.37}$$

The expected value and variance diverges for $\alpha \leq 1$ and $\alpha \leq 2$, respectively.

As a result of Equations (5.34) and (5.35), the hazard function of the Pareto distribution is given by

$$h(x) = \frac{\alpha}{x}.$$

We can use the built-in `numpy` and `scipy` methods to sample from a Pareto distribution and draw the corresponding PDF, as shown in Code Block 5.13. Note that the built-in `numpy.random.pareto` is actually a Pareto Type II distribution, or Lomax distribution, which we correct for by adding one to the output. The resulting output is shown in Figure 5.6.

5.3.6 The Weibull Distribution

A commonly used distribution in survival analysis is the Weibull distribution, which has a power-function hazard rate.

```
1  m = 1
2  alpha = 2
3  n_samples = 1000
4  samples = m * (1 + np.random.pareto(alpha, size=n_samples) )
5
6  x = np.linspace(1, 20)
7  y = scipy.stats.pareto.pdf(x, alpha, scale=m)
8  plt.hist(samples, bins=100, normed=True)
9  plt.plot(x,y)
```

Code Block 5.13: Samples from a Pareto distribution

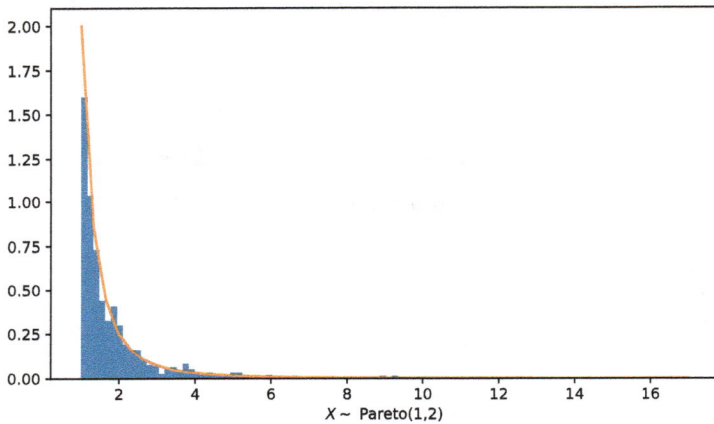

Fig. 5.6: Samples from Pareto(1,2).

Definition 5.10. *A random variable X is considered a* Weibull *random variable, denoted $X \sim$ Weibull(α, β), if it satisfies the* Weibull *distribution function*

$$f(x; \alpha, \beta) = \frac{\alpha}{\beta} \left(\frac{x}{\beta}\right)^{\alpha-1} e^{-(x/\beta)^\alpha}, \tag{5.38}$$

on the domain $x \in [0, \infty)$. The parameter α is called the shape parameter, *and the parameter β is called the* scale parameter.

The Weibull distribution has the following properties.

Proposition 5.7. *Let $X \sim$ Weibull(α, β). Then*

$$S(x) = e^{-(x/\beta)^\alpha}, \tag{5.39}$$

$$\mathbb{E}[X] = \beta \Gamma(1 + 1/\alpha), \tag{5.40}$$

$$\mathbb{V}(X) = \beta^2 \left[\Gamma(1 + 2/\alpha) + \Gamma(1 + 1/\alpha)^2\right], \tag{5.41}$$

where Γ represents the Gamma function (Definition 2.13).

As a result of Equations (5.38) and (5.39), the hazard function of the Pareto distribution is given by

$$h(x) = \frac{\alpha}{\beta} \left(\frac{x}{\beta}\right)^{\alpha-1}.$$

If a Weibull random variable $X \sim$ Weibull(α, β) represents the time-to-failure or lifetime, then the behavior can be described by three separate cases.

If $\alpha < 1$, the hazard rate decays over time, indicating the longer an item survives, the less its marginal probability of demise. This is sometimes referred to as the *Lindy effect*. As the item ages, its remaining life expectancy actually increases. Often certain nonperishable goods, such as ideas, books, or technology, exhibit this kind of reverse aging: many are killed in infancy, and the longer any one survives, the more its expected additional lifetime. If this seems counterintuitive, consider the expected remaining lifespan for a recently published mass paperback novel versus the works of Shakespeare. The fact that the works of Shakespeare have already survived half a millennium is a testament to their robustness as a staple of literature, whereas the romantic tripe published just yesterday might already be off the shelves by tomorrow.

If $\alpha = 1$, the hazard rate is constant over time, and the Weibull distribution reduces to an exponential distribution as a special case, in fact careful inspection reveals Weibull$(1, \beta) = $ Exp(β).

Finally, if $\alpha > 1$, the hazard rate increases with time, indicative of an aging process: the more an item ages, the more likely its impending doom.

We can use the built-in **numpy** and **scipy** methods to sample from a Weibull distribution and draw the corresponding PDF, as shown in Code Block 5.14. The resulting output is shown in Figure 5.7. Finally, the Weibull distribution is plotted for various parameter values in Figure 5.8.

```
alpha = 2
beta = 3
n_samples = 1000
samples = beta * np.random.weibull(alpha, size=n_samples)
x = np.linspace(0, 8)
y = scipy.stats.weibull_min.pdf(x, alpha, scale=beta)
```

Code Block 5.14: Samples from a Weibull distribution

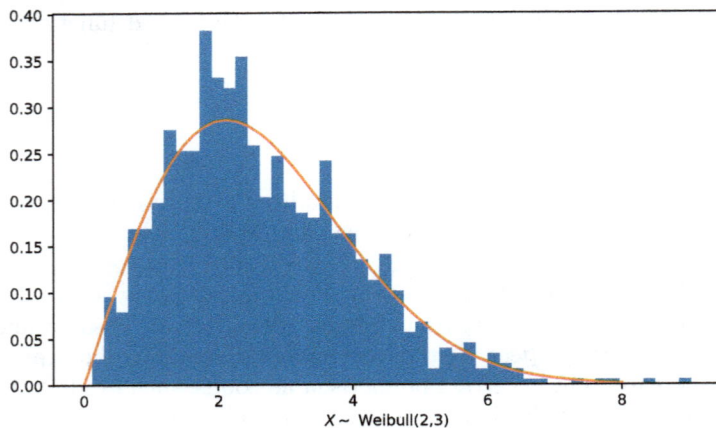

Fig. 5.7: Samples from Weibull(2,3).

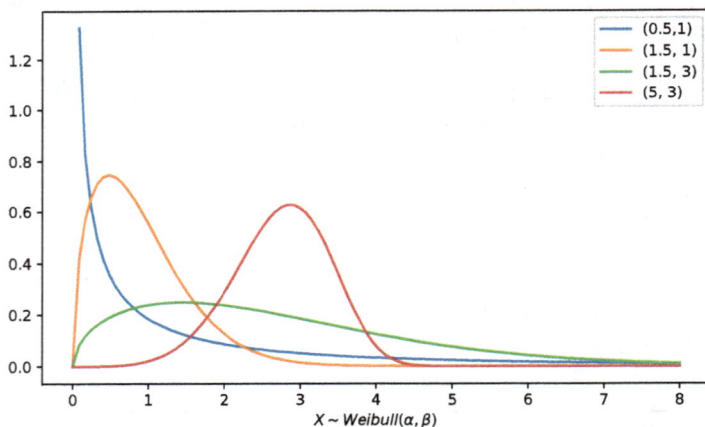

Fig. 5.8: Weibull(α, β) for various parameter values.

5.4 Online Processes

We conclude this chapter with a discussion on stochastic processes and, especially, a certain type of stochastic process we call online processes, which involve censored or truncated data. We will then discuss several empirical estimates for their underlying distribution function. We will not follow the traditional treatment of stochastic processes, for which we refer the reader to, e.g., Durrett [2016] or Hajek [2015]. Rather, we will develop our theory

based on a practical analysis of customer behavior for certain online products, such as websites and mobile apps. The idea is that user engagement
follows a distribution over time, which can be used to generate an accom-
panying stochastic process. The wrinkle is that, for online products, new
users can arrive each day, and the process begins anew for each user cohort.
Thus, we will consider a certain type of stochastic process that comes with
two indices, user cohort and the time that is measured within each cohort.
The online nature of the process causes the observed data to be diagonally
truncated or censored, depending on the specific application. The second
departure from traditional stochastic processes is that we will start with
the distributional nature of the process over time, using it to *generate* the
ensuing process.

5.4.1 Stochastic Processes

A stochastic process is defined over a sample space and over time as follows.
It is helpful to think of the probability space as consisting of the entire
process itself, as opposed to a single snapshot. That is, if we flip a coin
repeatedly, elements of the sample space are not $\{H, T\}$, but rather all
binary strings $\{HTHTTH\cdots, HHHTHT\cdots, \ldots\}$.

Definition 5.11. *A stochastic process (or* random process*) is a collection*
$\mathcal{P} = \{X_t\}_{t\in\mathcal{T}}$ *of random variables defined over a common probability space*
space $(\Omega, \mathcal{E}, \mathbb{P})$ *and indexed by a set* $\mathcal{T}$. *If* $\mathcal{T} = \mathbb{N}$, *then it is called a* discrete-
time random process. *If* $\mathcal{T} = \mathbb{R}$, *then it is called a* continuous-time random
process. *The range of the random variables* X_t *is called the* state space *and*
is denoted $\mathcal{S}$.

There are three equivalent ways to think of a stochastic process:

- the process X is a function over the Cartesian product $\mathcal{T} \times \Omega$, with
  value $X_t(\omega) = X(t, \omega)$ for any given $t \in \mathcal{T}$ and $\omega \in \Omega$;
- for each fixed $t \in \mathcal{T}$, the process $X_t(\omega)$ is a function over the sample
  space Ω;
- for each fixed sample outcome $\omega \in \Omega$, the process $X_t = X_t(\omega)$ is a
  function over the indexing set $\mathcal{T}$, known as a *realization* or *sample path*
  of the process.

Example 5.14. Let Ω be the sample space consisting of ten independent
flips of a fair coin. Then the set of random variables $\{X_t\}_{t=0}^{9}$, with state
space $\mathcal{S} = \mathbb{B} = \{0, 1\}$, defined so that $X_t = \mathbb{I}[i\text{th flip is heads}]$, constitutes
a stochastic process. An example outcome from the sample space is $\omega =$
$HTTHTHTTHT$, which would correspond to the sample path $X_0 = 1$,
$X_1 = 0$, $X_2 = 0$, $X_3 = 1$, etc. As this example shows, the sample space
for a stochastic process consists of all possible read-world outcomes of the
process; it is the set of all possible realizations of the process.   ▷

5.4.2 Common Types of Stochastic Processes

Before defining an online process, we first discuss some common flavors of
stochastic processes.

Markov Chains

A Markov chain, or Markov process, is a certain class of stochastic process
in which each random variable depends on its history only through the
immediately preceding random variable.

Definition 5.12. *A discrete-time stochastic process* $\{X_t\}_{t \in \mathbb{N}}$ *with state
space* $\mathcal{S}$ *satisfies the* Markov property *if*

$$\mathbb{P}(X_t \in S | X_0, X_1, \ldots, X_{t-1}) = \mathbb{P}(X_t \in S | X_{t-1}), \tag{5.42}$$

for all measurable $S \subset \mathcal{S}$.

A Markov process, *or* Markov chain, *is a stochastic process that satisfies
the Markov property.*

A Markov process is said to be homogeneous *if*

$$\mathbb{P}(X_{t+1} = x | X_t) = \mathbb{P}(X_1 = x | X_0);$$

*i.e., if the probabilistic relation of one state to the next is constant over
time. Moreover, the values* $p_{ij}(s,t)$ *defined by*

$$p_{ij}(s,t) = \mathbb{P}(X_t = j | X_s = i), \qquad for \ t \geq s,$$

are referred to as transition probabilities, *and which are independent of time
for homogeneous processes. For a finite state space, the set of transition
probabilities* $H(s,t) = \{p_{ij}(s,t) : s,t \in \mathcal{T}\}$ *constitutes a* state transition
matrix *between times s and t. For a homogeneous process, we will typically
refer to the state transition matrix as* $H = H(0,1)$.

Note 5.6. We limit our discussion to discrete-time Markov processes, as the
general definition relies on generated σ-algebras, a complexity unnecessary
for our present purpose. For completeness, however: A stochastic process is
said to satisfy the Markov property if

$$\mathbb{P}(X_t \in S | \mathcal{F}_s) = \mathbb{P}(X_t \in S | X_s), \qquad for \ t \geq s,$$

where $S \in \sigma(\mathcal{S})$ is in the σ-algebra over $\mathcal{S}$ and $\mathcal{F}_s = \sigma(\{X_h | h \leq s\})$ is
the σ-algebra generated by the history of the process X up until time s. In
other words, for $t \geq s$, the conditional probability of X_t given the entire
cumulative history of the process up until the point s is equivalent to the
conditional probability of X_t given the single state X_s. In yet even other
words: The probability distribution over a future state depends on its past
only through its present. ▷

Example 5.15. For fixed $p > 0$, a *random walk* is a discrete Markov process with state space $\mathcal{S} = \mathbb{Z}$ and transition probabilities

$$p_{ij} = \begin{cases} p & \text{if } j = i + 1 \\ (1 - p) & \text{if } j = i - 1 \\ 0 & \text{otherwise} \end{cases} .$$

That is, for each point in time, the state increments by one with probability p and decrements by one with probability $(1 - p)$.

```
n = 100
x = np.zeros(n)
x0 = 0
p = 0.5
for i in range(n):
    if i == 0:
        continue
    x[i] = x[i-1] + 2 * int(np.random.random() < p) - 1
```

Code Block 5.15: Random walk.

A random walk can easily be simulated, as shown in Code Block 5.15. The results of four random walks; i.e., four realizations or sample paths of this process; are shown in Figure 5.9.

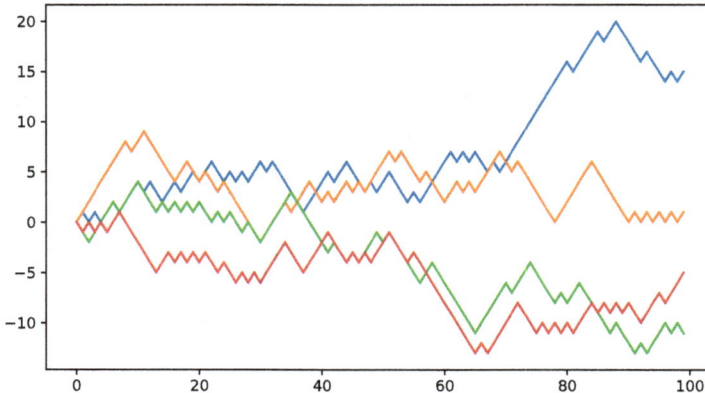

Fig. 5.9: Four random walks for Example 5.15.

A related problem known as *the gambler's ruin* is stated as follows: Let X_t be a random walk with $X_0 = s > 0$. Determine the probability that $X_t = 0$ for some $t > 0$. We leave this as an exercise for the reader.   ▷

Most of the time, however, we will consider discrete-time Markov processes with a finite (discrete) state space. In such cases, we can formulate the state transition matrices (since the enumeration of states is finite) and rely on the following two results.

Proposition 5.8. *Given a Markov process with a finite state space, the following* Chapman–Kolmogorov *equations hold*

$$H(s,t) = H(s,\tau)H(\tau,t),   \tag{5.43}$$

for any $s < \tau < t$.

Proof. This follows by considering the relation

$$\mathbb{P}(X_t = j | X_s = i) = \sum_{k \in \mathcal{S}} \mathbb{P}(X_t = j | X_\tau = k)\mathbb{P}(X_\tau = k | X_s = i),$$

which is equivalent to the result.   □

Proposition 5.9. *Let $\pi(t) = (\pi_i(t) : i \in \mathcal{S})$ be the* PMF *for the state of a homogeneous Markov process at time t, and let $H = H(0,1)$ be the single-step state transition matrix. Then*

$$\pi(t) = (H^T)^t \pi(0).   \tag{5.44}$$

Proof. Let e_i be the standard basis vector for state $i \in \mathcal{S}$. If the process is in state i at time t, i.e., if $X_t = i$, then the PMF over X_{t+1} is given by the transition probabilities

$$H_{ij} = \mathbb{P}(X_{t+1} = j | X_{t=i}),$$

which comprise the ith row of the matrix H. The PMF $\pi(t+1)$ is therefore the ith column of H^T, or $\pi(t+1) = H^T \cdot e_i$.

Next, suppose that the state at time t is not known, but a probability distribution $\pi(t)$. Note that this vector is equivalent to

$$\pi(t) = \sum_{i \in \mathcal{S}} \pi_i(t) e_i.$$

It follows that the probability distribution at time $t + 1$ is given by

$$\pi(t+1) = \sum_{i = \in \mathcal{S}} \pi_i(t) H^T e_i = H^T \pi(t);$$

i.e., by applying the transpose H^T to a probability distribution $\pi(t)$, we obtain the probability distribution over the states for the immediately following point in time. By successive application of this rule, we can transition a probability distribution over an initial state to the probability distribution t steps later:

$$\pi(t) = H^T \pi(t-1) = (H^T)^2 \pi(t-2) = \cdots = (H^T)^t \pi(0).$$

The result follows.   □

Counting and Arrival Processes

We have previously seen another example of a stochastic process, and that is the Poisson process. It turns out that the Poisson process is itself an example of an entire class of stochastic processes known as counting processes.

Definition 5.13. *A* counting process *is a stochastic process* $\{N(t)\}_{t \in \mathcal{T}}$ *with nonnegative-integer state space* $\mathcal{S} = \mathbb{N}$, *such that* $N(0) = 0$ *and such that, for any realization of the process, the sample function* N_t *is right-continuous and nondecreasing*[10].

A Poisson process is a special type of counting process that fits under the broader umbrella of *renewal processes*.

Definition 5.14. *Given a distribution* F *over* $\mathbb{R}_+ = (0, \infty)$[11], *known as* the interarrival distribution, *an* arrival process *is a discrete-time Markov process* $\{T_n\}_{n \in \mathbb{N}}$ *with* $T_0 = 0$, *such that the increments* $T_{n+1} - T_n \sim F$ *are* IID *over* F, *for* $n \in \mathbb{N}$. *The value* T_n, *for* $n \geq 1$, *is referred to as the* nth arrival time.

Note 5.7. The initial term $T_0 = 0$ is defined purely for convenience, so that the initial increment $T_1 - T_0 = T_1$ is well defined. Otherwise, we can disregard T_0.
Since the interarrival distribution is over $\mathbb{R}_+ = (0, \infty)$, it follows that the arrival times constitute a strictly increasing function of their index. ▷

Definition 5.15. *A* continuous-time counting process $\{N(t)\}_{t \in \mathbb{R}_*}$ *is a* renewal process *if there exists an arrival process* $\{T_n\}_{n \in \mathbb{N}}$ *such that*

$$N(t) = \sum_{i=1}^{\infty} \mathbb{I}[T_i \leq t]. \tag{5.45}$$

Definition 5.16. *A* Poisson process *is a renewal process with an exponentially distributed interarrival distribution.*

Thus, a renewal process is a generalization of a Poisson process where the interarrival distribution is not exponentially distributed.

---

[10] Right-continuous: $\lim_{t \to t_0^+} N_t = N_{t_0}$; nondecreasing: $N_{t+s} \geq N_t$, for any $s > 0$.
[11] I.e., F is the CDF of a positive random variable, so that $F(0) = 0$.

5.4.3 Aggregated Processes

We next turn to the case of a stochastic process that plays out independently over a collection of units, thereby generating a sample of realizations, which are then aggregated to form a more holistic picture.

Definition

The idea of an aggregate process is that it is just the sum of a sample of realizations from an underlying base process. Each sample outcome from the aggregation is therefore a set of sample outcomes from the base process. Therefore, a brief word is in order about the Cartesian product of a probability space.

Let $(\Omega, \mathcal{E}, \mathbb{P})$ be a probability space. Consider the Cartesian product of w copies of the sample space:

$$\Omega^w = \underbrace{\Omega \times \cdots \times \Omega}_{w \text{ times}}.$$

We can define the *product σ-algebra* as the σ-algebra generated[12] from the Cartesian product $\mathcal{E}^w$:

$$\sigma(\mathcal{E}^w) = \sigma\left(\{(E_1, \ldots, E_w) : E_1, \ldots, E_w \in \mathcal{E}\}\right).$$

Similarly, we can define a *product probability measure* $\mathbb{P}^w$ defined over $\sigma(\mathcal{E}^2)$. For the detailed definition of product measure, see any text on real analysis, such as DiBenedetto [2002] or Folland [1999]; or, for a definition in the context of probability theory, see Capinksi and Kopp [2005]. But essentially the product measure is a generalization of

$$\mathbb{P}^w((E_1, \ldots, E_w)) = \prod_{i=1}^{w} \mathbb{P}(E_i),$$

for $E_1, \ldots, E_w \in \mathcal{E}$, to the full σ-algebra $\sigma(\mathcal{E}^w)$.

Definition 5.17. *Given an underlying* base process $\{B_t\}_{t \in \mathcal{T}}$ *over* $(\Omega, \mathcal{E}, \mathbb{P})$, *an* aggregate process *with weight* w *is defined as the stochastic process* $\{X_t\}_{t \in \mathcal{T}}$ *over the product probability measure* $(\Omega^w, \mathcal{E}^w, \mathbb{P}^w)$, *such that*

$$X_t(\omega) = \sum_{i=1}^{w} B_t(\omega_i),$$

for any realization $\omega = (\omega_1, \ldots, \omega_w) \in \Omega^w$.

In other words: an aggregate process is simply the sum of a sample of realizations of an underlying base process. The sample mean of the realizations is therefore simply X_t/w.

---

[12] Let X be a set, and $\mathcal{S} \subset \mathcal{P}(X)$ a collection of subsets. Then the σ-algebra generated from $\mathcal{S}$, denoted $\sigma(\mathcal{S})$, is the smallest σ-algebra that contains the subsets $\mathcal{S}$.

Bernoulli Processes

Just how a binomial random variable is the sum of Bernoulli random variables, a binomial stochastic process is the sum of Bernoulli stochastic processes. It is therefore a perfect example of an aggregate process.

Definition 5.18. *A Bernoulli process is the stochastic process $\{X_t\}_{t\in\mathbb{N}}$ defined by $X_t \sim \text{Bern}(p)$, for some $p \in (0,1)$.*

A binomial process with weight w is the stochastic process $\{X_t\}_{t\in\mathbb{N}}$ defined by $X_t \sim \text{Binom}(w,p)$, for some $p \in (0,1)$.

If the parameter p is replaced by a function $p : \mathbb{N} \to (0,1)$, called the generator *of the process, either process is said to be* nonhomogeneous. *Futhermore, if the series $\sum_{t=0}^{\infty} p_t$ converges, the process is said to be* finite.

Clearly, the binomial process is simply an aggregate process, for which the underlying base process is a Bernoulli process.

We find the finite nonhomogeneous binomial process to be quite interesting. For example, consider a mobile app in which user engagement fluctuates over time. Each user's individual behavioral pattern constitutes a Bernoulli process. However, by aggregating the patterns over a collection of users, we have a binomial process. If enough users are aggregated together, we may infer something about the underlying generator p_t for the process; i.e., what is the distribution of engagement over time for the process.

If data for a binomial process do not exist beyond some $t > \tau$, we say the process is *truncated* at $t = \tau$. The process is truncated, and not censored, as we do not know the total count of events that would ultimately transpire, if the process were allowed to continue.

Survival Processes

Definition 5.19. *A unit survival process is a stochastic process $\{A_t\}_{t\in\mathcal{T}}$ with binary state space $\mathcal{S} = \mathbb{B} = \{0,1\}$ over the probability space $(\mathbb{R}_+, \mathcal{E}, \mathbb{P})$, defined for sample outcome $T \in \Omega = \mathbb{R}_+$:*

$$A_t = \mathbb{I}[T > t].$$

The sample outcome T is known as the survival time. *The probability distribution F_T over $\Omega = \mathbb{R}_+$ is called the* distribution of lifetimes, *and is said to* generate *the process.*

A survival process with weight w is an aggregation of w unit survival processes.

The aggregate survival process is simply the empirical survival function of the distribution F_T that generates the process.

If data for a survival process do not exist for $t > \tau$, we say the process is *censored* at $t = \tau$. The process is censored, and not truncated, as we know a priori the ultimate tally of survival events.

5.4.4 Online Processes

Many online products (apps, websites, etc.) are constantly acquiring new users who interact with them on a regular basis. A *user cohort* is a group of users who were acquired on the same day. It is often of interest to study certain random processes over users' lifecycles (e.g., engagement, churn, monetization, etc.). It is therefore useful to think of such online processes as a function of two indices: cohort membership and cohort time. *Cohort membership* is an index counting the various acquisition cohorts chronologically, and *cohort time* is a measure of each user cohort's time since acquisition. The idea is that each user cohort undergoes the same mathematical process once they are acquired; however, the acquisition of user cohorts happens in a staggered manner over time. Formally, we have the following.

Definition 5.20. *An* online process *is a stochastic process* $\{X_{ij}\}_{(i,j)\in\mathbb{N}^2}$ *with a two-dimensional discrete indexing set* $\mathbb{N}^2$ *and a set of weights* w, *such that*

1. calendar time *represents the total real calendar time since the first user cohort was acquired and the overall online process commenced,*
2. *index i represents the calendar time at which cohort i was acquired;*
3. *index j represents the* cohort time, *or the* time since acquisition, *within each cohort;*
4. *the isochrones* $t = i + j$ *represent the* calendar time *at which each random variable* X_{ij} *was observed;*
5. *the* process age τ *represents the current (cumulative) calendar time for the overall process;*
6. *the* cohort age $\tau - i$ *represents the current cohort time for cohort i;*
7. *the* cohort weights w *constitute a set of nonnegative weights assigned to each cohort, where the weight* w_i *typically represents the number of users belonging to cohort i;*
8. *the random variables* X_{ij} *are censored or truncated, depending on application, for* $t > \tau$.

Due to future-truncation across the isochrone $t > \tau$, the random variables X_{ij} are naturally expressed within an upper-left triangular matrix known as the *process matrix*. For example, the process matrix for an online process that is currently six days old (i.e., the process age is $\tau = 6$), is represented by

$$X = \begin{bmatrix} X_{00} & X_{01} & X_{02} & X_{03} & X_{04} & X_{05} \\ X_{10} & X_{11} & X_{12} & X_{13} & X_{14} & 0 \\ X_{20} & X_{21} & X_{22} & X_{23} & 0 & 0 \\ X_{30} & X_{31} & X_{32} & 0 & 0 & 0 \\ X_{40} & X_{41} & 0 & 0 & 0 & 0 \\ X_{50} & 0 & 0 & 0 & 0 & 0 \end{bmatrix}.$$

We can visualize the calendar time $t = i + j$, which represents the real, calendar times since the first cohort was acquired, by the series of diagonals represented below:

$$\begin{bmatrix} 0 & 1 & 2 & 3 & 4 & 5 \\ 1 & 2 & 3 & 4 & 5 & * \\ 2 & 3 & 4 & 5 & * & * \\ 3 & 4 & 5 & * & * & * \\ 4 & 5 & * & * & * & * \\ 5 & * & * & * & * & * \end{bmatrix}.$$

Notice that cohort 0 was acquired at calendar time 0; cohort 1 was acquired at calendar time 1, etc. The isochrones $t = i+j$ represent the actual calendar day on which the data were collected. The most recent day is the diagonal $t = \tau$. Data below this diagonal $(t > \tau)$ are missing (censored or truncated), as they represent days yet to come.

The more recently a cohort was acquired, the fewer data we have for the given cohort. It is useful to analyze online processes to determine whether the process is changing as a function of cohort age (e.g., recent cohorts are less engaged than cohorts from a year ago) and calendar-time isochrones (e.g., weekend or holiday effects, server outages, sale promotions, etc.).

Notwithstanding cohort and calendar effects, we typically like to model the process happening within each cohort. We therefore like to endow this array of random variables with some kind of additional, overarching structure that depends on the cohort time.

Definition 5.21. *Let $\mathcal{B}(\theta) = \{B_j | \theta\}_{j \in \mathbb{N}}$ represent a stochastic process that depends on a parameter θ. This process is the* base process *for an online process $\{X_{ij}\}_{(i,j) \in \mathbb{N}^2}$ with weights $\{w_i\}_{i \in \mathbb{N}}$ if, for each $i \in \mathbb{N}$, the process $\{X_{ij}\}_{j \in \mathbb{N}}$ constitutes an aggregate process over the base process $\mathcal{B}(\theta)$ with weight w_i.*

If the parameter θ depends on the cohort i, the online process is said to have a cohort effect. *If the parameter θ depends on the calendar time $t = i + j$, the online process is said to have a* calendar effect. *An online process with no cohort or calendar effect is said to be* pure.

Note 5.8. If a calendar effect depends on $t = i+j$ only through t mod 7, it is called a *day-of-week effect*. If a day-of-week effect corresponds to weekends, it is referred to as a *weekend effect*.   ▷

Definition 5.22. *An* online Bernoulli process *is an online process whose base process is the Bernoulli process (Definition 5.13).*

An online survival process *is an online process whose base process is the unit survival process (Definition 5.19).*

Example 5.16. Simulate the following online Bernoulli process, representing the counts of a certain engagement event within a mobile app:

1. cohort weights are normally distributed with mean 100 and standard deviation 20;
2. the random process $X_{ij} \sim \text{Binom}(w_i, \alpha f_T(j))$, where f_T is the probability mass function for a shifted geometric distribution with mean 20, and alpha is a scaling parameter;
3. the scaling parameter is typically 10, but after cohort 49, a sudden shift in user acquisitions causes this parameter to be reduced by 50%.
4. there was a bug in the app on days 30–15 resulting in a 60% drop in engagement,
5. there is a 50% lift in engagement on weekends; the first weekend occurred for calendar times t=5, 6.

All of these ingredients can easily be combined in a single simulation, as shown in Code Block 5.16. The result is plotted in Figure 5.10.

```
1   tau = 60
2   w = np.random.normal(loc=100, scale=20, size=tau).astype(int)
3   X = np.zeros((tau,tau))
4   alpha_0 = 10
5
6   for i in range(tau):
7       if i == 50: # cohort effect
8           alpha_0 *= 0.5
9       for j in range(tau-i):
10          t = i + j
11          alpha = alpha_0
12          if t % 7 in [5, 6]: # weekend effect
13              alpha *= 1.5
14          if t in range(30, 36): # calendar effect
15              alpha *= 0.4
16          p = alpha * scipy.stats.geom.pmf(j, 0.05, loc=-1)
17          X[i, j] = np.random.binomial(w[i], p)
18
19  plt.figure(figsize=(8, 9/2))
20  Y = (X / w.reshape((tau, 1))) # Divide each row by its weight
21  plt.pcolormesh(Y)
22  plt.gca().invert_yaxis()
23  pj_hat = X.sum(axis=0) / np.flip(np.cumsum(w))
24  se = np.sqrt(pj_hat * (1 - pj_hat) / np.flip(np.cumsum(w)))
```

Code Block 5.16: Simulation of an Online Process.

The horizontal axis in Figure 5.10 represents the cohort time j, whereas the vertical axis represents the user cohort i. Note that the fraction of engagements is decreasing from left to right, as users age within their individual cohorts. This is the structured binomial distribution at work. However,

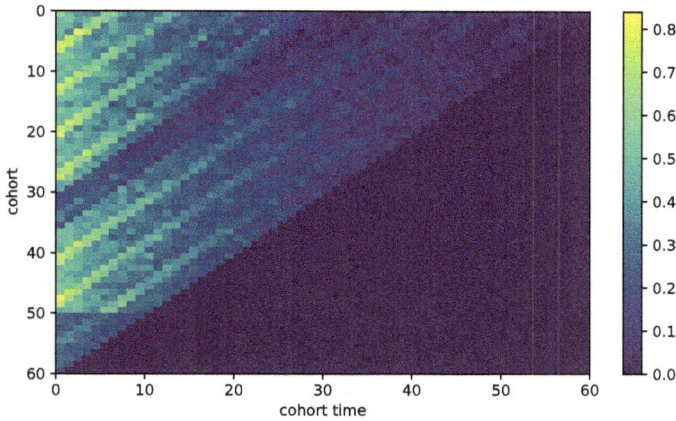

Fig. 5.10: Normalized process matrix for Example 5.16.

we can also easily spot some additional features: there is a dark blue diagonal strip starting with cohorts 30–35 and directed diagonally (north-east). These are the isochrones corresponding to calendar times 30–35, when there was a bug in the mobile app, causing diminished engagement across all cohorts for those particular calendar days. We further see a periodic sequence of bright diagonals, corresponding to the weekend effect. Finally, we notice that there is a darker shading for all cohorts beginning with cohort 50, indicating a downward shift in the type of user being acquired (and, perhaps, a cause for alarm).   ▷

Empirical Estimation of Online Bernoulli Process

Naturally, we are interested in approximating the distribution of events over cohort time, as this distribution tells us how each cohort individually is expected to age. The following theorem gives us an empirical estimate.

Theorem 5.5. *Let X_{ij} be a pure online Bernoulli process of age τ. Then an unbiased empirical estimate for the truncated generator function $p_j : \{0, \ldots, \tau\} \to (0, 1)$ is given as follows:*

$$\hat{p}_j = \frac{1}{\omega_{\tau-j}} \sum_{i=0}^{\tau-j} X_{ij}, \qquad (5.46)$$

where we define the cumulative weights ω_k as

$$\omega_k = \sum_{i=0}^{k} w_i, \qquad (5.47)$$

for $k = 0, \ldots, \tau$. Moreover, the variance in the estimate is given by

$$\mathbb{V}(\hat{p}_j) = \frac{p_j(1 - p_j)}{\omega_{\tau-j}}. \tag{5.48}$$

Proof. From the definitions, the random variables

$$X_{ij} \sim \text{Binom}(w_i, p_j),$$

which has expected value $\mathbb{E}[X_{ij}] = w_i p_j$. This processed is truncated for $t = i + j > \tau$, so that, for a fixed cohort time $j = 0, \ldots, \tau$, we only have data for cohorts $i = 0, \ldots, \tau - j$. Note that the probability depends only on the calendar time, so that

$$\sum_{i=0}^{\tau-j} X_{ij} \sim \text{Binom}(\omega_{\tau-j}, p_j),$$

for the cumulative weights ω_k defined in Equation (5.47). The expected value of this sum is therefore given by

$$\mathbb{E}\left[\sum_{i=0}^{\tau-j} X_{ij}\right] = \omega_{\tau-j} p_j.$$

The estimate provided in Equation (5.46) therefore constitutes an unbiased estimate for p_j. Moreover, the variance of a binomial random variable is given by

$$\mathbb{V}\left[\sum_{i=0}^{\tau-j} X_{ij}\right] = \omega_{\tau-j} p_j (1 - p_j),$$

which yields Equation (5.48). □

For Example 5.16, this empirical estimate and standard error are computed in lines 23–24 of Code Block 5.16. The results are shown in Figure 5.11, plotted concurrently with the geometric base distribution used for the simulation. (Note that the estimate works quite well, despite there being some cohort and calendar effects in the data.)

For comparison, Figure 5.12 was generated using the same simulation, except with the cohort and calendar effects removed.

Empirical Estimation of Online Survival Process

Next, we will consider online survival processes. We will rely on a result from survival analysis.

Theorem 5.6 (Kaplan-Meier). *Let $T_1, \ldots, T_n \sim F_T$ be a (possibly censored) sample of survival times from a distribution F_T over R_+. Let $d_j \geq 1$*

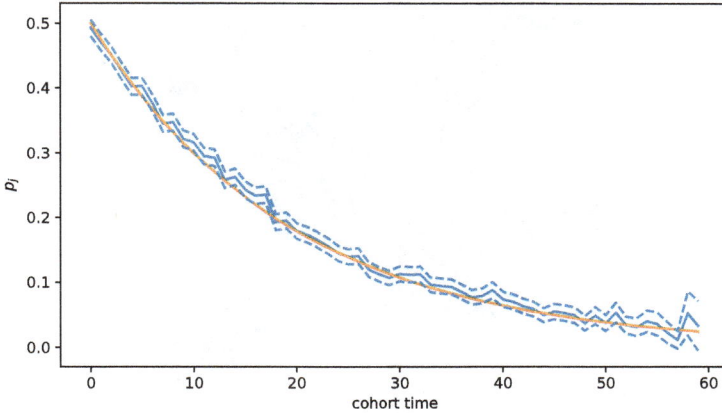

Fig. 5.11: Empirical estimate for $p_j = \alpha f_T(j)$ in Example 5.16, with 95% confidence bounds $(2\hat{se}(\hat{p}))$.

be the number of events occurring at time T_j, and let $Y(T_j)$ be the number of individuals at risk "just prior" to T_j. Then the Kaplan-Meier estimator

$$\hat{S}(t) = \prod_{T_j \leq t} \left(1 - \frac{d_j}{Y(T_j)}\right) \tag{5.49}$$

is an unbiased estimate for the survival curve $S(t) = 1 - F(t)$. Moreover, the variance in this estimator can be estimated by

$$\mathbb{V}\left(\hat{S}(t)\right) = \hat{S}(t)^2 \sum_{T_j \leq t} \frac{d_j}{Y(T_j)(Y(T_j) - d_j)}, \tag{5.50}$$

known as the Greenwood formula.

We will omit the proof. For discussion on this estimator, see Aalen, *et al.* [2008] or Klein and Moeschberger [2003]. For an online survival process, we recognize that $X_{i0} = w_i$, for each cohort i; i.e., each cohort starts its process with full survival. For $j > 0$, we therefore have

$$d_j = \sum_{i=0}^{\tau-j} \left[X_{i(j-1)} - X_{ij}\right]$$

$$Y(T_j) = \sum_{i=0}^{\tau-j} X_{i(j-1)}.$$

Of course, the number of units censored between $(j-1)$ and j is simply $X_{(\tau-j+1)(j-1)}$. Therefore, as a corollary to Theorem 5.6, we have the following:

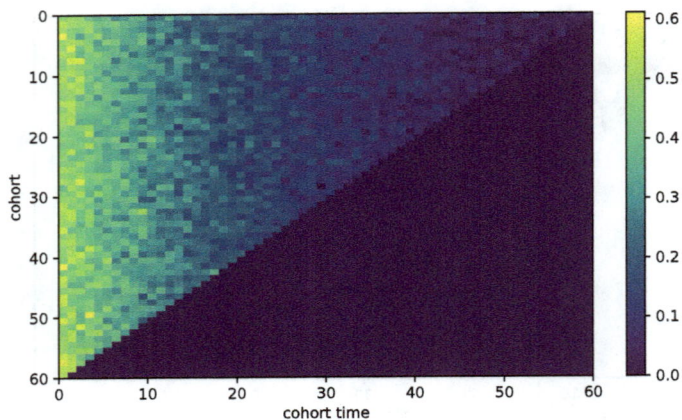

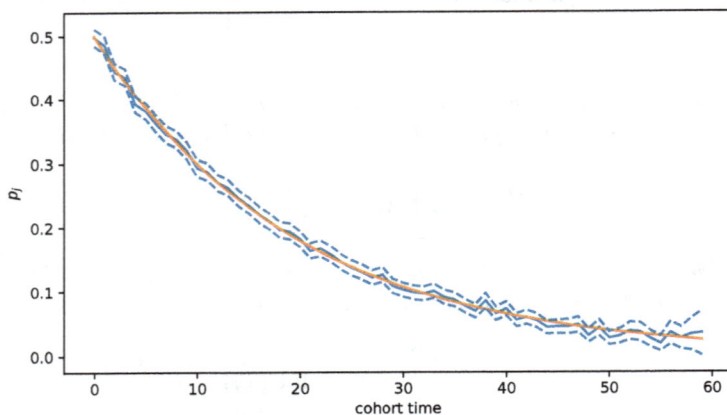

Fig. 5.12: Simulation with no calendar or cohort effects.

Theorem 5.7. *The Kaplan-Meier estimate for the survival function for an online survival process is given by*

$$\hat{S}_j = \prod_{k=1}^{j} \left(\frac{\sum_{i=0}^{\tau-k} X_{ik}}{\sum_{i=0}^{\tau-k} X_{i(k-1)}} \right). \tag{5.51}$$

Moreover, Greenwood's formula reduces to

$$\mathbb{V}\left(\hat{S}_j\right) = \hat{S}_j^2 \sum_{k=1}^{j} \left(\frac{1}{\sum_{i=0}^{\tau-k} X_{ik}} - \frac{1}{\sum_{i=0}^{\tau-k} X_{i(k-1)}} \right). \tag{5.52}$$

Proof. Simple arithmetic.   □

Example 5.17. Simulate a pure online survival process with base distribution $F_T = \text{Weibull}(2, 40)$ for $\tau = 60$. Plot the Weibull distribution with the Kaplan-Meier estimate from the simulated data.

The code for this is given in Code Block 5.17. The results are shown in Figure 5.13. (The process matrix was plotted using lines 19–22 from Code Block 5.16.)

```
1   alpha, beta, tau = 2, 40, 60
2   w = np.random.normal(loc=100, scale=20, size=tau).astype(int)
3   X = np.zeros((tau,tau))
4
5   for i in range(tau):
6       samples = beta * np.random.weibull(alpha, size=w[i])
7       for j in range(tau-i):
8           X[i, j] = np.sum( (samples > j) )
9
10  # Kaplan Meier estimate
11  col_sum = X.sum(axis=0)
12  diag_elements = np.flipud(X).diagonal()
13  col_at_risk = np.roll(col_sum - diag_elements, 1)
14  col_at_risk[0] = col_sum[0]
15
16  S_hat = np.cumprod(col_sum / col_at_risk)
17  S = scipy.stats.weibull_min.sf(np.arange(tau), alpha, scale=beta)
18  se = S_hat * np.sqrt( np.cumsum( 1 / col_sum - 1 / col_at_risk ) )
```

Code Block 5.17: Random walk.

As expected, the Kaplan-Meier estimate provides an accurate empirical estimate for the survival function of our online survival process.   ▷

Problems

5.1. Let $X_1, \ldots, X_n \sim \text{N}(\mu, \sigma^2)$. Show that the MLE estimates for μ and σ^2 are given by

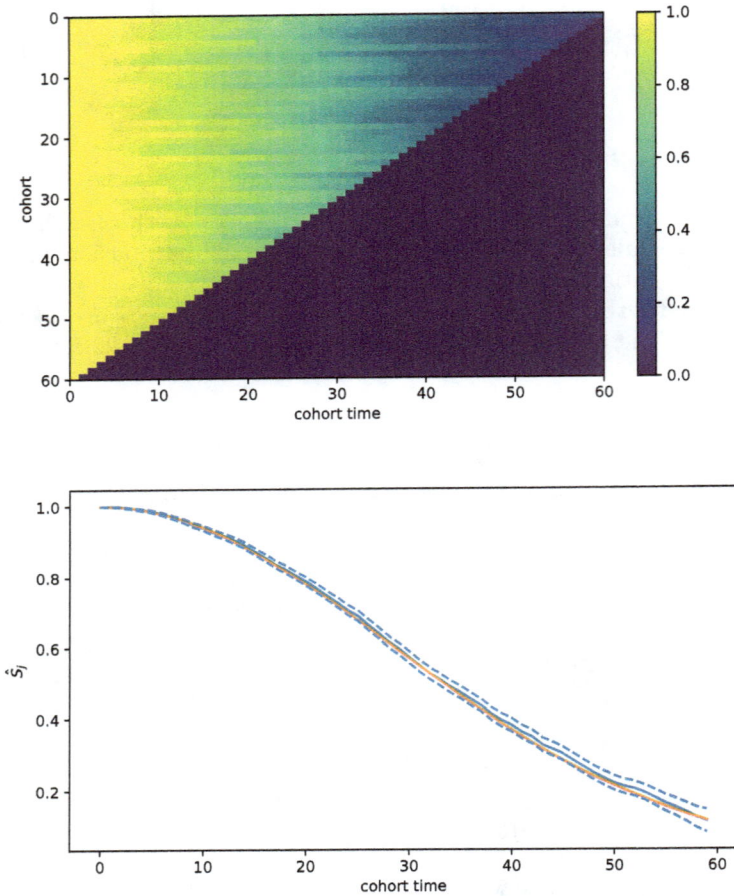

Fig. 5.13: Process matrix and Kaplan-Meier estimate for online survival process of Example 5.17.

$$\hat{\mu}_n = \frac{1}{n} \sum_{i=1}^{n} x_i,$$

$$\hat{\sigma}_n^2 = \frac{1}{n} \sum_{i=1}^{n} (x_i - \hat{\mu})^2.$$

5.2. Prove the assertion made in Note 5.4.

5.3. Show the missing steps between Equations (5.29) and (5.30).

5.4. Show that the definition of the hazard function, given by Equation (5.33), is equivalent to

$$h(x) = -d \log(S(x))/dx.$$

5.5. Let $T_1, \ldots, T_n \sim \text{Exp}(\beta)$ be IID exponential data that are left-truncated at $T = \tau$. Determine the likelihood and log-likelihood functions for the data. Show that the MLE is given by

$$\hat{\beta} = \left(\frac{1}{n} \sum_{i=0}^{n} t_i \right) - \tau.$$

Explain why this result makes sense intuitively.

5.6. Prove Proposition 5.7.

5.7. Determine the value of α for the Pareto distribution that results in the 80-20 rule.

5.8. Let $X \sim \text{Pareto}(1, \alpha)$, and define $Y = X | X > m$, for $m > 1$. Show that $Y \sim \text{Pareto}(m, \alpha)$.

5.9. Suppose that $X_1, \ldots, X_n \sim \text{Pareto}(1, \alpha)$ is IID Pareto-distributed data that represents the amounts of insurance claim payouts for a policy with a \$500 deductible and \$20,000 cap. Determine the likelihood function for the data.

References

Aalen, O.o., O. Borgan, H.K. Gjessing (2008) *Survival and Event History Analysis: A Process Point of View*, Springer.

Agresti, A. (2013) *Categorical Data Analysis*, 3rd ed., Wiley.

Agresti, A. (2019) *An Introduction to Categorical Data Analysis*, 3rd ed., Wiley.

Berger, P.D., R.E. Maurer, and G.B. Celli 2018 *Experimental Design: With Applications in Management, Engineering, and the Sciences*, 2nd ed., Springer.

Brier, G.W. (1950) Verification of forecasts expressed in terms of probability; *Monthly Weather Review* **78**(1)

Buitinck, *et al.* (2013) API design for machine learning software: experiences from the scikit-learn project, in *ECML PKDD Workshop: Languages for Data Mining and Machine Learning*, p. 108–122.

Capinksi, M. and E. Kopp (2005) *Measure, Integral, and Probability*, 2nd ed., Springer.

Casella, G., and R.L. Berger (2002) *Statistical Inference*, 2nd ed., Brooks/-Cole.

Chong, E.K.P. and S.H. Zak (2008) *An Introduction to Optimization*, 3rd ed., John Wiley & Sons.

Devroye, L. (1986) *Non-Uniform Random Variate Generation*, Springer.

DiBenedetto, E. (2002) *Real Analysis*, Birkhäuser.

Dobson, A.J. and A.G. Barnett (2018) *An Introduction to Generalized Linear Models*, 4th ed., CRC Press.

Dunn, P., and G.K. Smyth (2018) *Generalized Linear Models with Examples in R*, Springer.

Durrett, R. (2016) *Essentials of Stochastic Processes*, Springer.

Efron, B. (1971) Forcing a sequential experiment to be balanced, *Biometrika*, **58**, 403–417.

Fisher, R.A. (1935) *The Design of Experiments*, Oliver and Boyd.

Folland, G.B. (1999) *Real Analysis: Modern Techniques and their Applications*, 2nd ed., Wiley.

Hajek, B. (2015) *Random Processes for Engineers*, Cambridge University Press.

Härdle, W.K. and L. Simar (2019) *Applied Multivariate Statistical Analysis*, 5th ed., Springer.

Hastie, T, R. Tibshirani, and J. Friedman (2009) *The Elements of Statistical Learning: Data Mining, Inference, and Prediction*, Springer.

Hogg, R.V., E.A. Tanis, and D.L. Zimmerman (2015) *Probability and Statistical Inference*, 9th ed., Pearson.

Klein, J.P. and M.L. Moeschberger (2003) *Survival Analysis: Techniques for Censored and Truncated Data*, 2nd ed., Springer.

Kochenderfer, M.J. (2015) *Decision Making Under Uncertainty*, MIT Press.

Kohavi, R., D. Tang, and Y. Xu (2020) *Trustworthy Online Controlled Experiments: A Practical Guide to A/B Testing*, Cambridge University Press.

Kuhn, M. and K. Johnson (2013) *Applied Predictive Modeling*, Springer.

Maruskin, J. (2018) *Dynamical Systems and Geometric Mechanics: An Introduction*, 2nd ed., de Gruyter.

Mood, A.M. (1950) *Introduction to the Theory of Statistics*. McGraw-Hill.

McKinney, W. (2017) *Python for Data Analysis*, 2nd ed., O'Reilly.

Predicting good probabilities with supervised learning, In ICML-05 *International Conference on Machine Learning*, Aug. 2005, 625–632; https://doi.org/10.1145/1102351.1102430.

Olive, D. (2017) *Linear Regression*, Springer.

Pedregosa, *et al.* (2011) Scikit-learn: Machine Learning in Python, *Journal of Machine Learning Research* **12**: p. 2825–2830.

Platt, J. (2000) Probabilistic outputs for support vector machines and comparison to regularized likelihood models; In Bartlett B., B. Schölkopf, D. Schuurmans, A. Smola (eds.) *Advances in Kernel Methods Support Vector Learning*, p. 61–74, MIT Press.

Reinhart, A. (2015) *Statistics Done Wrong*, no starch press.

Robbins, H. (1952) Some aspects of the sequential design of experiments, *Bulletin of the American Mathematical Society* **58**(5): 528–535.

Rosenbaum, P.R. (2002) *Observational Studies*, 2nd ed., Springer.

Ross, S. (2012) *A First Course in Probability*, 9th ed., Prentice Hall.

Seber, G.A.F. and A.J. Lee (2003) *Linear Regression Analysis* (2nd ed.), John Wiley.

Selvamuthu, D. and D. Das (2018) *Introduction to Statistical Methods, Design of Experiments, and Statistical Quality Control*, Springer.

Shao, J. (2003) *Mathematical Statistics*, 2nd ed., Springer.

Sutton, R.S. and A.G. Barto (2018) *Reinforcement Learning: An Introduction*, 2nd ed., MIT Press.

Trussell, J. and D.E. Bloom (1979) A model distribution of height or weight at a given age, *Human Biology* **51**: 523-536.

Wasserman, L. (2004) *All of Statistics: A Concise Course in Statistical Inference*. Springer.

Wasserman, L. (2006) *All of Nonparametric Statistics*. Springer.

Withers, C.S. and S. Nadarajah (2014) The spectral decomposition and inverse of multinomial and negative multinomial covariances. *Brazilian Journal of Probability and Statistics*. Vol. 28, No.3, p. 376–380.

Zou, H., T. Hastie, and R. Tibshirani (2007) On the "degrees of freedom" of the lasso, *The Annals of Statistics* **35**(5): 2173–2192.

Index

ℓ_1 norm, 105, 127
ϵ-greedy, 225
σ-algebra, 4
p-value, 115
IID, 11

A/B test, 181
abstract class, 256
accept null hypothesis, 110
action, 110
action-value function, 225
ad tech, 226
aggregate process, 278
alternative hypothesis, 110
analysis of variance, 172
ANOVA, 172
ANOVA table, 176
arrival process, 277
assigment mechanism, 179

bandits, 224
batch gradient descent, 255
Bernouilli Trial, 39
Bernoulli process, 279
Bernoulli randomized experiment, 186
beta distribution, 82
beta function, 81
beta-binomial distribution, 89
beta-binomial random variable, 88
between-group sum of squares, 174,
     218
bias, 20
bias–variance tradeoff, 21
binary decision rule, 110

binomial bandits, 224
binomial coefficient, 41
binomial distribution, 40
binomial process, 279
binomial theorem, 41
block randomized A/B test, 196
blocking, 181
Bonferroni correction, 212
bootstrap sample, 28
bootstrap variance estimate, 30

calendar effect, 281
calendar time, 280
categorical data, 143
causal effect, 188
censored data, 262
central limit theorem, 68
chi-squared distribution, 73
chi-squared random variable, 70
     noncentral, 149
chi-squared test, 143
churn, 264
classic ANOVA null hypothesis, 173
click-through rate, 226
clicks, 226
cohort, 280
cohort age, 280
cohort effect, 281
cohort time, 280
completely randomized A/B test, 186
completely randomized design, 182,
     217
completely randomized experiment,
     202

compound distribution, 88
conditional distribution, 12
conditional expectation, 13
conditional probability, 5
confidence interval, 117
confidence level, 113
consistent estimator, 242
continuous random variable, 5
continuous-time random process, 273
convergence
    in distribution, 20, 68
    in probability, 20
convolution, 106
correlation, 11
counterfactual, 187
counterfactuals, 182
counting process, 277
covariance, 11
covariates, 179
CPC, 226
CPM, 226
creative, 226
critical value, 110
cumulative distribution function, 5
customer lifetime, 264

day-of-week effect, 281
decision rule, 110
delta function, 8
detectable effect space, 127
digamma function, 245
direct notation, 183
discrete random variable, 5
discrete-time random process, 273
disjoint, 4
distance
    between a point and a set, 127
dot notation, 173
Dvoretzky-Kiefer-Wolfowitz (DKW)
    inequality, 37

empirical distribution, 22
Euclidean norm, 127
event, 3
expected information, 239
experiment, 179
experimental analysis, 183
experimental notation, 183
experimentwise error rate, 211

exploration–exploitation tradeoff, 224
exponential distribution, 56

F statistic, 175
F-distribution, 78
    noncentral, 207
factors, 180
Fisher information, 239
Fisher's least significant difference
    test, 213
functional, 24

gambler's ruin, 275
gamma distribution, 83
gamma function, 73
generator, 256, 279
geometric distribution, 50
geometric random variable, 49
goodness of fit, 169
gradient, 241
gradient descent, 251
greedy, 225

harmonic mean, 218
hat operator, 218
hazard function, 268
Heaviside function, 8
Hessian matrix, 241
hierarchical model, 88
homogeneous
    Markov process, 274
homoscedasticity, 172
honestly significant difference, 215
HSD, 215
hypergeometric distribution, 51
hypergeometric random variable, 51
hyperparameters, 88
hypothesis of no effect, 183
hypothesis test, 109
    mean, known variance, 118
    mean, unknown variance, 120
    two means, equal variance, 134
    two means, unequal variance, 139

imbalance, 200
impressions, 226
independence, 5, 11
index notation, 183
indicator function, 4, 8
infimum, 26

infinity norm, 127, 204
interaction sum of squares, 218
interarrival time, 59
isochrone, 280

Jacobian, 34
joint probability density function, 10
joint probability mass function, 9

Kaplan-Meier esimtator, 284
kernel
  of objective function, 255

lambda function, 249
least significant difference, 213
left-tailed test, 132
likelihood function, 235
Lindy effect, 271
linear functional, 25
location parameter, 36
location–scale family, 36
LSD, 213

Mahalanobis distance, 93
Manhattan norm, 127
marginal distributions, 10
Markov process, 274
Markov property, 274
max norm, 127
maximum likelihood estimate, 236
  consistency, 242
  invariance, 242
mean squares, 174, 219
mean-squared error, 21
mean-squares, 218
memoryless, 50
method of scoring, 242
Method of Steepest Descent, 252
mini-batch gradient descent, 255
minimum detectable effect, 127
mixture distribution, 88
moment-generating function, 35, 85
multi-armed bandits, 224
multinomial coefficient, 100
multinomial distribution, 101
multinomial random variable, 99
multinomial random vector, 143
multinomial theorem, 101
multiple hypothesis tests, 211
multivariate Bernoulli trial, 99

multivariate Wald test, 124
mutually exclusive, 4

negative binomial distribution, 45
Newton–Raphson method, 241
noncentral chi-squared random
  variable, 149
noncentral F-distribution, 207
noncentrality effect, 208
noncentrality parameter
  chi-squared, 149
nonhomogeneous process, 279
normal distribution, 61
  multivariate, 91
nuisance factor, 181
null hypothesis, 109

objective function, 255
observed information, 239
observed outcome, 187
one-at-a-time simple random sampling,
  187
one-sided test, 110
one-sided tests, 193
oneway ANOVA, 172
online Bernoulli process, 281
online experiment, 200
online process, 280
online randomized block design, 200,
  203
online survival process, 281
order statistics, 166
outcome, 3

paired randomized experiment, 182
pairwise comparison, 211
Pareto distribution, 269
partition, 4
partitioning theorem, 219
partitioning theorem for sum of
  squares, 173
Pearson's chi-squared test, 143
plug-in estimator, 25
point estimator, 14
Poisson distribution, 54
Poisson process, 58
policy, 224
potential outcome, 187
potential outcomes, 182

power, 194
    minimum detectable effect, 128
power function, 112
probability density function, 6
probability distribution, 4
probability mass function, 6
probability simplex, 98
probability space, 4
process age, 280
process matrix, 280
product σ algebra, 278
product engineering, 179
product measure, 278
pure online process, 281
Python
    abstract class, 256
    generator, 256

quantiles, 26

random process, 273
random variable, 5
random walk, 275
randomized block design, 182
randomized block experiment, 203
realization, 273
reduced multinomial random vector,
    103
reduced multivariate Bernoulli random
    vector, 103
reject null hypothesis, 109
rejection region, 110
renewal process, 277
replication, 182
reward, 224
right-tailed test, 132

sample mean, 15
sample of realizations, 278
sample path, 273
sample quantile, 26
sample size, 128, 194
sample space, 3
sample variance, 15
sampling, 166
sampling distribution, 14
Satterthwaite's approximation, 138
scale parameter, 36
score equation, 237

score function, 237
score statistic, 237
significance level, 113
simplex, 98, 166
size of a test, 113
Snedecor's F distribution, 80
Snedecor's F random variable, 78
softmax, 226
spacings, 166
standard k-simplex, 98
standard error, 14
state space, 273
state transition matrix, 274
statistic, 14
statistical functional, 24
Steepest Descent, 252
stochastic gradient descent, 255
stochastic process, 273
strategy, 224
studentized range distribution, 215
sums of squares, 174, 218
supremum, 26
survival analysis, 268
survival process, 279

t-random variable, 77
t-test, 134
    for randomized block A/B test, 198
test statistic, 110
time since acquisition, 280
total sum of squares, 174, 218
training samples, 255
transfomration
    continuous random vector, 34
transformation
    discrete random vector, 33
transformations of random variables,
    31
transition matrix, 274
transition probabilities, 274
treatments, 179
truncated data, 262
Tukey's honestly significant difference
    test, 215
two-factor experiment, 217
two-level factorial deisgn, 181
two-sided test, 110
Type I and II Errors, 111

uncorrelated, 11

uniform random variable, 7
uniform spacings, 166
unit survival process, 279
user lifetime, 264

Wald test, 123

weekend effect, 281
Weibull distribution, 269
Welch's t-test, 139
with replacement, 97
within-group sum of squares, 174, 218

www.ingramcontent.com/pod-product-compliance
Lightning Source LLC
Chambersburg PA
CBHW071329210326
41597CB00015B/1387